GEOTHERMAL ENERGY AS A SOURCE OF ELECTRICITY

A Worldwide Survey of the
Design and Operation of Geothermal Power Plants

RONALD DiPIPPO

Professor of Mechanical Engineering
Southeastern Massachusetts University
North Dartmouth, Massachusetts   02747
and
Adjunct Professor of Engineering (Research)
Brown University
Providence, Rhode Island   02912

Prepared under the sponsorship of the
U.S. DEPARTMENT OF ENERGY
Assistant Secretary for Resource Applications
Division of Geothermal Energy

Books for Business
New York-Hong Kong

Geothermal Energy as a Source
of Electricity:
A Worlwide Survey of the
Design and Operation of Geothermal Power Plants

by
Ronald DiPippo

ISBN: 0-89499-153-1

Reprinted from the 1980 edition

Books for Business
New York - Hong Kong
http://www.BusinessBooksInternational.com

Preface

GEOTHERMAL ENERGY may not be destined to play a major role in solving the contemporary, cataclysmic energy crisis but it is sure to make a contribution. Just what percentage of this country's, or the world's needs for electricity will be covered from geothermal resources remains to be seen. The potential, however, is very encouraging. For example, the most recent summary by the United States Geological Survey contained in its Circular 790 estimates a potential of 21,000 MW of electrical output for thirty years from the already identified, hot-water, hydrothermal convection systems having temperatures exceeding 150°C (302°F), in the United States alone. The installed capacity in the whole world is now at a level of 1750 MW electrical, and growing.

Geothermal electricity, unlike fossil or nuclear, cannot be ordered: it must be developed, for there is nothing more hazardous than a premature order for conversion equipment. A hydrothermal field must be discovered, explored, tested and developed. At some point in this process it becomes possible to specify the characteristics of the conversion plant with good assurance of success, and to order the first set of turbines. There follows a long period of continuing development during which more and more electric capacity (up to a maximum limit) can be added. Thus, even though there are financial risks—as with any potentially profitable enterprise—these can be held to a minimum by harmonious cooperation between geologists and mechanical engineers.

To this day it is not possible to state with definite assurance whether, or which, aquifers are renewable, or whether, as assumed in U.S.G.S. Circular 790, geothermal energy must be "mined", like oil. One way or another, it is, nevertheless, possible to say that geothermal energy is here now, that it is waiting to be more fully exploited, and that it is sure to play a part in the elimination of our plight.

As part of an effort undertaken on behalf of the Division of Geothermal Energy of the Department of Energy, and under the overall management imaginatively performed by Mr. Clifton B. McFarland, the team at Brown University has undertaken the task of editing a *Sourcebook on the Production of Electricity from Geothermal Energy* which is scheduled to appear at about the same time as this volume. It was thought useful to provide the *Sourcebook* with a companion compendium which would contain in it a snapshot picture of all existing geothermal, electricity-producing installations in the world. This task was undertaken and imaginatively completed by my former student and now colleague, Professor Ronald DiPippo. He was able to assemble a vast set of documents describing geothermal installations, and to digest and organize them for the reader's easy access and use. Much of the information reported in this volume was obtained from direct conversations with equipment manufacturers, plant engineers, and utility managers during site visits to several installations. Both he and I believe that the resulting survey is essentially complete and accurate as of July 1979.

Work on this survey teaches that electricity from geothermal resources is viable, and that its commercial risks are no different from those associated with the established and more familiar technologies. Geothermal power has proved beneficial for a number of countries, and will play a particularly important role in such developing countries as El Salvador and the Philippines, primarily because of its simplicity of construction, operation, and maintenance. It has demonstrated the ability to achieve capacity factors (ratio of kilowatt-hours produced per annum to the maximum possible) exceeding 80%. The comparable figure for fossil plants is only about 50%. In those applications, geothermal power has proved to be economical, and clearly less expensive than installations using oil, coal or nuclear fuel.

As might have been expected, the 1750 MW now in use in the world have exploited the "easy" reservoirs in that roughly 60% of it is supplied with dry steam. The lowest temperature exploited at the wellhead is 190°C, and there are no commercial binary-fluid installations as yet (except for the 11 MW plant at East Mesa, California, now under construction). In short, further expansion must reckon with somewhat more challenging conditions for exploitation. These include mastery of low to moderate temperatures, avoidance of scaling and corrosion, and an ability to maintain a guaranteed flow of working fluid in an expanding installation. It appears that the development of binary-fluid plants will be essential to achieve these goals and to realize the potential that geothermal energy holds out to us.

Professor DiPippo and I hope that the present survey will have broad appeal, particularly among energy planners. This book should convince them that geothermal electric power generation is already very important in some parts of the world, that the unrealized potential of geothermal energy should be recognized, and that it must constitute an integral part of any plan to solve the energy problems facing the United States.

J. KESTIN
Principal Project Investigator.

July 1979
Providence, R.I.

Acknowledgements

THE AUTHOR IS INDEBTED to many people who have helped in the process of accumulating the vast amount of information in this volume. While I cannot name all of those who helped, special thanks must be extended to the following individuals: K. Aikawa, Mitsubishi Heavy Industries, Ltd.; D. N. Anderson, Geothermal Resources Council; J. H. Barkman, Republic Geothermal, Inc.; R. S. Bolton, Ministry of Works and Development, New Zealand; B. J. Cossette, Pacific Gas and Electric Company; G. Cuéllar, Comisión Ejecutiva Hidroeléctricia del Río Lempa (CEL), El Salvador; K. Danno, Mitsubishi Heavy Industries, Ltd.; J. H. Eskesen, General Electric Company; J. P. Finney, Pacific Gas and Electric Company; T. C. Hinrichs, Imperial Magma; T. Kobori, Nissho-Iwai American Corporation; J. T. Kuwada, Rogers Engineering Company, Inc.; N. C. McLeod, Ministry of Works and Development, New Zealand; H. Nakamura, Japan Metals & Chemicals Co., Ltd.; R. H. Sheehan, World Bank; B. Tanaka, Mitsubishi International Corporation.

The bulk of the writing was done during a sabbatical leave from Southeastern Massachusetts University, North Dartmouth (July 1977–August 1978). The author wishes to acknowledge the excellent technical support of the Division of Engineering staff at Brown University, especially the secretarial assistance of Mrs. Leslie A. Giacin, and the work of the Drafting Department headed by Mrs. Eleanor E. Tartaglia, including Mrs. Muriel S. Anderson, Mrs. Mildred L. Brown, and Michael L. Waldygo.

Clifton B. McFarland, Division of Geothermal Energy, Department of Energy, provided encouragement and support throughout the project, which was funded under contract No. AS02-76ET28230 to Brown University.

I wish to acknowledge, with warmest gratitude, Dr. Joseph Kestin, Professor of Engineering at Brown University and Principal Investigator on this project. It was he who first revealed to me the mysteries of thermodynamics and offered guidance and encouragement during my graduate studies.

Finally, I am grateful to my wife. Lois, and my children, Debora, John and Michael, for their patience and understanding during the writing of this book, and to all those who have provided me with encouragement and inspiration.

RONALD DiPIPPO

July 18, 1979
Barrington, R.I.

Contents

Chapter 6. Japan—Continued

Chapter 13. Countries Planning Geothermal Power Plants   *Page*

CHAPTER 1

OVERVIEW OF GEOTHERMAL POWER GENERATION

THE GEOTHERMAL ENERGY stored within the interior of the earth amounts to a vast quantity which, if fully exploited, could supply the energy needs of man for millennia. If one considers the United States, for example, and includes only those geothermal resources with temperatures greater than 150°C (300°F), it has been estimated that about 450 million kilowatts (kW) could be generated. This is only slightly less than the total installed electric generating capacity of the country from all energy sources [EPRI, 1977].

The main difficulty in putting this enormous energy resource to use lies in its diffuse nature. Over the course of a year about 4×10^{17} kilojoules (kJ) or 100 billion megawatt-hours of heat energy is conducted to the surface from the interior of the earth [Bowen and Groh, 1977]. This is about twenty times the electric energy used by the entire world.

The rest of this chapter will present a short description of the various types of geothermal energy resources, a historical account of geothermal energy usage, a survey of geothermal energy conversion technology along with a clarification of terms used to characterize the operations and reliability of central power stations and, finally, a summary of current geothermal power plants and those planned for the near future.

1.1 General nature of geothermal resources

Even though the quantity of heat reaching the surface of the earth is colossal, it is distributed more or less evenly over nearly the entire surface, resulting in a heat flux of only about 63 milliwatts per square meter, 0.063 W/m². Were it not for the fact that there are a number of regions where anomalously large geothermal heat fluxes occur, it is unlikely that geothermal energy would be very useful as a practical means of supplying electrical energy.

These anomalous regions generally coincide with the boundaries of the tectonic plates that comprise the earth's crust. Figure 1.1. shows the major plates and their associated boundaries. These may be of the ridge, transform, or subduction type. It is clear that vulcanism and seismic activity are abundant along these boundaries and in the areas adjacent to them. Not

1

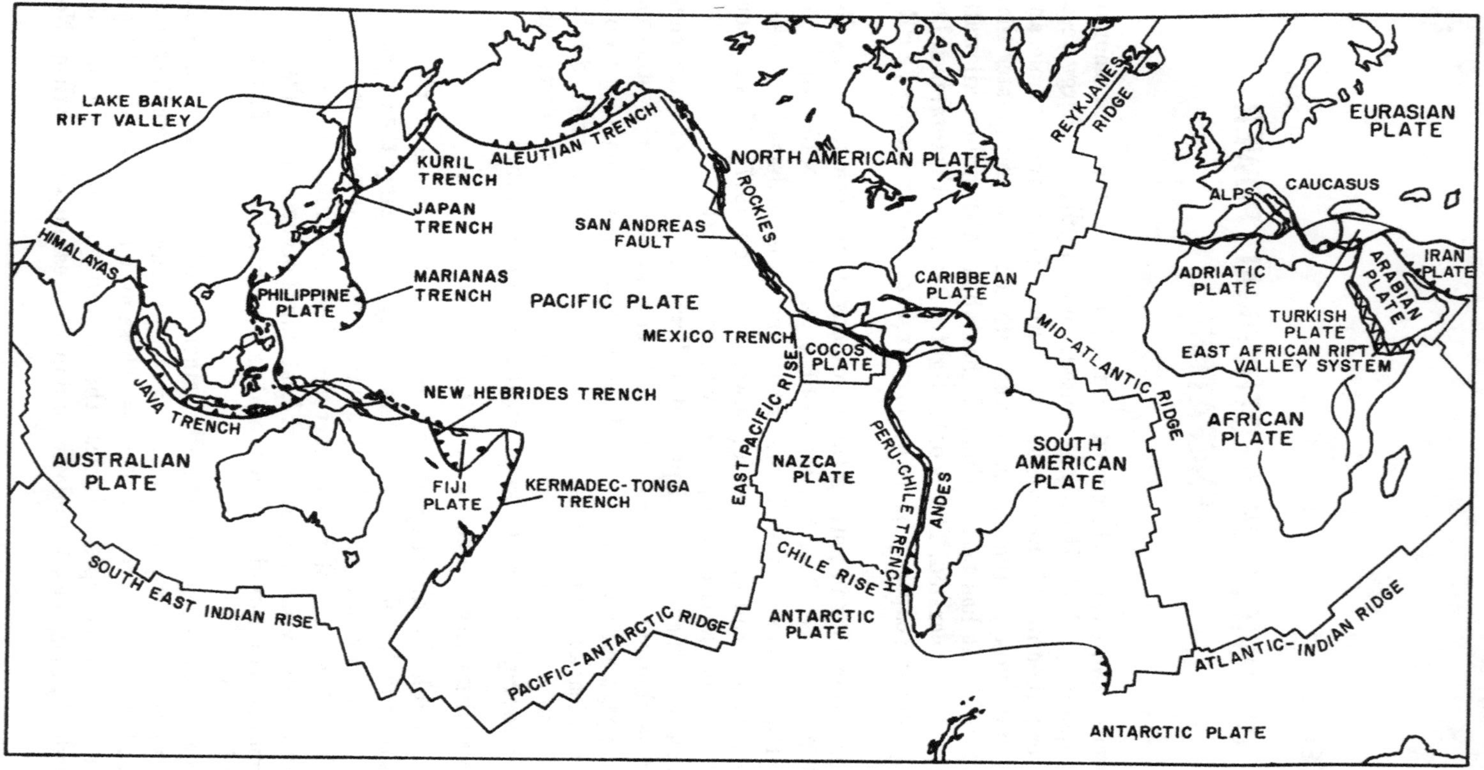

FIGURE 1.1—Plate tectonics. Lithosphere or crust consists of many large plates in relative motion. Boundaries are of three types: ridge axes (plates diverging); transforms (plates sliding); subduction zones (plates converging, with one diving beneath the other).

surprisingly, the most favorable regions for the exploitation of geothermal energy are characterized by volcanic and earthquake activity.

Geothermal energy resources may be classified into five categories: (1) normal-gradient, (2) radiogenic, (3) high heat flow areas, (4) geopressured, and (5) point heat sources. The first three are of relatively low practical value at the level of present-day technology, the fourth holds considerable promise, particularly in the Gulf Coast region of the United States, and the last is already being exploited for electric power in a number of countries around the world.

1.1.1 Normal-gradient regions

In normal regions, i.e., areas outside the anomalous zones near plate boundaries, the temperature gradient in the crust averages about 2.5°C/100 m (1.4°F/100 ft). If one sets a temperature of 150°C (300°F) as the minimum for feasible electric power generation, one must drill to a depth of about 5000 m (16,400 ft) before encountering temperatures as high as this. Irrespective of the detailed nature of the resource, it is uneconomical at present to attempt to exploit such geothermal systems.

Nevertheless there is a program underway aimed at the development of techniques to fracture hot but dry rock formations, and to introduce water from the surface into the factures in order to extract the thermal energy from the rocks. The Los Alamos Scientific Laboratory has demonstrated that such a scheme is technically feasible [Pettitt, 1978]. It is conceivable that continued success in the development of hot dry rock technology will eventually open up many regions where geothermal energy is not now practical or economical.

1.1.2 Radiogenic regions

The geothermal energy flowing to the land surface of the earth has as its origin the radioactive decay of a number of elements contained within the crust [Bullard, 1973]. The principal ones are uranium (U), thorium (Th), and potassium (K). Granite is one of the main constituents of the crust and contains about 3.4% K, 20 parts per million (ppm) Th, and 4.7 ppm U. The energy flow rate caused by radioactivity in granite amounts to about one-billionth of a watt per kilogram of granite (10^{-9} W/kg).

In regions with above-average concentrations of granite, i.e., the White Mountains of New Hampshire in the United States, there will be a stronger geothermal energy heat flux. However, this energy is quite diffuse and a suitable medium may not be readily available to permit its large-scale extraction. That is, there may not be sufficient permeability in the granite to allow for circulation of water, or water may not be present at all.

1.1.3   High heat flow areas

Heat transfer through the earth's crust takes place by means of conduction when there are no subterranean water reservoirs. In such cases
the heat flow rate is proportional to the thermal conductivity of the rock
and the temperature gradient. In some areas where abnormally low thermal
conductivities are found in combination with high heat flows, very high
thermal gradients will exist. As a consequence relatively shallow wells
should be able to encounter high-temperature sources. This situation exists,
for example, in the Hungarian Basin where thermal gradients range from
$40°$ to $75°C/km$ ($2.2°$ to $4.1°F/100$ ft), or as much as three times the
normal gradient [Boldizsár, 1970]. Such high heat flow areas may be
associated with regions in which the crust is abnormally thin, thus allowing the mantle to come into closer proximity to the surface, or in which
a large, deep-seated magma chamber is enclosed within the earth's crust.
From a practical point of view resources of this type have not proved
feasible for electric power production because of the diffuse nature of the
energy contained in them.

1.1.4   Geopressured resources

There are many basins that have been formed by the steady deposition
of sedimentary rocks containing within their pores fluids at pressures in
excess of the normal hydrostatic pressure. When the vertical pressure
gradient exceeds about 10.5 kPa/m (0.465 lbf/in^2 · ft), the energy contained within such reservoirs is called geopressured-geothermal energy
[Wallace et al., 1978]. Normal hydrostatic pressure gradient is 9.78 kPa/m
(0.433 lbf/in^2 · ft). The attractiveness of this form of resource lies in the
fact that three types of energy may be extracted: hydraulic energy (by
virtue of the high pressure), thermal energy (by virtue of the high temperature), and fossil energy (by virtue of dissolved methane gas).

Several questions must be answered satisfactorily, however, before the
exploitation of geopressured-geothermal energy will take place on an
economically sound basis. These relate to the temperature of the pore fluid
(it must be sufficiently high), to the quantity of dissolved methane present
(it must be enough to justify its extraction at current gas prices), to the
thickness and permeability of the sand zone (it must be sufficiently large
and fractured to allow large volumes of fluid to circulate), and to the
fault-bounded nature of the zone (it must be bounded by faults but not
excessively fractured). Of even greater importance, since these regions are
usually associated with coastal areas, is the possibility of significant land
subsidence resulting from the withdrawal of large quantities of high-pressure fluid and the consequent flooding of low-lying coastal plains. Reinjection of waste liquid may alleviate this problem to some extent.

In the United States, the Gulf Coast of Texas and Louisiana is currently being explored with deep wells, although no power plants have

been built to utilize this potentially huge source of energy. Additional details on geopressured resources are given in section 12.9.4.

1.1.5 Point heat sources

The most easily exploited geothermal energy resources are those classified as point heat sources. The source of the thermal anomaly is a hot magma intrusion or pluton trapped in the crust of the earth, typically at a depth of between 7 and 15 km (23,000–49,000 ft). Figure 1.2 illustrates schematically how magma chambers can be formed, for example, in subduction zones where one plate is diving beneath an adjacent plate. A region of partial melting along the edges of the moving lithosphere gives rise to molten basalt (magma), which makes its way upward toward the surface. Volcanoes are usually created this way in geologically young zones, whereas in older areas it is not uncommon to find pockets of magma embedded in the upper crust.

Some attention is being given to the direct tapping of such magma intrusions [Eaton, 1975], but the most practical way to convert this geothermal energy into electricity is by extraction of steam and hot water from subterranean reservoirs heated by the magma intrusions. Such systems are usually called hydrothermal geothermal resources.

A simplified schematic cross-sectional view of a hydrothermal system is shown in figure 1.3. The top of the magma intrusion heats the surrounding crystalline rock by conduction and by gases that are released and travel through fissures and pores. One of the key features of a useful hydrothermal system is a permeable layer that allows the circulation of water. This water, which falls as rain or reaches the area as runoff on the surface, eventually percolates down to the reservoir through fractures in the overlying rock. Enough fractures must be present to permit adequate recharge of the permeable layer, but too many would allow much of the geothermal energy to escape as geysers, fumaroles, or hot springs. Thus a layer of cap rock is essential to provide a lid for the natural "boiler."

In summary, the five elements necessary for a practical hydrothermal geothermal resource are: (1) a high-temperature magma chamber at a depth of about 10 km (32,800 ft); (2) fault boundaries to delimit the actual reservoir; (3) a permeable layer; (4) a thick layer of cap rock; and (5) a mechanism for recharge of water to the reservoir during exploitation.

Many techniques have been developed to carry out exploration for suitable hydrothermal reservoirs [Combs and Muffler, 1973]. These include geological, hydrological, geochemical, and geophysical surveys. Among the geophysical surveys usually performed are temperature gradient and heat flux measurements, electrical resistivity, electromagnetic and gravity surveys, and active and passive seismic techniques. Often more reliable predictions can be made by interpreting the results of several different types of

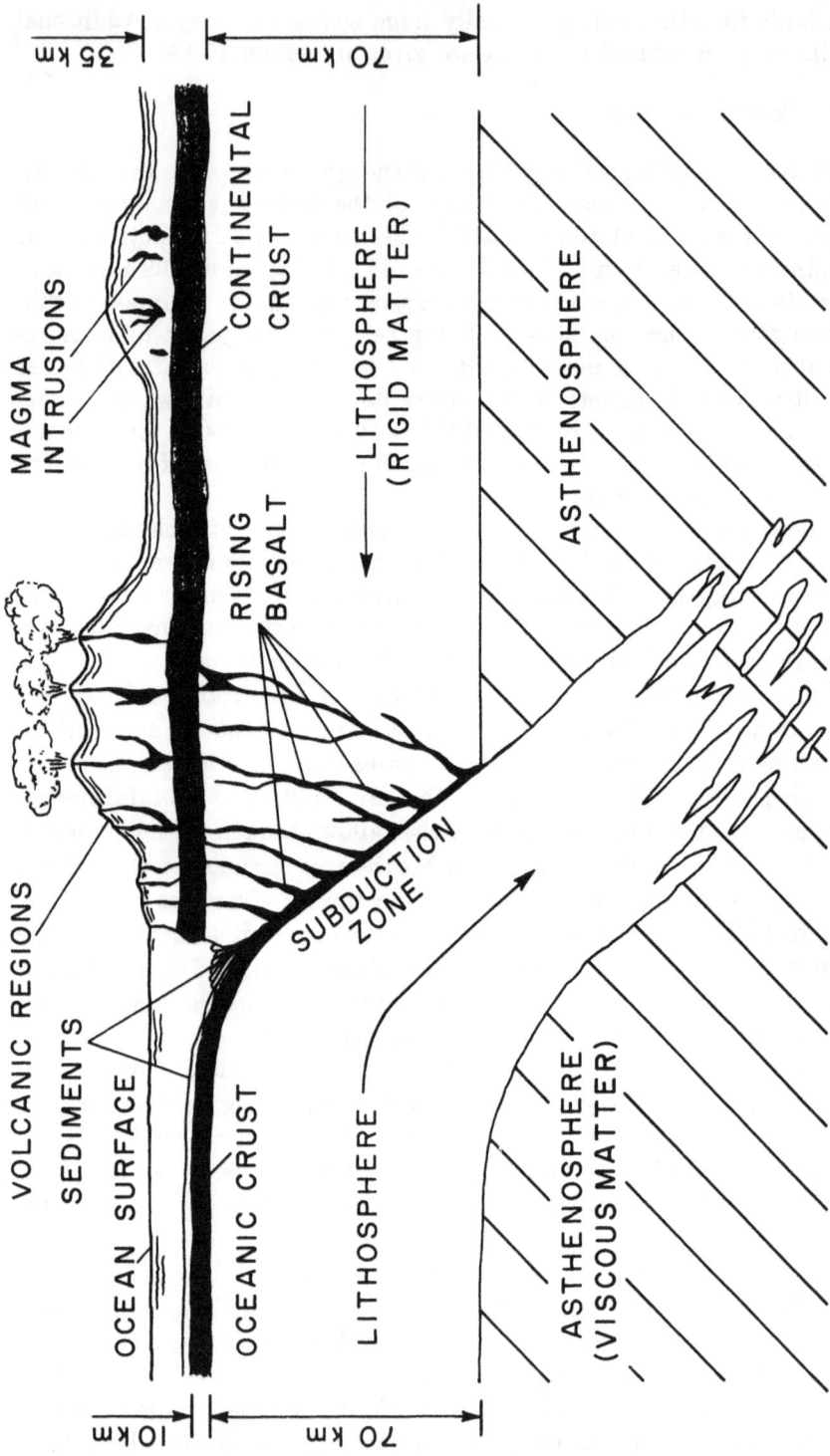

FIGURE 1.2—Subduction zone. Diving plate forms zone of partial melting, giving rise to molten basalt. Young regions are highly volcanic; other regions may contain magma intrusions.

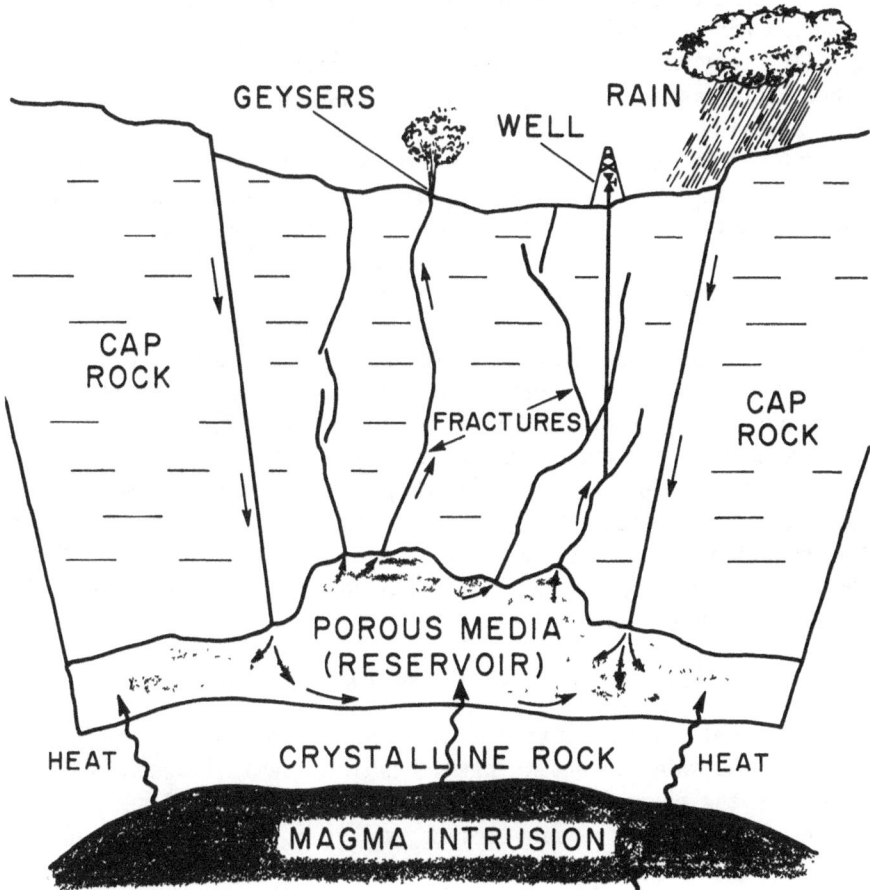

FIGURE 1.3—Hydrothermal geothermal reservoir. Simplified representation showing major elements: magma intrusion, porous layer, impermeable cap rock, fractures, and recharge mechanism.

surveys, since there are times when indications from a particular technique may be misleading [Meidav and Tonani, 1975].

In the end, it will be necessary to drill exploratory wells to determine the precise nature of the resource. Depending on the relative amount of steam and hot water produced at the surface, hydrothermal systems are usually classified as follows: (1) dry steam (saturated or slightly superheated water vapor), (2) vapor-dominated (high percentage of water vapor with relatively small amounts of liquid), or (3) liquid-dominated (high percentage of liquid with relatively little steam).

The fluids produced from hydrothermal reservoirs are complex and contain a large number of dissolved minerals and gases. These "waters" have been classified as: (1) alkali chloride, (2) acid sulfate, (3) acid sulfate-chloride, and (4) bicarbonate. Some of the gases that may be released from the fluid when the pressure is lowered include carbon dioxide (CO_2), hydrogen sulfide (H_2S), methane (CH_4), hydrogen (H_2), nitrogen (N_2), oxygen (O_2), ammonia (NH_3), argon (Ar), radon (Rn), and other noble gases such as helium (He), neon (Ne), krypton (Kr), and xenon (Xe) [Ellis and Mahon, 1977].

Exploitation of hydrothermal resources is hampered by the corrosive and/or scaling properties of geothermal "brines." In particular silica and calcium tend to precipitate in wells and pipes, forming plugs that restrict flow. It is not uncommon for geofluids to contain 10,000–25,000 ppm of dissolved solids, and concentrations as large as 250,000–300,000 ppm are found in the brines of the Salton Sea geothermal area in the Imperial Valley of southern California. For comparison, sea water contains about 30,000 ppm. A concerted effort is being made to solve the problems of processing such hostile geothermal fluids [Austin et al., 1977].

More than sixty countries possess hydrothermal geothermal resources. Those resources being used for electric power generation are some of the hottest and cleanest now known, but current developments are leading to the technology to exploit even some of the most hostile resources. These developments will be covered in detail in the individual chapters of this book.

1.2   Historical evolution of geothermal power technology

Geothermal energy was known two thousand years ago to the Romans who used the waters of hot springs to heat their baths [Goguel, 1976] and bathed in the natural hot springs on the banks of the Danube River [Stone, 1978]. Direct heating uses of geothermal energy have a long history. Towns such as Chaudes Aygues, Dax, and Ax-les-Thermes in France have been distributing geothermal water at 80°C (176°F) for domestic purposes since the Middle Ages [Goguel, 1976]. In the 1700's, a number of balneological institutes were founded in Hungary to capitalize on the therapeutic virtues of geothermal waters, and some of them are still in operation today [Stone, 1978]. The commercial potential of the mineral-laden geothermal waters in the Larderello region of Italy led to wars between the Tuscan republics during the Middle Ages [ENEL, 1970].

It was at Larderello in 1904 that Prince Piero Ginori Conti first harnessed the power of natural geothermal steam to produce electricity [Conti, 1924]. Conti's system used a reciprocating engine to receive steam that had been separated from the water. The engine was of the noncondensing type, exhausting to the atmosphere, and produced about 15 kW of electricity from a DC generator to provide lighting for the boric acid factory at Larderello. The original engine was replaced by a turboalternator of 250-kW capacity in 1913, thus marking the beginning of the production of electricity from geothermal energy on a commercial scale.

In the United States, a major dry steam geothermal resource was discovered in 1847 in northern California. Known as The Big Geysers, this thermal anomaly was exploited as a hot-springs resort until the mid-1920's. The Geysers, as it is known today, was then used to power a 250-kW electric generator [Siegfried, 1925]. This tiny plant was not competitive with other sources of power, and soon fell into disuse. It was not until 1960 that a large-scale commercial geothermal power plant was built and operated at The Geysers (see section 12.1).

The geothermal fields at Larderello and The Geysers are dry steam resources, and as such are relatively easy to exploit as a source of electric power. Most geothermal reservoirs, however, are not as amendable to use. The vast majority of them are liquid dominated, i.e., they yield a mixture of steam and hot water at the wellhead.

A liquid-dominated geothermal resource on the Japanese island of Kyushu was tapped in 1925 to produce 1.12 kW. This very small unit operated off the Tsunumi geothermal well in the famous hot-springs city of Beppu in Oita Prefecture, and was followed in 1951 by a 30-kW test plant, the Hakuryu geothermal power station, also in Beppu. This plant was operated by the Kogyo Gijutsu Institute of the Japanese Ministry of International Trade and Industry (MITI), but was eventually abandoned.

Exploitation of liquid-dominated hydrothermal reservoirs did not take place on a large scale until the Wairakei plant was built in New Zealand in 1958 with an installed capacity of 192,600 kW. Since that time many other countries have begun to use their geothermal fields for electric power generation, including the United States, Mexico, the Soviet Union, Japan, El Salvador, the People's Republic of China, the Philippines, and Iceland. Furthermore, several countries are on the verge of developing their geothermal resources. The geothermal power plants of these countries will be discussed in the remaining chapters of this book.

1.3 Energy conversion systems

A variety of energy conversion systems are in use for geothermal power generation throughout the world, and several new systems are in the research and development stage. Present-day geothermal power plants may be categorized as follows:

    Dry (or superheated) steam plants
    Separated-steam (or "single-flash") plants
    Separated-steam/hot-water-flash (or "double-flash") plants
    Separated-steam/multiple-flash (or "multiflash") plants
    Single-flash steam plants with pumped wells
    Double-flash steam plants with pumped wells
    Binary cycle plants with a secondary working fluid
    Combined flash/binary plants
    Hybrid fossil/geothermal plants
    Total flow systems.

Plants of the first four types are in commercial operation. Flash plants with pumped wells and binary plants are either under construction or in the pilot plant stage. Combined cycle, hybrid, and total flow plants are in the conceptual design phase and are being considered seriously for certain applications. In the following subsections we will present simple illustrative flow diagrams and describe these systems briefly.

1.3.1 Dry (or superheated) steam plants

Of all types of geothermal resources, the simplest to exploit for electrical power production is dry steam. Power plants at The Geysers in California; Larderello, Monte Amiata, and Travale in Italy; and Matsukawa in Japan operate with direct dry steam. Figure 1.4 is a highly simplified flow diagram for such a system. The sketch shows a condensing turbine with a direct-contact condenser and a mechanically-induced-draft cooling tower. All Italian geopower plants and the plant at Matsukawa use natural-draft cooling towers, and some of the Italian plants use noncondensing, exhausting-to-atmosphere turbines. When it is necessary to segregate the noncondensable gases in the geothermal steam from the cooling water (e.g., for environmental protection), then shell-and-tube type surface condensers are usually employed. State 1 is either dry saturated vapor or slightly superheated steam; state 2 is wet steam with a dryness fraction of about 0.90 (i.e., 90% vapor and 10% liquid by weight).

As can be seen from figure 1.4, the working fluid does not undergo a cycle in the usual sense of the word. It is admitted to the turbine at 1, condensed at 3, and either evaporated from the stack of the cooling tower or reinjected into the reservoir. Thus the usual definition of power plant

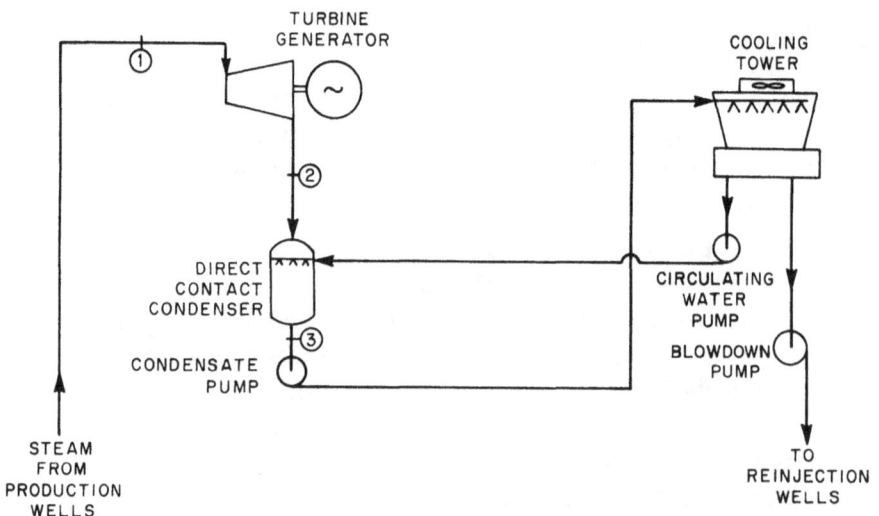

FIGURE 1.4—Dry-steam geothermal power plant.

cycle efficiency is not appropriate for assessing the overall thermodynamic performance of this type of geothermal plant. In that definition, the efficiency was given as

$$\eta = \dot{W}_{net}/\dot{Q}, \tag{1.1}$$

where $\dot{W}_{net}$ is the net plant output (turbine output minus all pump work), and $\dot{Q}$ is the rate at which heat is supplied to the working fluid. Since $\dot{Q}$ in this case is produced geothermally and not by the combustion of fuel, the determination of its value becomes ambiguous. Furthermore, equation 1.1 is valid only for cyclic operations.

The appropriate measure of plant performance in the thermodynamic sense is the geothermal resource utilization efficiency, η_u, which compares the net output of the plant to the maximum theoretically obtainable output, i.e.,

$$\eta_u = \dot{W}_{net}/\dot{\mathcal{E}}, \tag{1.2}$$

where $\dot{\mathcal{E}}$ is called the exergy and is defined as the difference between the thermodynamic availability of the entering working fluid at state 1 and that of the working fluid at the ambient, sink condition or the dead state 0 [Kestin, 1978]. It can be shown that the exergy is given by

$$\dot{\mathcal{E}} = \dot{m}[h_1 - h_0 - T_0(s_1 - s_0)], \tag{1.3}$$

where 　　$\dot{m}$ = mass flow rate of geothermal fluid;
　　　　h_0, h_1 = enthalpy of geofluid at states 0, 1;
　　　　s_0, s_1 = entropy of geofluid at states 0, 1;
　　　　T_0 = absolute temperature at the dead state.

Sometimes it is convenient to speak about the specific fluid consumption, SFC, of the plant:

$$SFC = \dot{m}/\dot{W}_{net}. \tag{1.4}$$

The figures of merit, η_u and SFC, can be used to assess the thermodynamic performance of any type of geothermal power plant, not only dry steam plants. Well-designed dry steam geothermal power plants with condensing turbines operate with utilization efficiencies of between 50% and 60%, and with steam consumptions of about 18 lbm/kW·h (8 kg/kW · h).

1.3.2 Separated-steam (or "single-flash") plants

It is common in the case of unpumped geothermal wells for the well-head product to consist of a two-phase mixture of liquid and vapor. The quality of the mixture (i.e., the mass fraction of the vapor phase) depends on reservoir properties and wellhead pressure. It is not difficult to separate the phases, say, at each wellhead, at centrally located stations, or at the powerhouse. Plants using a single stage of steam separation are called separated-steam plants.

In all likelihood the fluid condition in the reservoir is that of a compressed liquid at elevated temperature. As the fluid comes to the surface under a falling pressure, it flashes into steam and attains a wellhead quality ranging from about 10% to 50% for individual wells. Because of the flashing process, such plants are often called flash-steam or single-flash steam plants, even though the flashing takes place in the well or in the reservoir formation. In figure 1.5, the plant equipment is essentially

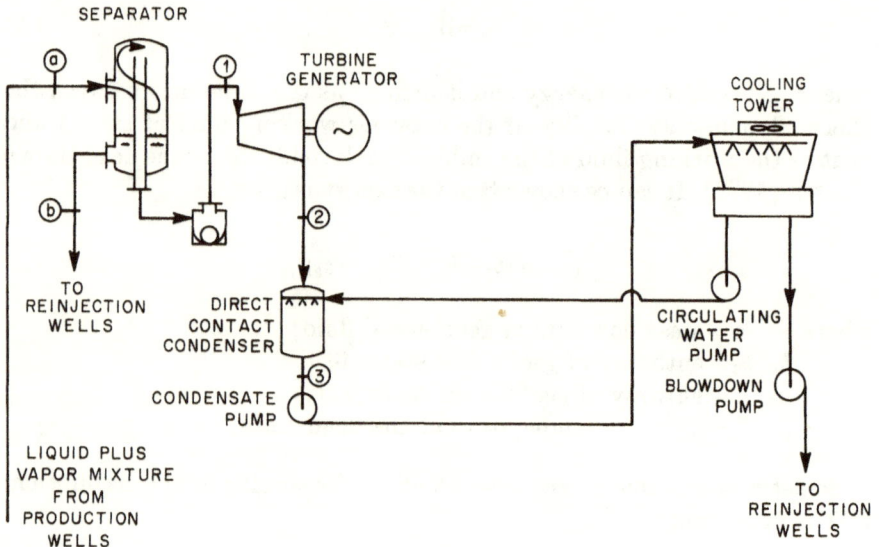

FIGURE 1.5—Separated-steam or "single-flash" geothermal power plant.

the same as for the dry steam system. The differences include the addition of the separator and a float-ball check valve to prevent massive ingestion of liquid by the turbine in the event of a backup in the separator.

Utilization efficiency for this type of plant is inferior to that of the dry steam plant because a portion of the geofluid mixture, the hot water fraction, is discarded from the plant (state b) without being used. Examples of this type of power plant are found in Mexico (Cerro Prieto), Japan (Otake, Onuma, Onikobe, Kakkonda), El Salvador (Ahuachapán Units 1 and 2), and the Soviet Union (Pauzhetka), to name a few.

1.3.3 Separated-steam/hot-water-flash (or "double-flash") plants

The flow diagram in figure 1.6 is similar to that in figure 1.5, except that a flash vessel is included to generate additional steam from the hot

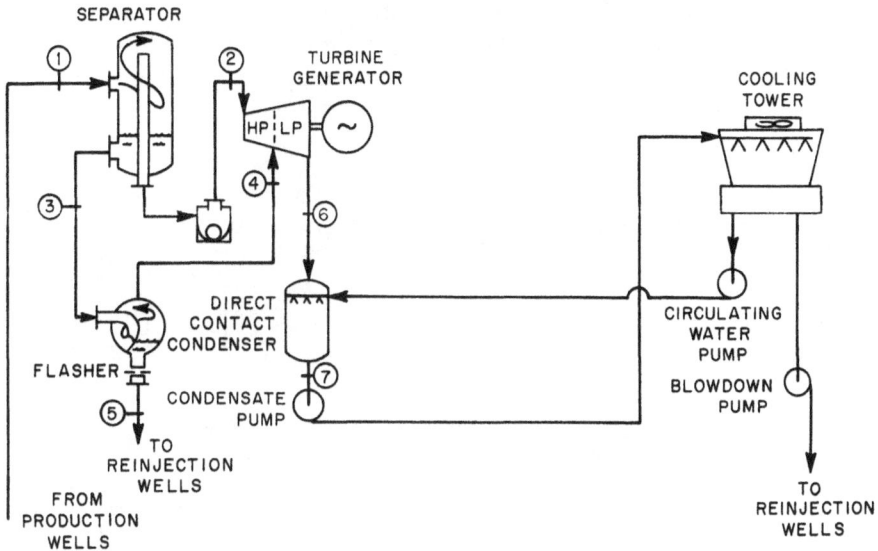

FIGURE 1.6—Separated-steam/hot-water-flash or "double-flash" geothermal power plant.

water separated from the wellhead mixture. The added steam at state 4 is admitted to the turbine via a plenum, where it mixes with the primary steam before expanding through the low-pressure stages. The term "double-flash" arises from the fact that two flashes occur, one below the surface and one above the surface in the specially designed flash tank. The main point of interest is that a two-stage process of steam generation and utilization is employed, the second stage capturing a portion of the energy otherwise wasted in a one-stage system. Plants of this type are operating in Japan (Hatchobaru) and Iceland (Krafla). An ambitious effort is underway to construct a large number of these units in the Philippines.

1.3.4 Separated-steam/multiple-flash (or "multiflash") plants

A multiflash plant employs steam at three or more levels of pressure at turbine entry points. The only known commercial plant of this type is in New Zealand (Wairakei). Multiflash plants are not economically attractive at the present time. The arrangement of the energy conversion system at Wairakei (see section 8.2) resulted from the requirements of an auxiliary chemical processing plant that was to have used a portion of the geothermal steam from the power station.

1.3.5 Single-flash plants with pumped wells

It may be desirable to place pumps down inside the production wells in order to increase the flow rate of the geothermal fluid, or to prevent flashing which may lead to precipitation and clogging of the wells. The pressurized fluid could then be flashed in surface equipment, as shown in figure 1.7, to generate steam for use in the turbine. A plant of this type is being

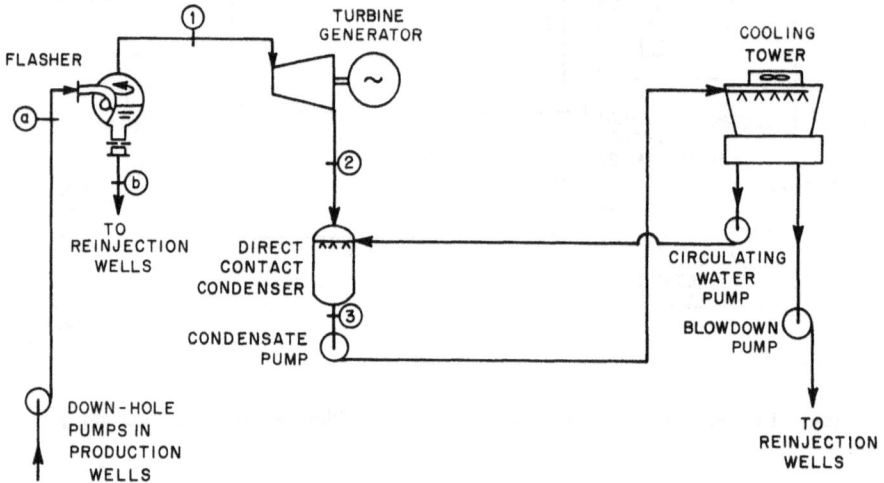

FIGURE 1.7—Single-flash geothermal power plant with pumped wells.

built at the East Mesa geothermal area in the Imperial Valley of California by Republic Geothermal, Inc. (see section 12.4).

1.3.6 Double-flash plants with pumped wells

If an additional flash vessel is employed to generate secondary steam from the liquid at state b shown in figure 1.7, the resulting double-flash plant will be more efficient than a single-flash plant. Either a dual-admission turbine, such as the one shown in figure 1.6, or two separate tandem-compound turbines may be used. The second stage of the Republic Geothermal plant at East Mesa will be of the tandem-compound type.

1.3.7 Binary cycle plants

These systems are called binary cycles because a secondary working fluid, typically a fluorocarbon or hydrocarbon, is used in a Rankine cycle with the geothermal fluid serving as the source for the required thermal energy to vaporize and superheat the secondary working fluid. A flow diagram for a simple binary plant is shown in figure 1.8, and the correspond-

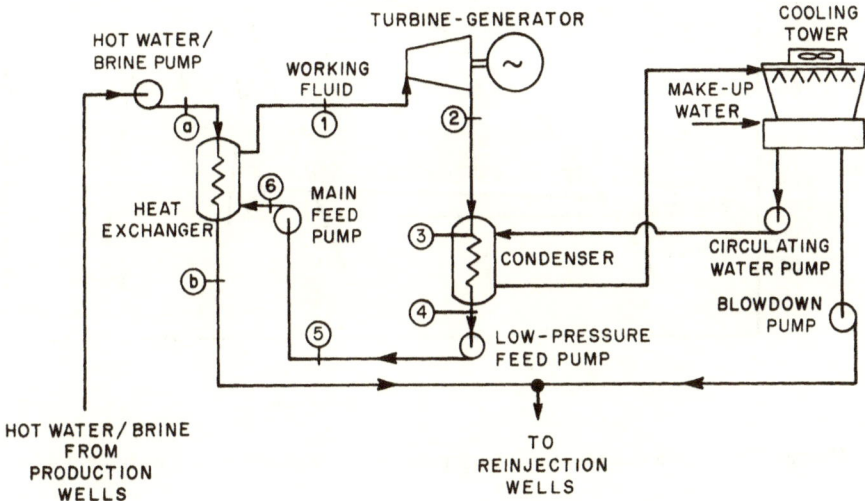

FIGURE 1.8—Binary geothermal power plant.

ing cycle diagram in pressure-enthalpy coordinates is given in figure 1.9. The latter is drawn for isobutane as the working fluid in the power loop with a supercritical boiler pressure (process 6 to 1).

A number of advantages have been claimed [Anderson, 1973] for a binary cycle compared to a flash-steam system, including the following:

> More suited to low-temperature hydrothermal resources
> Smaller turbine size for given output
> Less expensive turbine for given output
> High-pressure operation throughout, eliminating vacuum operation
> No problems of air in-leakage, etc.
> Non-corrosive working fluid in turbine
> Higher isentropic turbine efficiencies
> Completely dry expansion, eliminating erosion problems
> Condensing temperatures can be lower for better cycle efficiency.

Some of the disadvantages include:

> Suitable secondary working fluids are expensive
> No leaks can be permitted

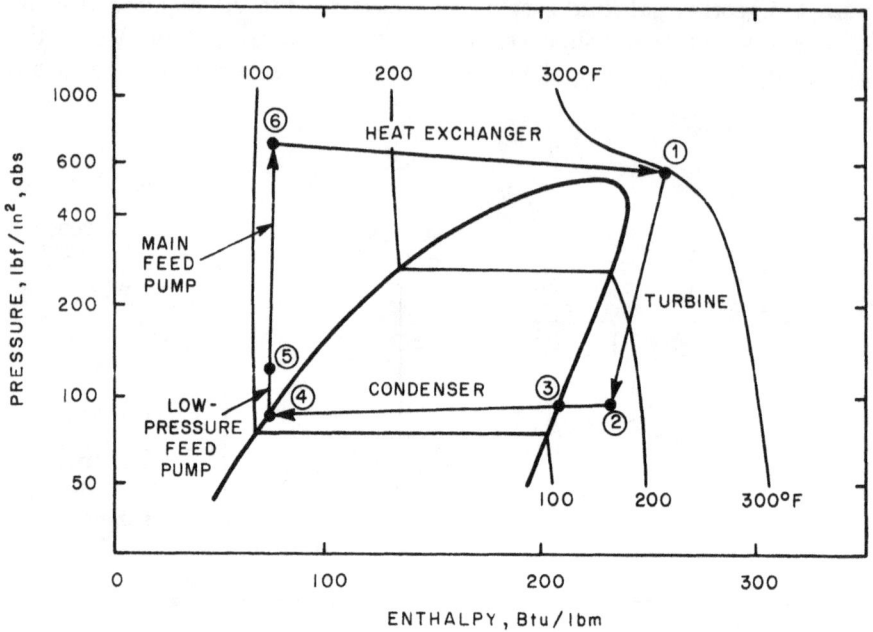

FIGURE 1.9—Cycle diagram for binary plant with isobutane as the working fluid.

  Heat exchangers are major elements and are costly
  Huge brine flow rates are needed for a reasonable-size plant, leading
    to disposal problems
  Flammability of hydrocarbon working fluids requires fire protection
    design.

A test plant using Refrigerant-12 was built and operated at Paratunka on the Kamchatka Peninsula of the Soviet Union. The plant was constructed in 1967, ran for a number of years, and has since been dismantled. The Japanese tested two plants of the binary type, one at Otake and one at Mori, and one plant is operating at East Mesa in the Imperial Valley of California. Each will be described in detail in later sections of this book.

It is interesting to note that the first commercial geothermal power plant in Italy used endogenous steam as a heating medium to boil pure water in a secondary power loop. In this way the direct, corrosive steam did not come in contact with the turbine, but was used instead in vertical-tube heat exchangers. In modern terminology, such systems might be called "endogenous-steam/pure-water binary cycles." Eventually, as turbine materials technology improved and troubles with heat exchanger performance arose, the binary cycle was abandoned in Italy in favor of reaction turbines using direct, natural steam [Hahn, 1923].

1.3.8   Combined flash/binary plants

One of many possible combined flash and binary cycle configurations is shown in figure 1.10, which indicates only the major components of the

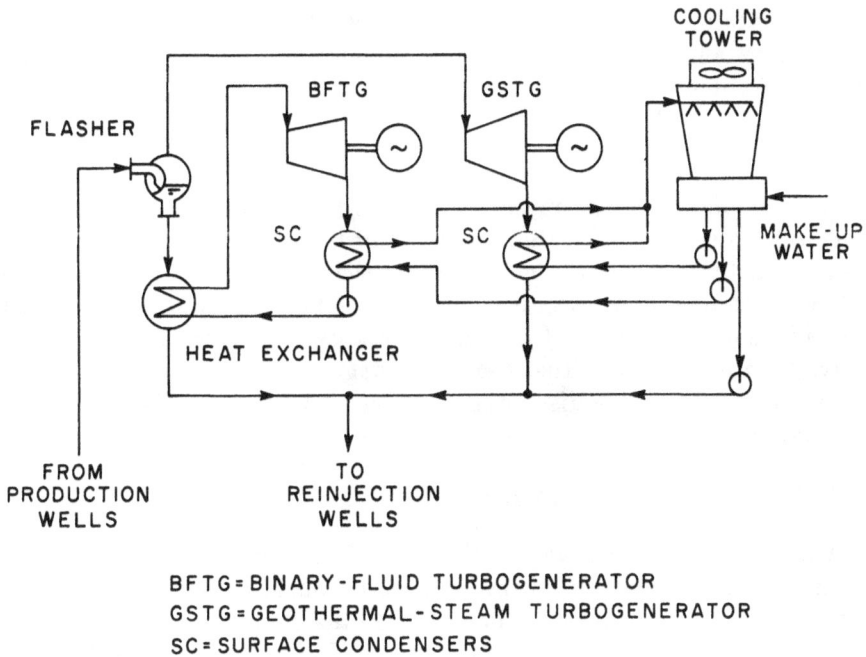

BFTG = BINARY-FLUID TURBOGENERATOR
GSTG = GEOTHERMAL-STEAM TURBOGENERATOR
SC = SURFACE CONDENSERS

FIGURE 1.10—Combined flash-steam and binary-fluid geothermal power plant.

plant. Other conceivable systems include those with more than one stage of flashing for the geofluid, and those in which the flashed steam as well as the liquid is used for heating the secondary working fluid instead of being employed in a steam turbine. Variations on these combined cycles have been proposed from time to time in order to suit particular resources, such as the hypersaline brines of the Salton Sea [SDG&E, 1977] or geofluids burdened with excessive amounts of noncondensable gases [Eskesen, 1977]. So far no combined cycles of this general type have been built for commercial power generation.

1.3.9   Hybrid fossil/geothermal plants

A hybrid fossil/geothermal power plant combines a fossil-energy fuel with a geothermal resource in a single plant in such a way as to take advantage of the synergistic possibilities of this configuration. Several different types of hybrid plants have been proposed and thermodynamically analyzed by the group at Brown University [Kestin et al., 1978; Khalifa et al., 1978a]. Generally such plants fall into one of two categories: geothermal-preheat systems [Khalifa et al., 1978b] or fossil-superheat systems [DiPippo et al., 1978]. Compound hybrid systems combine the features of these two systems, resulting in improved utilization and more flexibility [DiPippo and Avelar, 1979]. These are mentioned in section 12.9.2 and figures 12.24–12.26 in connection with their possible use at any of a number of resources in the western United States.

1.3.10   Total flow systems

The simplest energy conversion concept for geothermal application is the total flow system. In this case the full two-phase flow from the well is admitted into the expander. The theoretical advantage derives from the elimination of the losses associated with the flashing or separation processes in more conventional geothermal plants. A number of such devices have been developed at the laboratory stage, including expanders of the axial-flow and positive-displacement types [Austin et al., 1977], and field tests are being conducted for the latter type.

1.4   Power plant performance factors

One of the most outstanding features of the geothermal power plants that have been in operation throughout the world is their reliability. As each of the power stations is described in the chapters that follow, such terms as capacity factor, load factor, availability factor, and others will be used. Since there are sometimes differences in the basic definitions of these terms among power plant people, we will give our defining equations for these terms at the outset.

1.4.1   Operating factors

Several factors are commonly used to describe the operations and reliability of a power plant. There are three power plant operating factors which, when taken together, indicate the manner in which the plant is used to meet variations in demand. These are:

$$\text{Load factor, } F_L: \qquad F_L = \overline{L}/L^*, \qquad\qquad (1.5)$$

$$\text{Capacity factor, } F_C: \qquad F_C = \overline{L}/C, \qquad\qquad (1.6)$$

$$\text{Utilization factor, } F_U: \qquad F_U = L^*/C, \qquad\qquad (1.7)$$

where the terms used have the following meanings:
$\overline{L}$ = average load for a given period,
L^* = peak load for a given period, and
C = rated capacity of the plant or unit.

The average load $\bar{L}$ is found from

$$\bar{L}=G/h, \tag{1.8}$$

where $G=$total electrical generation for a given period, and
$\quad$ $h=$number of hours in the period (usually 8760 h$=1$ year).
It is clear that

$$F_\text{C}=F_\text{L}\times F_\text{U}. \tag{1.9}$$

and that whenever the utilization factor equals unity (i.e., pure base-load operation), then the load factor becomes identical numerically to the capacity factor.

1.4.2 Reliability factors

There are three commonly used power plant reliability factors:

$$\text{Availability factor, } F_\text{A}: \qquad F_\text{A}=P_\text{S}/h, \tag{1.10}$$

$$\text{Forced outage factor, } F_\text{FO}: \qquad F_\text{FO}=P_\text{FO}/h, \tag{1.11}$$

$$\text{Scheduled outage factor, } F_\text{SO}: \; F_\text{SO}=P_\text{SO}/h, \tag{1.12}$$

where
$\quad$ $P_\text{S}=$service period, i.e., the number of hours that the unit operated with breakers closed to the station bus,
$\quad$ $P_\text{FO}=$forced outage period, i.e., hours of downtime caused by equipment failure or malfunction, and
$\quad$ $P_\text{SO}=$scheduled outage period, i.e., hours of downtime for planned maintenance.

It is clear that

$$P_\text{S}=h-P_\text{FO}-P_\text{SO} \tag{1.13}$$

The material presented in the following chapters will show geothermal power plants are characterized generally by high capacity factors, typically greater than 80%, and very high availability factors, often exceeding 95%.

1.5 Summary of present and planned geothermal power plants

The total worldwide installed geothermal electric power capacity is about 1750 MW (as of July 1979), of which about 1100 MW is from dry steam plants. The number of individual power units is 85, for an average capacity of about 20 MW per unit. This is indicative of the generally small size of a typical geothermal plant. The largest single unit is rated at 135 MW and is scheduled to be installed at The Geysers in northern California during 1980.

Table 1.1 is a capsule summary of the active geothermal power capacity in the world as of July 1979. A detailed breakdown of the various power plants in each country is given in table 1.2. Table 1.3 summarizes actual installed capacity, capacity under construction or in planning for 1982, and projected capacities for the near-term beyond 1982. About 3600 MW will likely be delivered by geothermal power plants by 1982, and for the near-term future beyond 1982 almost 6000 MW may be generated from geothermal resources around the world. Note that these projections do not include the full, proven reserves of The Geysers or the assessed potential of the Imperial Valley in California. It may be expected that geothermal power will continue to grow throughout the rest of the twentieth century and beyond the year 2000.

Table 1.1—*Worldwide geothermal electric power capacity in operation as of July 1979*

	No. of units in operation	Installed capacity, MW	Future capacity, MW [a]
China	1	1	([b])
El Salvador	2	60	35
Iceland	3	32	30
Italy	37	420. 6	([b])
Japan	6	165	55
Mexico	4	150	30
New Zealand	14	202. 6	([b])
Philippines	3	59. 2	710
Turkey	1	0. 5	([b])
U.S.S.R.	1	5	([b])
United States	13	663	967. 55
Total	85	1758. 9	1827. 55

[a] Under construction or in planning for 1982. Additional capacity may come from countries not presently using geothermal energy for electricity, such as Costa Rica, Kenya, Nicaragua, and others (see table 1.3).
[b] Estimates not available.

Table 1.2—*Geothermal power units in operation or in planning*

Country/Plant Site	Year of installation	Capacity, MW
Azores		
São Miguel	1979	3. 0
Chile		
El Tatio	Future	15. 0
China		
Yang-pa-ching	1978	1. 0
Costa Rica		
Miravalles	Future	40. 0

Table 1.2—*Geothermal power units in operation or in planning*—Continued

Country/Plant Site	Year of installation	Capacity, MW
El Salvador		
Ahuachapán:		
Auxiliary Unit..........................	1975	1. 1
Unit 1.................................	1975	30. 0
Unit 2.................................	1976	30. 0
Unit 3.................................	1980	35. 0
Berlín...................................	Future	100. 0
Chinameca.	Future	100. 0
Chipilapa...............................	Future	50. 0
San Vicente.............................	Future	100. 0
Guatemala		
Amatitlán...............................	Future	50. 0
Honduras		
Pavana or San Ignacio....................	Future	50. 0
Iceland		
Námafjall (dismantled)...................	1969	3. 0
Krafla:		
Unit 1.................................	1978	30. 0
Unit 2.................................	Future	30. 0
Svartsengi (Grindavik)...................	1978	2. 0
Indonesia		
Kawah Kamojang.........................	1979	0. 25
Kawah Kamojang.........................	Future	100. 0
Italy		
Larderello 2 (5 units).....................	(ᵃ)	69. 0
Larderello 3 (6 units).....................	1969	120. 0
Gabbro..................................	1969	15. 0
Castelnuovo-V.C. (4 units)...............	(ᵃ)	50. 0
Serrazzano (5 units)......................	(ᵃ)	47. 0
Lago 2 (3 units)..........................	(ᵃ)	33. 5
Sasso Pisano (2 units)....................	(ᵃ)	15. 7
Monterotondo...........................	(ᵃ)	12. 5
Sant'Ippolito-Vallonsordo................	1972	0. 9
Lagoni Rossi 1...........................	1962	3. 5
Lagoni Rossi 2...........................	(ᵃ)	3. 0
Sasso 1.................................	(ᵃ)	7. 0
Capriola................................	(ᵃ)	3. 0
Molinetto...............................	(ᵃ)	3. 5
Travale.................................	1973	15. 0
Bagnore 1...............................	1959	3. 5
Bagnore 2...............................	(ᵃ)	3. 5
Piancastagnaio..........................	1969	15. 0
(unspecified sites).......................	Future	400. 0
Japan		
Matsukawa..............................	1966	20. 0
Otake...................................	1967	10. 0
Onuma..................................	1973	10. 0

ᵃ Not available.

Table 1.2—*Geothermal power units in operation or in planning*—Continued

Country/Plant Site	Year of installation	Capacity, MW
Japan—Continued		
Onikobe......	1975	25. 0
Hatchobaru... ...	1977	50. 0
Kakkonda (Takinoue)......	1978	50. 0
Otake (pilot binary; dismantled)...	1978	1. 0
Mori (pilot binary; dismantled)...	1978	1. 0
Mori...	1979	55. 0
Kuzuneda.	Future	50. 0
Kenya		
Olkaria...	Future	15. 0
Mexico		
Cerro Prieto I:		
Unit 1...	1973	37. 5
Unit 2...	1973	37. 5
Unit 3.....	1979	37. 5
Unit 4...	1979	37. 5
Unit 5.....	1982	30. 0
Cerro Prieto II..	Future	110. 0
New Zealand		
Kawerau.....	1961	10. 0
Wairakei, Station A:		
Machine 1 (IP)...	1959	11. 2
Machine 2 (HP)...	1958	6. 5
Machine 3 (HP)...	1959	6. 5
Machine 4 (IP)...	1959	11. 2
Machine 5 (HP)...	1962	11. 2
Machine 6 (HP)...	1962	11. 2
Machine 7 (LP)...	1959	11. 2
Machine 8 (LP).	1959	11. 2
Machine 9 (LP)...	1960	11. 2
Machine 10 (LP)..	1960	11. 2
Wairakei, Station B:		
Machine 11 (MP).....	1962	30. 0
Machine 12 (MP)...	1963	30. 0
Machine 13 (MP).....	1963	30. 0
Ohaki (Broadlands):		
Unit 1......	1983	50. 0
Unit 2...	1984	50. 0
Unit 3...	Future	50. 0
Nicaragua		
Momotombo...	Future	30. 0
Panama		
Cerro Pando...	Future	(ᵃ)
Philippines		
Tongonan (Leyte):		
Portable Unit...	1977	3. 0
Unit 1...	1980	55. 0
Unit 2...	1981	55. 0

ᵃ Not available.

Table 1.2—*Geothermal power units in operation or in planning*—Continued

Country/Plant Site	Year of installation	Capacity, MW
Philippines—Continued		
Makiling Banahaw (Los Baños):		
Wellhead Unit............................	1977	1. 2
Unit 1....................................	1979	55. 0
Unit 2....................................	1979	55. 0
Unit 3....................................	1981	55. 0
Unit 4....................................	1982	55. 0
Tiwi:		
Unit 1....................................	1979	55. 0
Unit 2....................................	1979	55. 0
Unit 3....................................	1981	55. 0
Unit 4....................................	1981	55. 0
Southern Negros:		
Unit 1....................................	1981	55. 0
Unit 2....................................	1982	55. 0
Manat-Masara:		
Unit 1....................................	1981	55. 0
Unit 2....................................	1982	55. 0
All sites.................................	Future	555. 0
Turkey		
Kizildere:		
Wellhead Unit.............	1975	0. 5
Unit 1.............................	Future	14. 0
U.S.S.R.		
Paratunka (dismantled)................	1967	0. 7
Pauzhetka............................	1967	5. 0
Bolshoye-Bannoye.......	Future	8. 0
All sites....	Future	70. 0
United States		
The Geysers:		
PG&E Unit 1..........	1960	11. 0
PG&E Unit 2...............	1963	13. 0
PG&E Unit 3...............................	1967	27. 0
PG&E Unit 4........	1968	27. 0
PG&E Unit 5....	1971	53. 0
PG&E Unit 6............	1971	53. 0
PG&E Unit 7..............................	1972	53. 0
PG&E Unit 8........................	1972	53. 0
PG&E Unit 9............	1973	53. 0
PG&E Unit 10.........	1973	53. 0
PG&E Unit 11............................ ..	1976	106. 0
PG&E Unit 12.......................	1979	106. 0
PG&E Unit 13......................•......	1980	135. 0
PG&E Unit 14.........................	1980	110. 0
PG&E Unit 15.............................	1979	55. 0
PG&E Unit 16......	1983	110. 0
PG&E Unit 17....................	1982	110. 0
PG&E Unit 18....	1982	110. 0

Table 1.2—*Geothermal power units in operation or in planning*—Continued

Country/Plant Site	Year of installation	Capacity, MW
United States—Continued		
The Geysers—Continued		
PG&E Unit 19......	Future	110. 0
NCPA Unit 1.............................	1981	110. 0
East Mesa:		
Magmamax Dual Binary...............	1979	11. 2
Republic Geothermal........................	1980	48. 0
Brawley, So. Cal. Edison..............	1980	10. 0
Puna, HGP Unit.............................	1980	5. 0
Raft River, Binary Unit.....................	1980	3. 35
Valles Caldera, Baca No. 1..................	1982	50. 0
Salton Sea, SDG&E.......	1982	50. 0
Heber:		
So. Cal. Edison....	Future	45. 0
SDG&E.........................	Future	. 50. 0
Westmorland...............................	Future	50. 0
Roosevelt Hot Springs.......................	Future	52. 0
Desert Peak................................	Future	20. 0

Table 1.3—*Summary of actual, planned and projected geothermal power capacity in the world as of July 1979*

		MW
(1)	Total installed capacity........	1758. 9
(2)	Planned or under construction............	1720. 8
(3)	Future projections, near-term.......... ...	2474. 0
	Total (1) + (2)................	3589. 7
	Total (1) + (2) + (3).................	5953. 7

REFERENCES

NOTE: The United Nations has sponsored three worldwide conferences on new sources of energy, including geothermal. Papers from these conferences are cited throughout this book. The full citations for these conferences are listed below; hereafter a shorthand notation will be used.

Proceedings of the United Nations Conference on New Sources of Energy: Solar Energy, Wind Power and Geothermal Energy, Rome, Aug. 21–31, 1961, Vols. 2 and 3, "Geothermal Energy," United Nations, New York, 1964. (Referred to as *Rome, 1961* hereafter.)

Proceedings of the United Nations Symposium on the Development and Utilization of Geothermal Resources, Pisa, Sept. 22–Oct. 1, 1970; *Geothermics*, Spec. Issue 2,

Vols. 1 and 2, 1970, Pergamon Press, Inc., New York, 1970. (Referred to as *Pisa, 1970* hereafter.)

Proceedings of the Second United Nations Symposium on the Development and Use of Geothermal Resources, San Francisco, May 22–29, 1975, Vols. 1–3, U.S. Government Printing Office, Washington, 1976. (Referred to as *San Francisco, 1975* hereafter.)

Anderson, J. H., 1973. "The Vapor-Turbine Cycle for Geothermal Power Generation," in *Geothermal Energy: Resources, Production, Stimulation*, P. Kruger and C. Otte, eds., Stanford Univ. Press, Stanford, CA., pp. 163–175.

Austin, A. L., Lundberg, A. W., Owen, L. B., and Tardiff, G. E., 1977. "The LLL Geothermal Energy Program Status Report, January 1976–January 1977," UCRL-50046-76, Lawrence Livermore Laboratory, Livermore, CA.

Boldizsár, T., 1970. "Geothermal Energy Production from Porous Sediments in Hungary," *Pisa, 1970*, Vol. 2, pp. 99–109.

Bowen, R. G. and Groh, E. A., 1977. "Geothermal Energy," in *Energy Technology Handbook*, D. M. Considine, ed., McGraw-Hill, New York, pp. 7.4–7.13.

Bullard, Sir E., 1973. "Basic Theories," in *Geothermal Energy: Review of Research and Development*, H. C. H. Armstead, ed., UNESCO, Paris, pp. 19–29.

Combs, J. and Muffler, L. J. P., 1973. "Exploration of Geothermal Resources," in *Geothermal Energy*, P. Kruger and C. Otte, eds., Stanford Univ. Press, Stanford, CA, pp. 93–128.

Conti, Prince P. G., 1924. "The Larderello Natural Steam Power Plant," *Proc. First World Power Conf.*, June 30–July 12, London.

DiPippo, R. and Avelar, E. M., 1979. "Compound Hybrid Geothermal-Fossil Power Plants", *Geothermal Resources Council Trans.*, Vol. 3, pp. 165–168.

DiPippo, R., Khalifa, H. E., Correia, R. J., and Kestin, J., 1978. "Fossil Superheating in Geothermal Steam Power Plants," *Proc. 13th Intersociety Energy Conv. Engin. Conf.*, Vol. 2, pp. 1095–1101.

Eaton, W. W., 1975. *Geothermal Energy*, ERDA-TIC, Oak Ridge, TN, p. 27.

Ellis, A. J. and Mahon, W. A. J., 1977. *Chemistry and Geothermal Systems*, Academic Press, New York.

ENEL, 1970, *Larderello and Monte Amiata: Electric Power by Endogenous Steam*, Ente Nazionale per l'Energia Elettrica, Compartimento di Firenze, Direzione Studi E Richerche, Roma. (In English.)

EPRI, 1977. "Geothermal Energy, The Hot Prospect," *EPRI Journal*, Vol. 2, No. 3, pp. 6–13.

Eskesen, J. H., 1977. "Cost and Performance Comparison of Flash Binary and Steam Turbine Cycles for the Imperial Valley, California," *Proc. 12th Intersociety Energy Conv. Engin. Conf.*, Vol. 1, pp. 842–849.

Goguel, J., 1976. *Geothermics*, S. P. Clark, Jr., ed., A. Rite, trans., McGraw-Hill, New York.

Hahn, E., 1923. "Some Unusual Steam Plants in Tuscany," *Power*, Vol. 57, No. 23, pp. 882–885.

Kestin, J., 1978. "Available Work in Geothermal Energy," Brown Univ. Rep. No. CATMEC/20, DOE No. COO/4051-25, Providence, RI.

Kestin, J., DiPippo, R., and Khalifa, H. E., 1978. "Hybrid Geothermal-Fossil Power Plants," *Mech. Engineering*, Vol. 100, No. 12, pp. 28–35.

Khalifa, H. E., DiPippo, R., and Kestin, J., 1978a. "Hybrid Fossil-Geothermal Power Plants," *Proc. 5th Energy Tech. Conf.*, pp. 960–970.

Khalifa, H. E., DiPippo, R., and Kestin, J., 1978b. "Geothermal Preheating in Fossil-Fired Steam Power Plants," *Proc. 13th Intersociety Energy Conv. Conf.*, Vol. 2, pp. 1068–1073.

Meidav, T. and Tonani, F., 1975. "A Critique of Geothermal Exploration Techniques," *San Francisco, 1975,* Vol. 2, pp. 1143–1154.

Pettitt, R. A., 1978. "Hot Dry Rock: A New Potential for Energy," *Geothermal Energy Magazine,* Vol. 6, No. 11, pp. 11–19.

SDG&E, 1977. "Expression of Interest for a Geothermal Demonstration Power Plant," presented to ERDA on June 20 by San Diego Gas and Electric Co., San Diego.

Siegfried, H. N., 1925. "The Geysers," in *Geothermal Exploration in the First Quarter Century,* D. N. Anderson and B. A. Hall, eds., Geothermal Resources Council Spec. Rep. No. 3, Davis, CA, 1973, pp. 59–88.

Stone, A. M., 1978. "Geothermal in Hungary," *Geothermal Energy Magazine,* Vol. 6, No. 11, pp. 27–28.

Wallace, R. H., Jr., Kraemer, T. F., Taylor, R. E., and Wesselman, J. B., 1978. "Assessment of Geopressured-Geothermal Resources in the Northern Gulf of Mexico Basin," in *Assessment of Geothermal Resources of the United States, 1978.* L. J. P. Muffler, ed., Geological Survey Circular 790, U.S. Dept. of Interior, pp. 132–155.

CHAPTER 2

CHINA

2.1 Overview

With the recent emphasis on education, science, and technology in the People's Republic of China, geothermal energy is expected to be developed for practical purposes. At least one geothermal electric power plant is in operation in China, but there are indications that the geothermal resources in the Himalayan mountain area will be developed rapidly for generation of electricity and for other beneficial purposes [Wu, 1978].

2.2 Yang-pa-ching

2.2.1 Geologic setting

The Yang-pa-ching geothermal power plant is located high in the Himalayan mountains of Tibet within a vast geothermal field. It is believed that this resource is only one of many such fields in China [Peking, 1978]. The Tibetan field is situated on a plateau about 4300 m (14,000 ft) above sea level, in the center of Yang-pa-ching Basin, west of the city of Lhasa. The basin is bounded on the south by the Himalayan Mountains and on the north by the Nyenchin Tangla Range, glacier-covered peaks rising to elevations of 5500–6000 m (18,000–19,700 ft). The basin extends for 2000 km (1250 mi) to the east and then turns toward the south at the Transverse Mountains, and extends to western Yunnan Province. Along this Himalayan geothermal belt are more than 400 sites of geothermal surface activity such as hot springs and fumaroles. Even rare phenomena such as geothermal water explosions and intermittent springs or geysers have been observed [Peking, 1978].

In November 1975 a colossal geothermal water explosion occurred at the Chupu area in Punan Prefecture in Tibet. A hugh volume of dark gray steam erupted to a height of 800–900 m (2600–2950 ft), leaving a crater of several tens of meters (~130–140 ft). This crater is still spewing hot water and steam. In the same area another crater caused by a geothermal eruption has a diameter of 80 m (260 ft).

The largest geyser found so far in China is the Da Chia Ger intermittent hot springs in Mao Nen Prefecture. This geyser shoots a hot water jet of

2 m (6.6 ft) diameter to a height of 20 m (66 ft) ; an accompanying steam jet shoots to 40–50 m (130–165 ft). Another well-known geothermal field lies east of the Himalayan in Ten-Chung, where more than 50 volcanic cones are in evidence. About 13 km (8 mi) from this city is a large geothermal field called Erh Hai, which means "hot sea" [Peking, 1978].

The geothermal resource at Yang-pa-ching is of the liquid-dominated type that produces a mixture of liquid and vapor at the wellhead as a result of flashing in the wellbore as the geothermal liquid rises from the reservoir to the surface. The site is favored by a ready supply of cooling water for the power plant owing to the presence of the Tsangpo River cutting across the region from southwest to northeast. The water temperature is quite low since the river is fed by melting snow from the surrounding mountain peaks.

2.2.2 Energy conversion system

The geothermal plant at Yang-pa-ching is of the separated-steam (or "single-flash") variety and has an installed capacity of 1000 kW. Constructed within three months by Chinese and Tibetan workers, engineers, and managers, the plant received special recognition for technical achievement by the Tibet Regional Committee of the Communist Party.

Figure 2.1 shows a simplified flow diagram for the plant. Geofluid from a single well is processed in a cyclonic separator which produces a stream of hot water that is led to a disposal area through a silencer, and a stream of vapor that is used to drive the turbogenerator. A barometric, direct-contact condenser is installed on the upper part of a hill and is supplied with cold water from the Tsangpo River. Noncondensable gases, which include carbon dioxide, hydrogen sulfide, oxygen, and nitrogen, are removed from the condenser by means of a high-pressure, water jet ejector [Wu, 1978].

Test runs of the plant indicate that power can be produced continuously without any adverse effects from the extremely high altitude of the site. The plant is now serving the inhabitants of the Yang-pa-ching Basin with electricity. No other technical specifications for the plant are available at this time.

2.3 Plans for the future

There are plans to construct geothermal power plants of the separated-steam/hot-water-flash (or "double-flash") variety that take advantage of additional (or secondary) steam released by flashing the hot water separated at the wellhead [Wu, 1978]. A number of small binary plants in the 50–200 kW range have been constructed and tested, but technical information on these is difficult to obtain [Finn, 1979]. Multipurpose utilization of geothermal energy including agricultural, industrial, and residen-

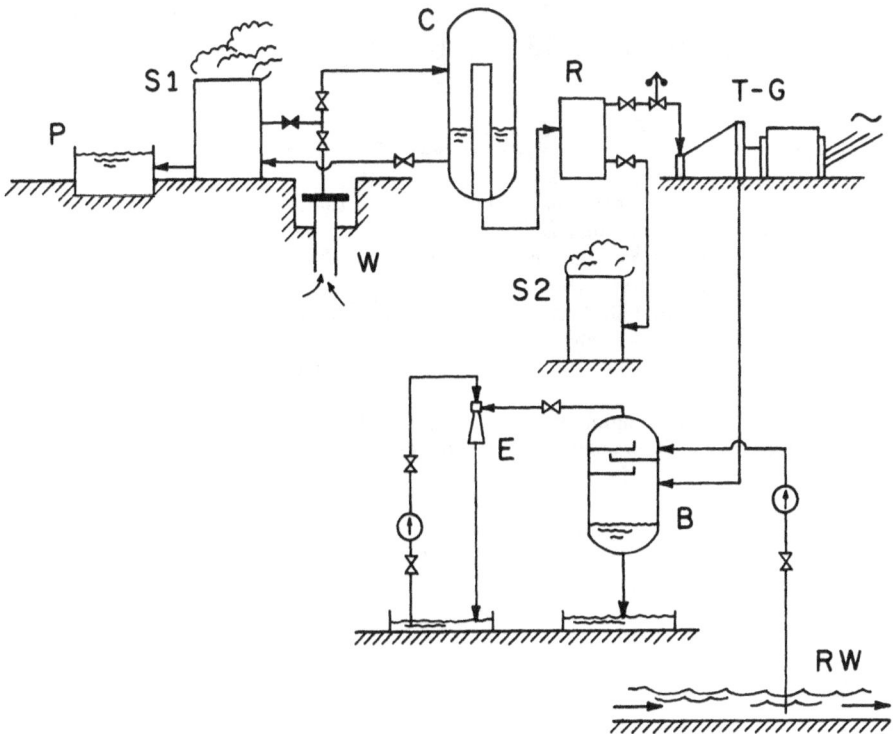

FIGURE 2.1—Flow diagram of 1000-kW Chinese geothermal power plant near Yang-pa-ching. Key: P=water-holding pond; S1=wellhead silencer; W=geothermal well; C=cyclonic separator; R=steam receiver; T-G=turbogenerator; S2=station silencer; B=barometric condenser; E=water jet ejector; RW=river water [after Wu, 1978].

tial space heating as well as electric power generation is also being contemplated.

Attention is being paid to possible problems associated with geothermal energy utilization such as pollution of streams and rivers by discharge of mineral-laden waste waters, corrosion caused by the discharge of H₂S to the atmosphere, and clogging of wells and pipelines from deposition of salts from the geothermal fluid. Reinjection and chemical treatment are being considered in this regard.

Since the Cultural Revolution, a number of geothermal projects have been launched throughout the provinces of China. A large-scale geothermal power plant is reportedly under construction at this time [Peking, 1978]. It has been estimated that the eastern, central, and southern parts of China also hold great geothermal potential and may eventually be put to practical use.

It seems reasonable to assume that more details about China's geothermal energy program will soon become available as a result of China's new foreign policy announced in early 1979.

REFERENCES

Finn, D. F. X., 1979. "Geothermal Developments in the People's Republic of China", *Geothermal Resources Council Trans.*, Vol. 3, pp. 209–210.

Peking, 1978. "Geothermal Resources in China," Geothermal Research Group, Geology and Geography Unit, University of Peking; Reported in the weekly *Chinese Industry News,* No. 439, May 29, pp. 6–7. (In Chinese.)

Wu, F. T., 1978. Yang-pa-ching Geothermal Experimental Power Plant," *Chinese Industry News,* No. 439, May 29, pp. 2–6. (In Chinese.)

NOTE: The author is indebted to Takao Sato (Brown University) and Dr. C. H. Chen (Southeastern Massachusetts University) for their assistance in preparing this chapter.

CHAPTER 3

EL SALVADOR

3.1 Overview

El Salvador is the first of the Central American countries to construct and operate a geothermal electric generating station. Exploration began in the mid-1960's at the geothermal field near Ahuachapán in western El Salvador. The first power unit, a separated-steam (or "single-flash") plant, was started up in June 1975 and was followed a year later by an identical unit. The 60 MW of geothermal capacity presently constitutes 14% of the total electric generating capacity of El Salvador, but during 1977 the Ahuachapán plant produced nearly one-third of the electricity generated in the country.

The Comisión Ejecutiva Hidroeléctrica del Río Lempa (CEL) is in the process of installing the third unit at Ahuachapán, a dual-pressure ("double-flash") unit that will be rated at 35 MW. In addition, CEL is actively pursuing several other promising sites for additional geothermal plants. Eventually geothermal energy could contribute as much as 450 MW of electric generating capacity. In any event, it appears that by 1985 El Salvador will be able to meet its domestic needs for electricity by means of its indigenous geothermal and hydroelectric power plants, thus eliminating any dependence on imported petroleum for power generation.

The map of El Salvador in figure 3.1 shows the location of the geothermal sites being explored, together with the existing power plants, both geothermal and conventional.

3.2 Ahuachapán

The Ahuachapán geothermal field is located in the western portion of El Salvador, 18 km (11 mi) from the Rio Paz, which forms a portion of the international boundary between El Salvador and Guatemala. The power plant is sited on moderately sloping terrain on the northern side of the coastal volcanic mountain chain that extends the length of the country. An aerial view of the power plant and the bore field is shown in figure 3.2.

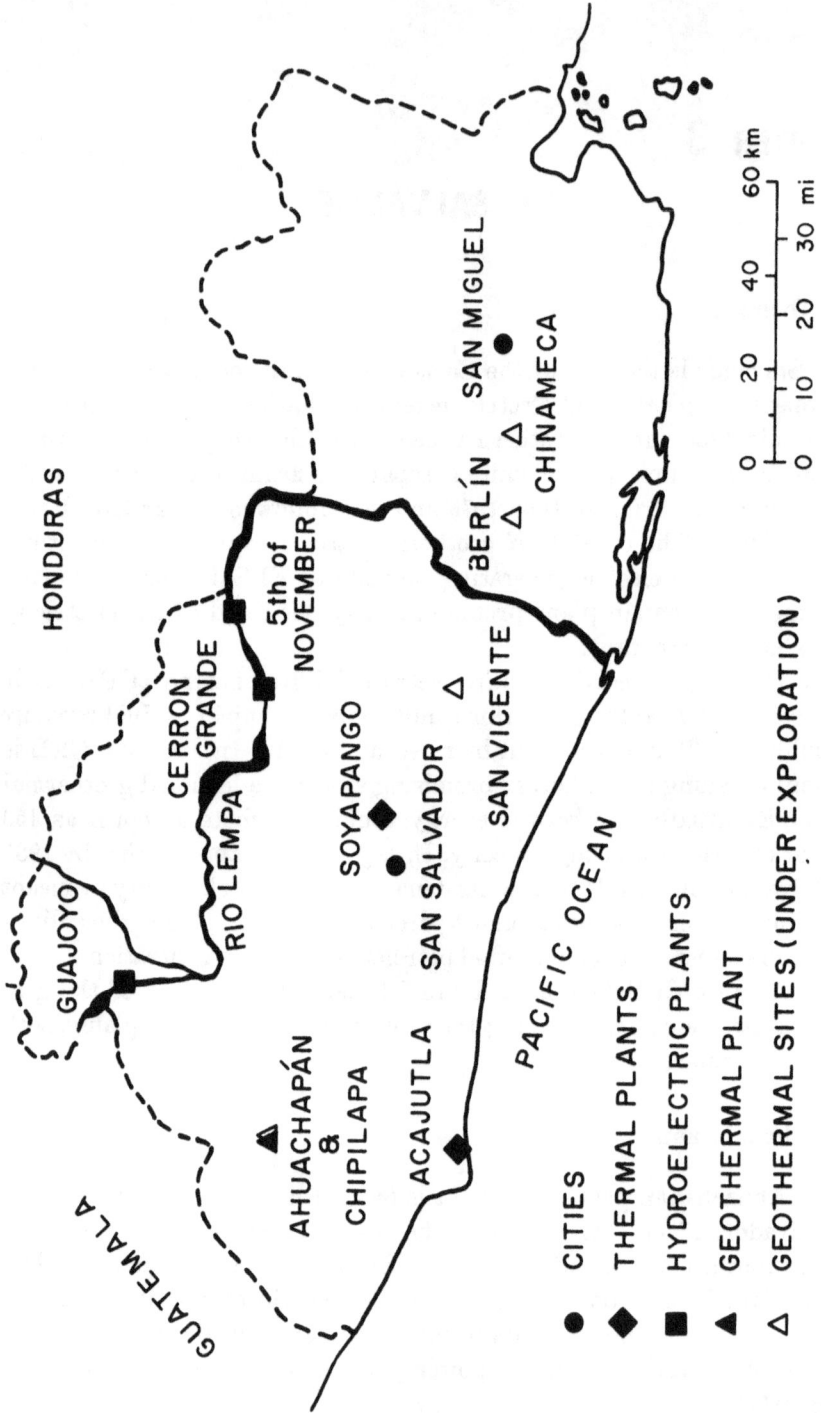

FIGURE 3.1—Map of El Salvador showing geothermal sites and existing power plants [after CEL, 1976].

FIGURE 3.2—The Ahuachapán 60-MW geothermal power plant and bore field [CEL, 1976].

3.2.1 Geology

Information gathered since 1965, when exploration and deep drilling
began at Ahuachapán, indicates that the geothermal formation consists
essentially of the following layers:

Brown tuff and pyroclastics (top 50 m)
Andesites (next 50 m)
Agglomerated tuff and pyroclastics (next 20–150 m)
Andesites (next 50–100 m)
Young agglomerates (next 100–250 m)
Ahuachapán andesites (next 10–300 m, absent in parts)
Ancient agglomerates (basement rock).

The young agglomerates serve as the reservoir cap, the Ahuachapán
andesites as the aquifer.

The lithology of the formation is not known precisely, but three faults
cut across the field trending north-northwest. It is believed that the aquifer
splits into two layers toward the eastern portion of the field, and that at the
western side the aquifer thins out to a contact surface about 500 m (1640
ft) below the surface [G. Cuéllar, personal communication].

3.2.2 Well programs and gathering system

At the time of this writing 28 wells have been completed, arranged as
shown in figure 3.3. The casing programs for typical production and reinjection wells are shown in figure 3.4. Since the formation is relatively hard,
it is not necessary to install a slotted production liner; for about half of
the wells an open hole is sufficient. For reinjection wells, the inner casing
is hung from the surface (not cemented) to allow for easy removal in the
event that the well may someday be used as a producer. Table 3.1 summarizes information on well completions. "Dual-purpose" wells such as
AH-8 and AH-17 are producing wells that may also be used for reinjection.

The drilling program for well AH-26 is given in figure 3.5. Completed
in 49 days to a depth of 804 m (2638 ft), the penetration rate averaged
about 2 m/h (6.6 ft/h). Drilling mud was used until the aquifer was
reached at about 400 m (1312 ft); drilling proceeded with water to the full
depth of the well.

The separation between wells is not less than about 150 m (490 ft). Over
the entire field the average spacing is roughly 23 ha (57 acres) per well,
although in the central portion of the field the wells are more densely
spaced, 11 ha (28 acres) per well. Figure 3.6 shows the layout of production and reinjection wells and the location of the powerhouse. The area to
the south of the plant is the site of numerous surface thermal manifestations such as steam vents, hot springs, boiling pools, and mud pools (see
figure 3.7). Pipelines shown schematically in figure 3.6 do not depict the
actual configurations, which include expansion bends (see figure 3.2). The

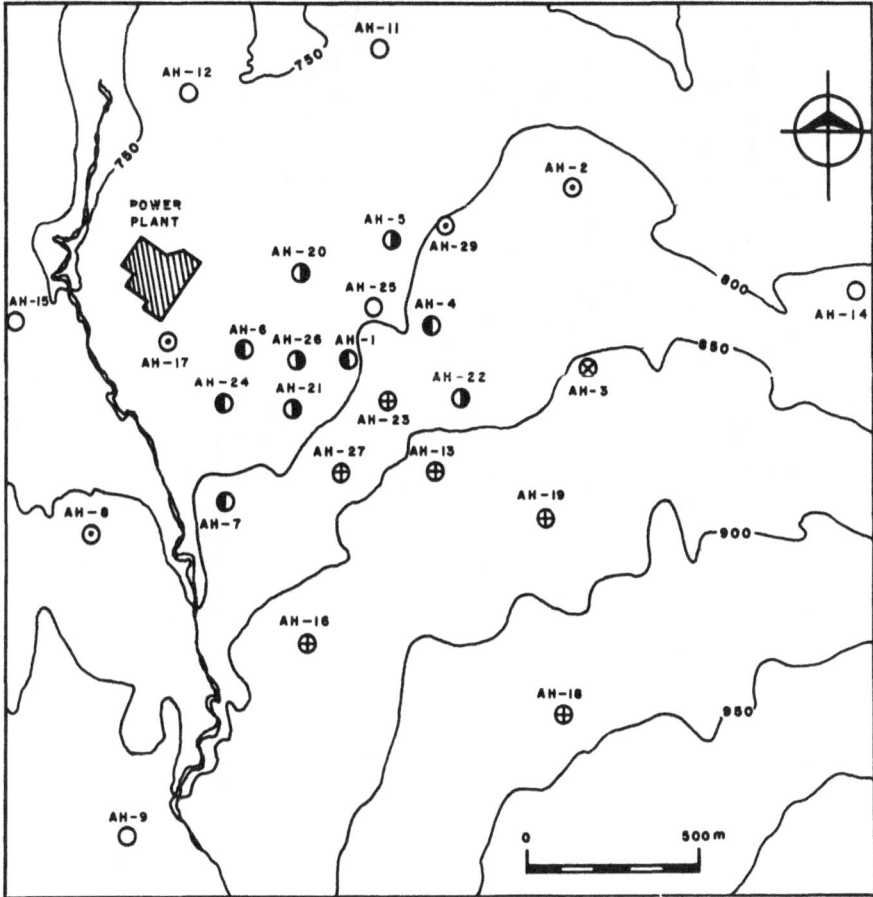

FIGURE 3.3—Arrangement of wells at Ahuachapán. ◐ = Wells for Unit 1; ◑ = Wells for Unit 2; ⊙ = Reinjection wells; ○ = Nonproductive wells; ⊗ = Collapsed well; ⊕ = Stand-by wells.

true lengths and diameters of the various steam and liquid reinjection lines are given in table 3.2.

Table 3.3 gives production data for the 10 wells supplying Units 1 and 2. Wellhead separator pressure, liquid flow rate, steam flow rate, and total flow rate are listed for two time periods, October 1976 and April 1978. Average wellhead quality was about 17% in April 1978, whereas it was nearly 19% in October 1976. The highest quality occurs at well AH-26, which delivers 35% steam. This anomalous behavior is probably caused by a relatively tight formation in the neighborhood of this well that may lead to flashing in the formation. The power potential of well AH-4 is 17 MW, but only 13 MW is being extracted because the size of the wellhead separator limits flow. Well AH-21 is also an excellent well, with a 9-MW po-

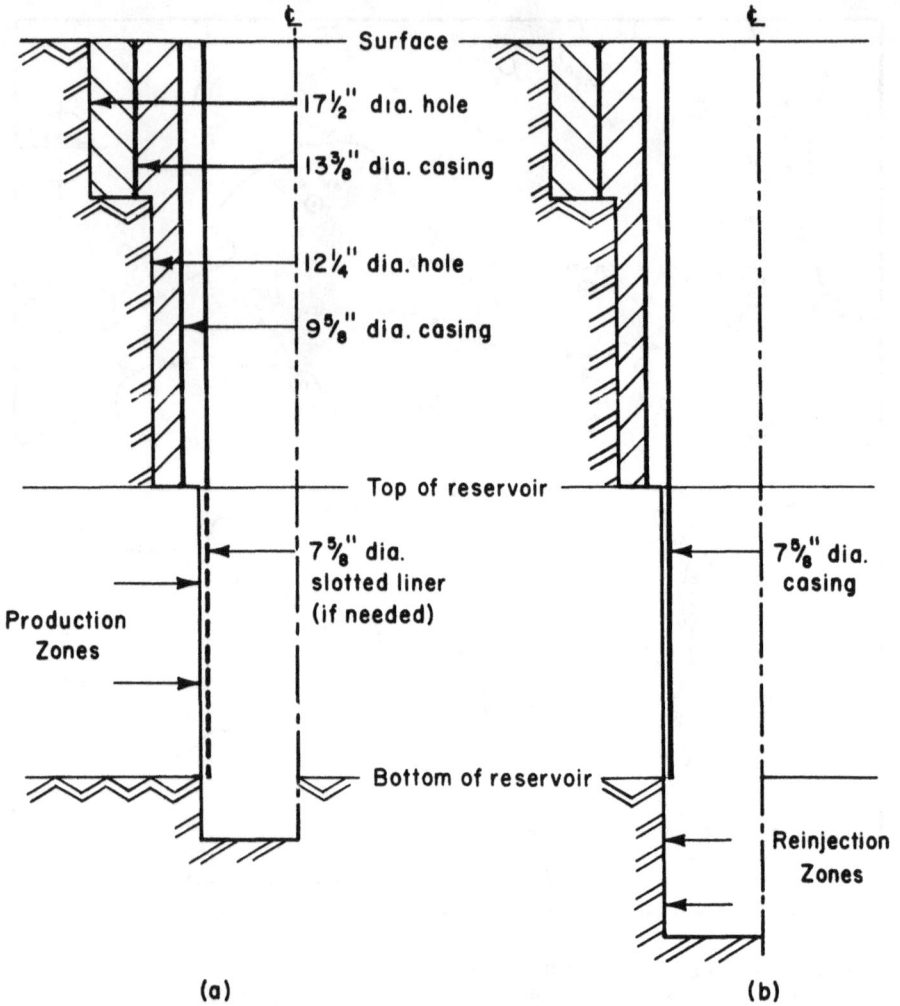

FIGURE 3.4—Typical well casing programs for (a) production and (b) reinjection wells at Ahuachapán.

tential and 7-MW actual utilization. Figure 3.8 gives the overall mass balance for the wells serving Units 1 and 2, both with and without reinjection. Figure 3.9 shows the wellhead equipment at well AH-20.

3.2.3   Geofluid characteristics

Total dissolved solids in the liquid at the wells average about 18,400 ppm or 1.84%. The principal constituents are chloride (10,430 ppm), sodium (5690 ppm), and potassium (950 ppm). Table 3.4 lists the concentration of all impurities in the liquid from 9 of the 10 wells that supply Units 1

Table 3.1—*Well information at Ahuachapán* [a]

Well No.	Elevation		Depth		Comments
	m	ft	m	ft	
AH-1..........	802.8	2634	1205	3954	Producer for Unit 1.
AH-2..........	808.0	2651	1200	3937	Reinjector for AH-4.
AH-3..........	855.5	2807	802	2631	Collapsed during drilling.
AH-4..........	812.2	2665	640	2100	Producer for Unit 1.
AH-5..........	789.5	2590	952	3124	Producer for Unit 2.
AH-6..........	783.0	2569	591	1939	Producer for Unit 1.
AH-7..........	804.8	2641	950	3117	Producer for Unit 1.
AH-8..........	811.0	2661	988	3242	Dual-purpose; reinjector for AH-7.
AH-9..........	871.3	2859	1424	4672	Dry hole, beyond the field.
AH-10........	723.8	2375	1524	5000	Dry hole, beyond the field.
AH-11........	759.3	2491	943	3094	Dry hole, beyond the field.
AH-12........	758.8	2490	1003	3291	Dry hole, beyond the field.
AH-13........	859.6	2820	860	2822	Producer, on standby.
AH-14........	822.0	2697	1053	3455	Dry hole, but highest temperature.
AH-15........	772.7	2535	704	2310	Dry hole, beyond the field.
AH-16........	869.0	2851	1006	3301	Producer, on stand-by.
AH-17........	773.0	2536	1200	3937	Dual-purpose; reinjector for AH-6.
AH-18........	926.3	3039	1256	4121	Newly drilled, not in equilibrium yet.
AH-19........	~880	~2887	[b]	[b]	Newly drilled.
AH-20........	792.9	2602	600	1969	Producer for Unit 2.
AH-21........	795.0	2608	849	2786	Producer for Unit 2.
AH-22........	842.0	2763	660	2165	Producer for Unit 2.
AH-23........	825.4	2708	924	3032	Producer, on standby.
AH-24........	783.1	2569	850	2789	Producer for Unit 1.
AH-25........	798.5	2620	943	3094	Dry hole in middle of field.
AH-26........	791.1	2596	804	2638	Producer for Unit 2.
AH-27........	~830	~2723	[b]	[b]	Producer, on standby.
AH-28........	[b]	[b]	[b]	[b]	To be sited and drilled.
AH-29........	794.8	2608	1200	3937	Reinjector for AH-1.

[a] Source: Cuéllar, 1978.
[b] Not available.

and 2. (Data on well AH-24 were not available.) Table 3.5 gives average concentrations for the constituents of these 9 wells. Noncondensable gases amount to 0.05% by weight of the total well flow or 0.2% by weight of the steam flow. Composition of the noncondensable gases is shown in table 3.6, where the percentages are volumetric.

Prior to entering the powerhouse, the steam passes through a final moisture separator. Table 3.7 gives the concentration of the various elements found in the liquid that settles in the receiver trap at the moisture separator. Also included in table 3.7 is the composition of the steam condensate at the condenser hot well.

Table 3.2—*Lengths and diameters of stream transmission lines and liquid reinjection lines at Ahuachapán*

		Pipe Diameter		Length from Wellhead to Receiver	
Unit No.	Well No.	mm	in	m	ft
1............................	AH-1	406	16	560	1840
1............................	AH-4	508	20	820	2690
1............................	AH-6	406	16	280	920
1............................	AH-7	305	12	695	2280
1............................	AH-24	305	12	303	995
2............................	AH-5	305	12	740	2430
2............................	AH-20	406	16	420	1380
2............................	AH-21	ᵃ 406	16	256	840
2............................	AH-22	508	20	900	2955
2............................	AH-26	ᵃ 406	16	100	330
Reinjection..................	AH-2R	305	12	600	1970
Reinjection..................	AH-8R	305	12	350	1150
Reinjection..................	ᵇAH-17R	305	12	250	820
Reinjection..................	AH-29R	305	12	500	1640

ᵃ Joined into a 508-mm (20-in) line that runs 470 m (1540 ft).
ᵇ Values given are for the connection between AH-6 and AH-17R; a 254-mm (10-in) line runs from AH-21 to the line joining AH-6 and AH-17R.

Table 3.3—*Characteristics of well production for Units 1 and 2*

	October 1976				April 1978			
	p^a	$\dot{m}_l$	$\dot{m}_v$	$\dot{m}_t$	p^a	$\dot{m}_l$	$\dot{m}_v$	$\dot{m}_t$
Well No.	kPa	kg/s	kg/s	kg/s	kPa	kg/s	kg/s	kg/s
			Unit 1					
AH-1..............	665. 3	81. 70	13. 20	94. 90	670. 3	76. 39	14. 16	90. 55
AH-4..............	699. 4	102. 97	23. 66	126. 63	660. 2	131. 73	23. 69	155. 42
AH-6..............	670. 5	44. 97	17. 65	62. 62	651. 1	61. 80	15. 18	76. 98
AH-7..............	660. 5	53. 89	9. 17	63. 06	591. 5	44. 32	6. 94	51. 26
AH-24.............	—	—	—	—	602. 0	54. 01	7. 82	61. 83
Total...........		283. 53	63. 68	347. 21		368. 25	67. 79	436. 04
			Unit 2					
AH-5..............	631. 2	47. 72	6. 15	53. 87	601. 8	55. 09	7. 69	62. 78
AH-20.............	626. 3	44. 72	10. 74	55. 46	611. 6	48. 67	14. 87	63. 54
AH-21.............	650. 9	81. 29	12. 51	93. 80	655. 8	59. 63	12. 50	72. 13
AH-22.............	635. 3	54. 24	16. 47	70. 71	591. 2	48. 48	13. 66	62. 14
AH-26.............	640. 9	19. 55	12. 37	31. 92	601. 7	19. 44	10. 26	29. 70
Total...........		247. 52	58. 24	305. 76		231. 31	58. 98	290. 29
Plant total.......		531. 05	121. 92	652. 97		599. 56	126. 77	726. 33

ᵃ Pressure at wellhead separator.

Table 3.4—*Chemical analysis of liquid from wells at Ahuachapán* [a]

Substance	Well Number								
	AH-1	AH-4	AH-5	AH-6	AH-7	AH-20	AH-21	AH-22	AH-26
Cl.........	10600	9050	9110	10900	12500	10900	11500	9217	10130
Na........	5800	5000	5000	6000	6600	6000	6100	5080	5600
K.........	1000	740	680	1050	1280	1040	1140	710	900
SiO₂.......	577	534	470	500	610	556	535	552	500
Ca.........	425	400	440	439	486	450	480	437	430
B..........	147	144	138	156	178	155	169	127	142
Br.........	46. 0	35. 7	38. 1	43. 0	50. 4	45. 9	48. 4	39. 8	47. 7
HCO₃......	34. 4	32. 5	28. 1	46. 5	26. 6	31. 6	40. 3	31. 5	36. 3
SO₄........	30. 8	34. 2	40. 8	29. 7	23. 3	30. 2	40. 5	32. 7	44. 2
Li.........	18. 5	15. 7	14. 8	18. 8	20. 0	18. 5	19. 1	15. 0	17. 5
As.........	10. 8	10. 0	9. 9	11. 6	14. 0	12. 0	11. 8	10. 2	11. 0
I..........	8. 2	6. 5	7. 0	8. 5	9. 2	8. 5	10. 1	6. 9	8. 5
Rb........	7. 9	5. 6	5. 1	8. 3	8. 5	7. 9	8. 3	5. 2	7. 3
Cs.........	5. 8	4. 6	4. 1	6. 3	6. 6	6. 0	6. 3	4. 5	5. 8
Sr.........	4. 5	4. 6	5. 7	4. 5	4. 8	4. 5	4. 7	4. 7	4. 5
Sb.........	2. 3	1. 9	1. 8	2. 1	2. 5	—	—	—	—
F..........	1. 5	1. 3	1. 2	1. 8	1. 8	1. 6	1. 7	1. 3	1. 4
Mg........	0. 09	0. 13	0. 24	0. 09	0. 08	0. 07	0. 01	0. 16	0. 30

[a] All values in ppm.

Table 3.5—*Average chemical composition of liquid from wells at Ahuachapán* [a]

Substance	Concentration (ppm)
Chloride, Cl....................................	10430
Sodium, Na....................................	5690
Potassium, K...................................	950
Silica, SiO₂....................................	537
Calcium, Ca....................................	443
Boron, B.......................................	151
Bromide, Br....................................	43. 9
Bicarbonate, HCO₃.............................	34. 2
Sulfate, SO₄....................................	34. 0
Lithium, Li.....................................	17. 5
Arsenic, As.....................................	11. 3
Iodide, I.......................................	8. 1
Rubidium, Rb..................................	7. 1
Cesium, Cs.....................................	5. 5
Strontium, Sr...................................	4. 7
Antimony, Sb...................................	2. 1
Fluoride, F.....................................	1. 5
Magnesium, Mg................................	0. 13

[a] Samples taken from wells AH-1, 4, 5, 6, 7, 20, 21, 22, and 26.

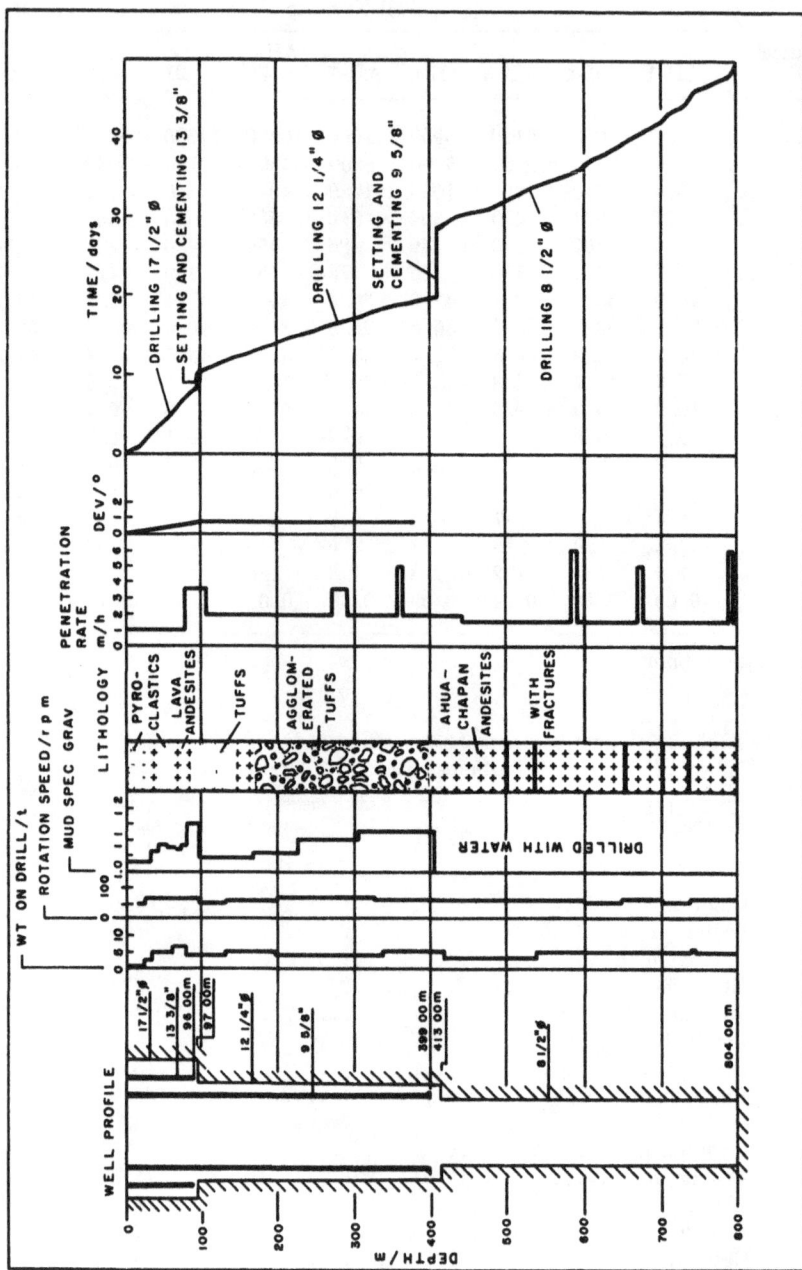

FIGURE 3.5—Drilling program for well AH-26 at Ahuachapán [G. Cuéllar, personal communication].

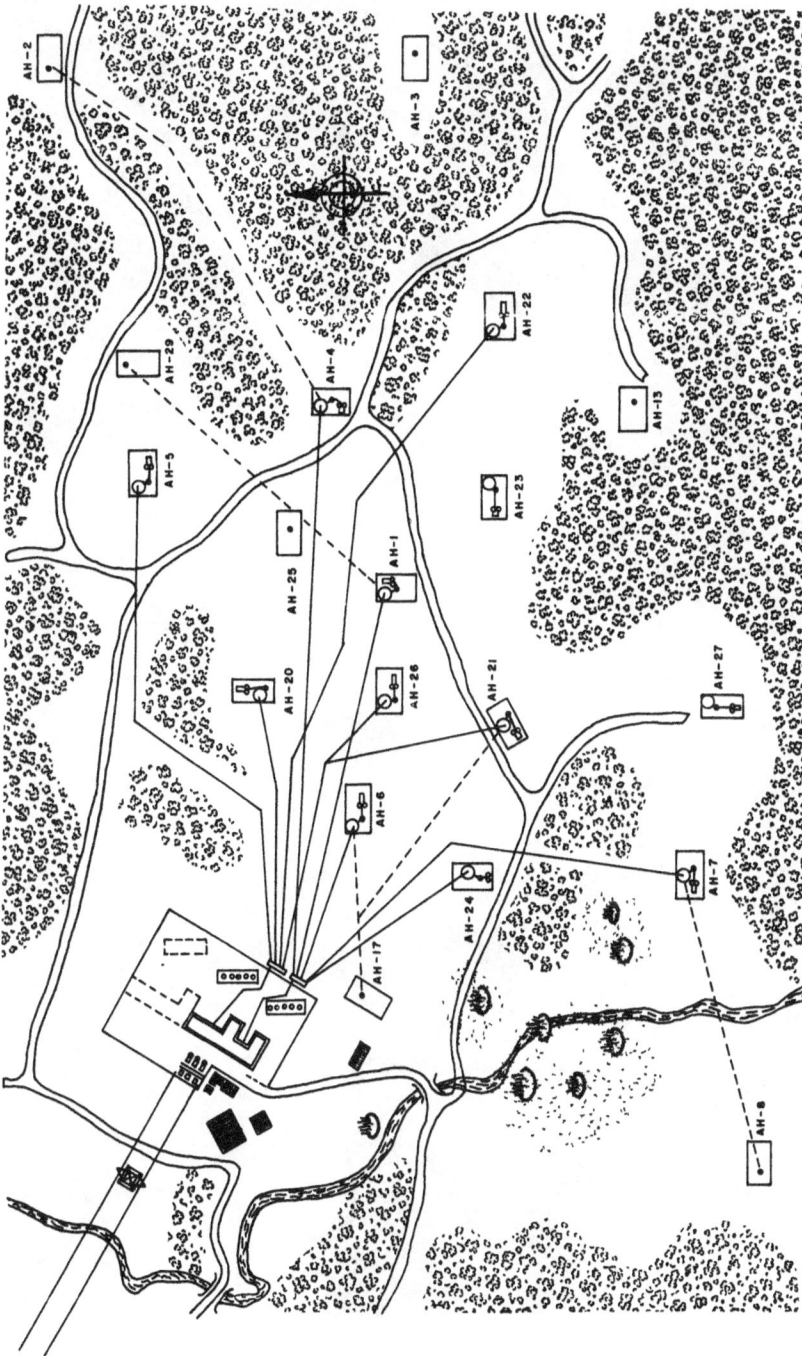

FIGURE 3.6—Layout of production and reinjection wells for Units 1 and 2 at Ahuachapán [G. Cuéllar, personal communication].

FIGURE 3.7—Surface thermal manifestations at Ahuachapán: boiling mud pools and
steam vents in foreground, well AH-21 in background [photo by R. DiPippo].

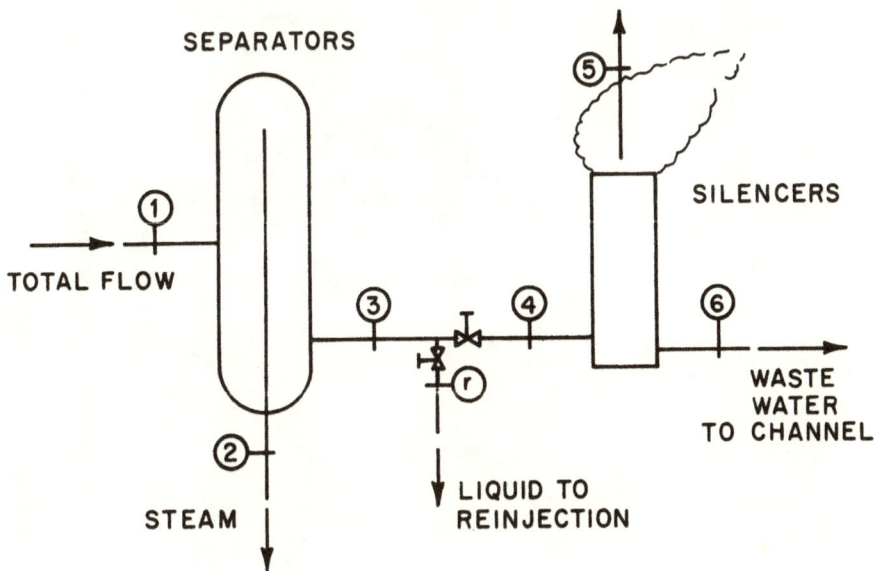

FIGURE 3.8—Overall mass balance for 10 wells serving Units 1 and 2 at Ahuachapán

Table 3.6—*Composition of noncondensable gases in Ahuachapán geothermal steam*

Carbon dioxide, CO_2	86. 8%
Hydrogen sulfide, H_2S	12. 1%
Hydrogen, H_2	0. 126%
Nitrogen, N_2	0. 05%
Ammonia, NH_3	1%
Methane, CH_4	

FIGURE 3.9—Wellhead equipment for well AH-20 at Ahuachapán [photo by R. DiPippo].

Table 3.7—*Composition of steam condensate at receiver and hot well*

Liquid at Receiver Trap		Condensate in Hot Well	
Constituent	Concentration (ppm)	Constituent	Concentration (ppm)
Cl.	13. 6	Cl.	59. 3
Na.	7. 15	Na.	130.
K.	1. 25	K.	0. 35
SO₄.	0. 6	SO₄.	181. 5
SiO₂.	1. 61	SiO₂.	0. 86
Ca.	0. 80	NH₄.	0. 36
HCO₃.	2. 90	Fe.	0. 10
B.	0. 75	CaCO₃.	6. 49
		S.	0. 15
pH.	5. 28	pH.	7. 22

3.2.4 Energy conversion systems for Units 1 and 2

The layout of the powerhouse, switchyard, and cooling towers is shown in figure 3.10. Cooling towers are oriented so that the prevailing wind carries the plumes away from the powerhouse and transformers. Two steam receivers (one for each unit) are located between the cooling towers for Units 1 and 2 at the end farther from the plant. Collected steam travels by means of elevated pipelines to the final moisture separators and thence to the turbines. A cross-sectional view of the turbine, a product of Mitsubishi Heavy Industries, Ltd., is shown in figure 3.11. It is a five-stage, double-flow, impulse-reaction machine housed in a single cylinder. The last-stage blade height is 520 mm (20.5 in). Identical machines are used for each of the first two units.

A schematic flow diagram for each unit is given in figure 3.12. Five wells supply each unit; steam is separated by means of a simple Webre-type cyclone separator with the steam passing through a ball-check valve prior to entering the steam transmission pipeline. Each turbine exhausts to a low-level, direct-contact condenser fitted with a slanted barometric pipe. This feature allows the condenser hot well to be located adjacent to the turbine building for ease of accessibility to circulating pumps, water treatment equipment, etc.

A two-stage steam ejector gas extraction system is connected to the gas cooler section of the condenser. For each unit there are two sets of extractors arranged in parallel; one serves as a standby system. Cooling towers are of the crossflow, mechancially-induced-draft type. Each has five cells and uses wood packing with fiberglass for the exterior of the stacks.

Technical particulars for Units 1 and 2 are listed in table 3.8. The geothermal energy resource utilization efficiency η_u may be computed using the data from table 3.3 for April 1978 and an output of 60 MW, relative

Table 3.8—*Technical specifications for Ahuachapán geothermal power plant*

	Units 1 and 2 1975, 1976 [a]	Unit 3 1980 [a]	Auxiliary Unit 1979 [a]
Turbine data:			
Type..	Single-cylinder, double-flow, impulse, 5×2	Single-cylinder, double-flow, impulse-reaction, (3, 4) ×2	Single-cylinder, one Curtis stage, noncondensing, geared
Rated capacity, MW	30, each	35	1.1
Maximum capacity, MW	35, each	40	1.3
Speed, rpm	3600	3600	7129/1800
Main steam pressure, lbf/in²	81.1	79.5	80.2
Secondary steam pressure, lbf/in²	(None)	21.8	(None)
Main steam temperature, °F	313.0	311.6	313.0
Secondary steam temperature, °F	(None)	232.6	(None)
Exhaust pressure, in Hg	2.46	2.46	28.4
Main steam flow rate, 10³ lbm/h	507, each	377	46.3
Secondary steam flow rate, 10³ lbm/h	(None)	320	(None)
Last-stage blade height, in	20.5	22.2	([b])

Condenser data:

Type.. Low-level, direct-contact type with slanted barometric pipe (None)

Cooling water temperature, °F............ 80.6 80.6 —

Outlet water temperature, °F............. 104.5 104.5 —

Cooling water flow rate, 10^6 lbm/h.... 19.1 27.0 —

Gas extractor data:

Type.. Two-stage, steam jet ejector with inter- and after-condenser (None)

Suction pressure, in Hg...................... 2.32 (b) (b)

Gas capacity, ft³/min......................... 6,886 (b) (b)

Steam consumption, 10^3 lbm/h.......... 9.04, each

Cooling tower data:

Type.. Crossflow, mechanically-induced-draft with vertical axial fans (None)

Number of cells................................. 5, each 5 —

Design wet-bulb temperature., °F....... 71.6 71.6 —

Fan motor power, kW/fan.................... 80 80 —

a Year of startup.
b Not available.

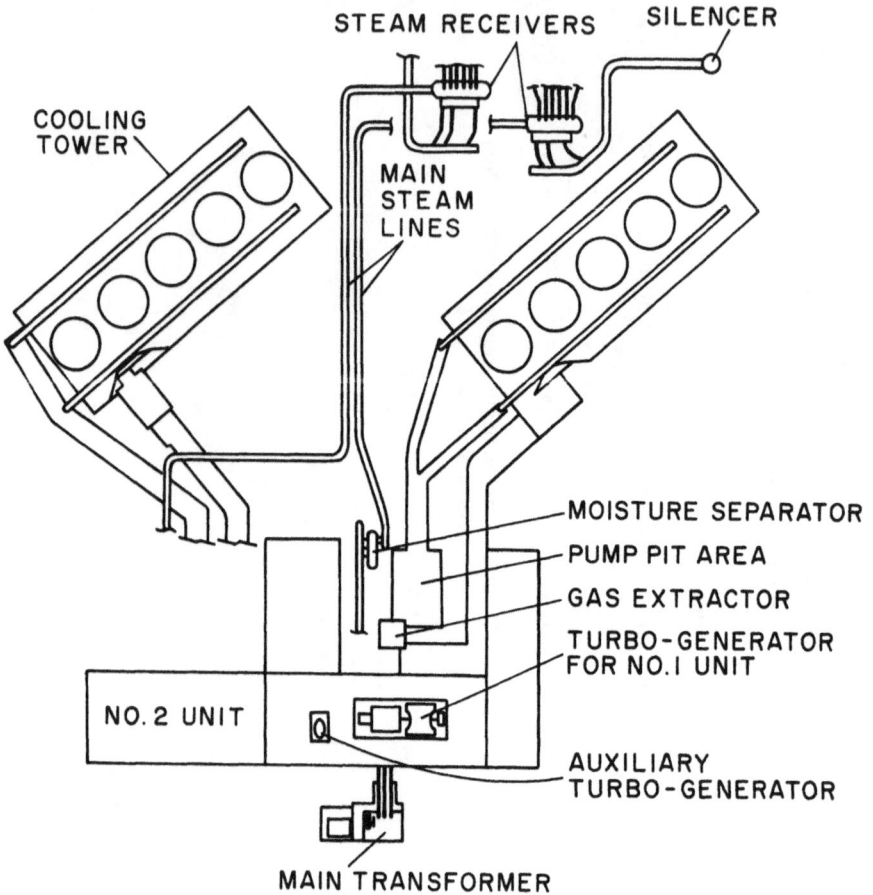

FIGURE 3.10—Power plant site arrangement: Ahuachapán Units 1 and 2 [MHI, 1977].

to a sink condition at 22.0°C (71.6°F), the design wet-bulb temperature. For these conditions $\eta_u = 37\%$. Overall steam consumption for the plant is about 7.6 Mg/MW·h (16.8 lbm/kW·h).

3.2.5 Proposed energy conversion system for Unit 3

The third unit for Ahuachapán was originally planned as a 30-MW low-pressure unit operating on steam flashed from separated bore liquid. As the field became more developed and confidence in the steam supply grew, it was decided to install a dual-pressure unit that would be supplied with medium-pressure (MP) steam from wells together with low-pressure (LP) steam from flashed liquid. The unit was upgraded to 35 MW as well.

A highly simplified flow diagram for Unit 3 is given in figure 3.13. Hot water at the wellhead separator pressure is drawn from eight wells into

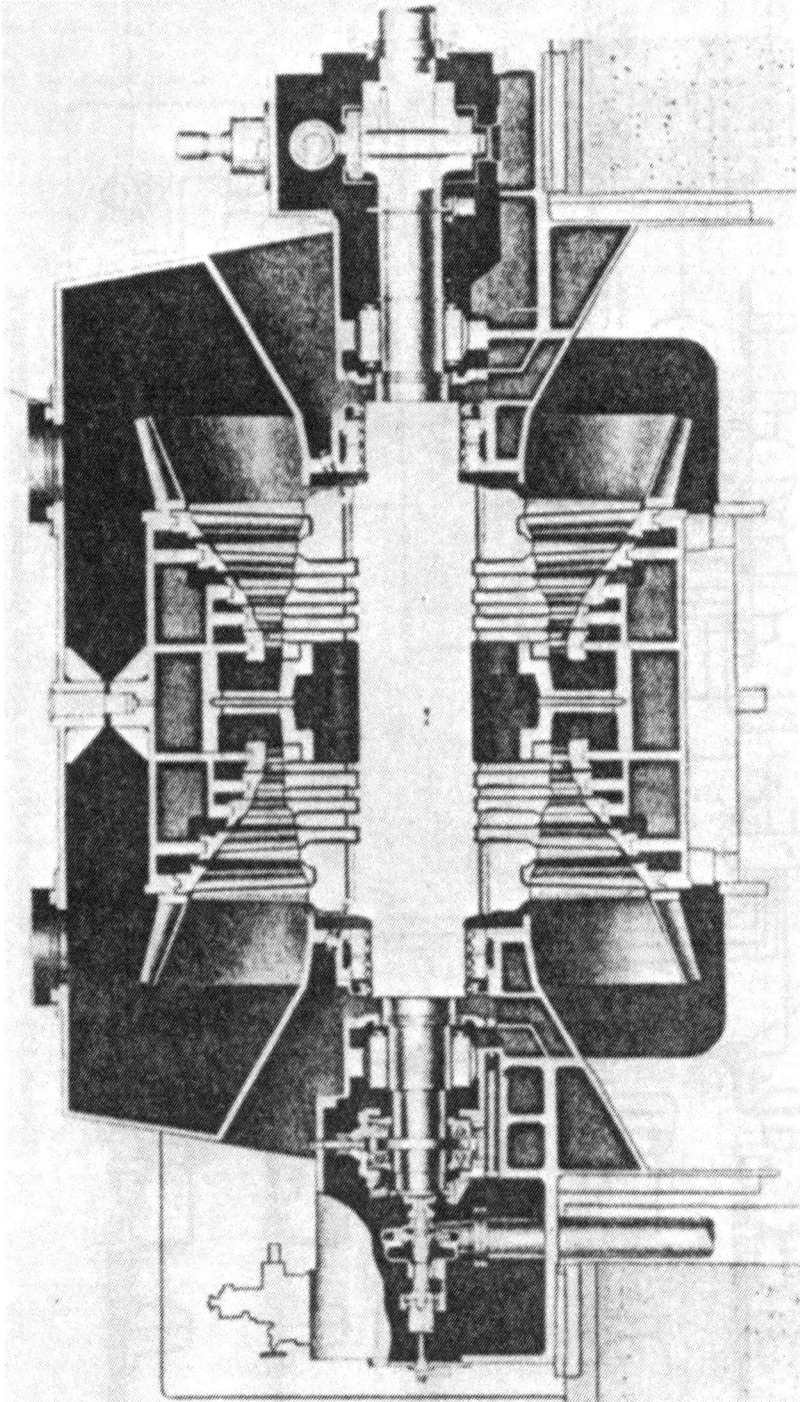

FIGURE 3.11—Cross-section of turbine for Units 1 and 2 at Ahuachapán [MIII, 1977].

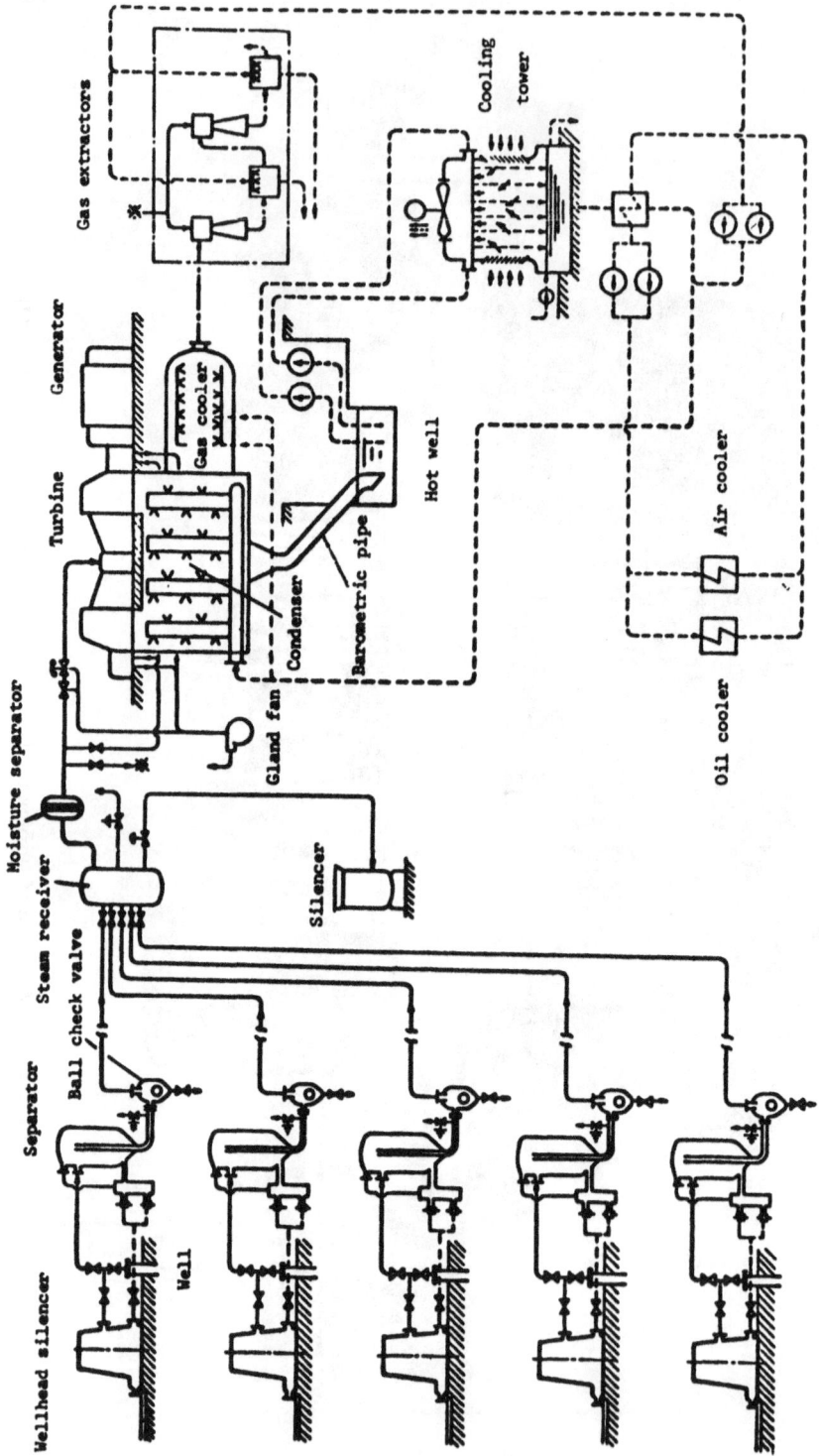

FIGURE 3.12—Flow diagram for Units 1 and 2 at Ahuachapán [after MHI, 1977].

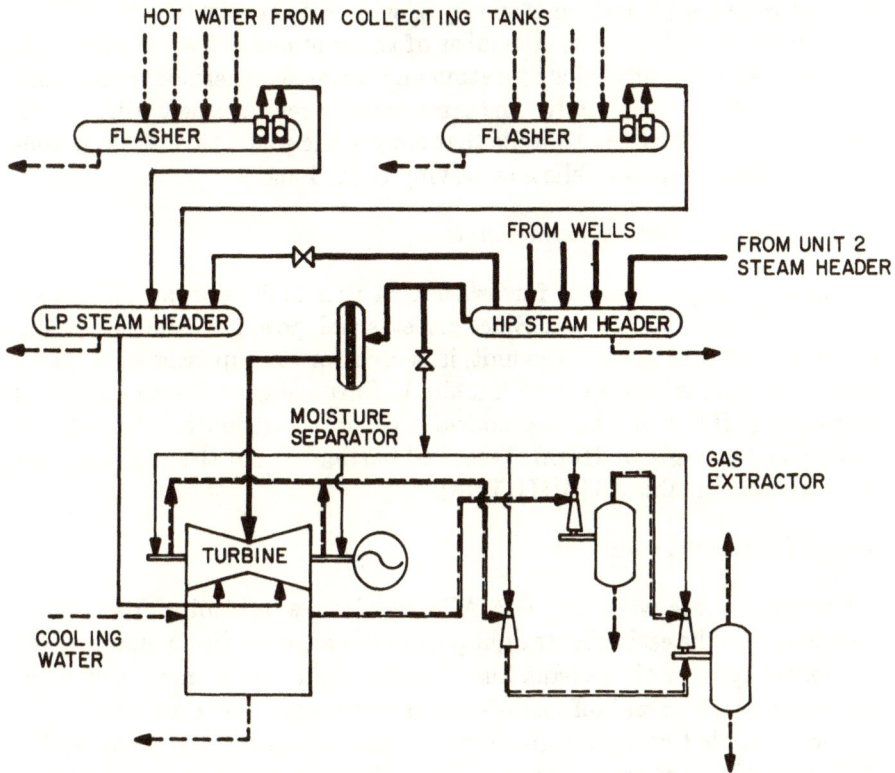

FIGURE 3.13—Simplified flow diagram for Unit 3 at Ahuachapán [after Fuji, 1977]

two horizontal flash vessels. Steam thus generated leaves each vessel through a pair of ball-check valves and flows to an LP steam header. In addition, MP steam produced by separation at three wells is mixed with MP steam from the header for the existing Unit 2 in a new MP steam header. Provision is made for flashing a portion of the MP steam from the MP header down to the LP header, if necessary. The MP steam is admitted to the turbine entrance after being scrubbed of moisture; the LP steam is admitted to the turbine at a pass-in section. Gas extraction is carried out by a conventional two-stage steam ejector; a single steam ejector is used to purge the turbine gland seals.

The turbine will be of the dual-pressure, double-flow type in a single cylinder with the medium-pressure section consisting of three stages (essentially impulse blading) followed by the low-pressure section of four impulse-reaction stages. The generator will be air-cooled, rated at 40,000 kVA, 13.6 kV, at 60 Hz with a 0.875 (lagging) power factor.

Table 3.8 lists the currently projected technical specifications for the third unit, which is under construction and is expected to begin generating electricity early in 1980. According to the data now available, the geothermal resource utilization efficiency η_u will be approximately 42%. This value has been estimated on a number of assumptions and should be recalculated after the unit begins operating and the actual thermodynamic conditions are known accurately. The three units, taken together, will have an efficiency $\eta_u \cong 43\%$ if the 13 wells that supply the plant have average conditions matching the 10 wells now serving Units 1 and 2.

3.2.6 Auxiliary turbogenerator unit

The Ahuachapán plant is furnished with a 1.1-MW noncondensing auxiliary turbine-generator set. Since no external power source or cooling water is needed to operate the unit, it is used for startup from cold conditions. The unit is equipped with a single Curtis stage with an air-cooled lubricating oil system. All mechanical, electrical, and control elements are mounted on a single platform. Technical particulars for the auxiliary unit may be found in table 3.8 [MHI, 1978].

3.2.7 Construction materials

Casings for the wells are J-55 API standard-weight pipe. The cement used to secure the casing is straight portland cement. Drilling mud is of the bentonitic type, with coconut husks, coffee bean shells, mica, and other materials added to seal off loss-of-circulation zones during drilling.

The pipes that carry the geothermal steam are fabricated from ASTM A-53 grade B seamless carbon steel pipe. The velocity of the steam is kept below 50 m/s (164 ft/s). Blocks of calcium silicate are used to insulate the steam pipes. These are wired onto the pipes, covered with composite kraft paper-aluminum sheet (which acts as a vapor barrier), and enclosed within a jacket of galvanized steel. Thickness of insulation depends on pipe size (see table 3.9). Pipes carrying waste liquid to the reinjection wells are not insulated.

Circulating water in the plant is carried by 304 stainless steel pipes; steam pipes in the plant are made of 316 stainless. Turbine blades are of 13 Cr alloy steel with stellite inserts where needed. Cooling towers contain redwood packing, and sodium hydroxide is used for pH control.

Table 3.9—*Thickness of insulation on steam pipes*

Pipe diameter		Insulation thickness	
mm	in	mm	in
305	12	80	3. 15
406	16	80	3. 15
508	20	100	3. 94

3.2.8 Effluent and emissions handling systems

Two methods are used for disposal of the waste liquid from the plant. One method is reinjection; the other is discharge to the Pacific Ocean through a covered concrete channel 1 m² (11 ft²) in cross-sectional area and 75 km (47 mi) in length.

Four wells currently serve as reinjectors (see table 3.2 and figure 3.3). The total amount of liquid being reinjected (for 60-MW capacity) is 368.8 kg/s (5846 gal/min), and constitutes about 63% of the waste liquid discharged from the plant. The liquid to be reinjected is taken directly from the wellhead separators, at the pressure of the separator, and piped to the reinjection wells without the aid of booster pumps and without chemical treatment or exposure to the atmosphere. Because the temperature of the liquid is not lower than 150°C (302°F), mineral deposition has been avoided in the reinjection lines and the wells [Einarsson et al., 1975]. More than 13×10^6 Mg (29×10^9 lbm) have been returned to the formation since rejection was started in 1975.

The remainder of the liquid waste, including steam condensate from the turbines, is carried through a discharge channel from the plant—around, over, and through the mountains, eventually to the ocean. The channel passes close to the Rio Paz on the western border of El Salvador, and for a time (until the channel was completed) a short connection between the channel and the Rio Paz allowed the waste water to be disposed of in that river. This was a temporary and not wholly satisfactory solution. During this period it was necessary at times to curtail the output of the Ahuachapán plant in order not to exceed the allowable limits on arsenic and boron in the Rio Paz, which is used for irrigation and other purposes. The completed aqueduct contains 16 siphons (made from sections of pipe) to carry the fluid over the hills and through valleys across the rugged terrain that lies between the plant and the Pacific Ocean. About 90% of the length of the channel is concrete and 10% is pipe. The route was chosen to minimize the average slope of the channel.

Two labyrinth retention tanks in the bore field provide a settling time of 50 to 60 minutes before the waste water enters the channel. One of these is located between wells AH-1 and AH-25 (see figures 3.3 and 3.6); it receives all the liquid from well AH-22 and part of the liquid from AH-4. As can be seen from the photograph in figure 3.14, the open tank is considerably easier to clean than the channel, which is covered with removable concrete slabs. The other settling tank is located just north of well AH-6 (see figure 3.6). Settling tanks have proved to be a very effective means of controlling the deposition of silica in the effluent lines used for surface disposal of the waste liquid [Cuéllar, 1975].

There are no emissions controls on hydrogen sulfide, which reaches concentrations of 1–4 ppm at the boundary of the plant site. Downwind of the powerhouse and cooling towers the odor of H_2S is noticeable but not objectionable. A minimum of about 100 kg/h (220 lbm/h) of nonconden-

FIGURE 3.14—Baffled retention tank for waste liquid at Ahuachapán. Well AH-1 is in background [photo by R. DiPippo].

sables is ejected at full output. This does not include air that enters at the turbine gland seals and in the direct-contact condenser from the cooling water. Of this amount, roughly 95 kg/h) (210 lbm/h) of H_2S is emitted, or 1580 g/MW·h (3.5 lbm/MW·h), on a specific power basis.

The designers of this plant have done an admirable job of minimizing the adverse environmental effects possible from such a plant. Although the plant site is sparsely populated, the city of Ahuachapán is only a few kilometers away and some families actually live on the plant property. These inhabitants use the hot springs for domestic purposes and graze their cattle among the wellheads and steam pipes of the bore field. The use of a covered waste liquid channel and the careful design and arrangement of the wellhead equipment have prevented despoliation of the countryside and effected a reasonable harmonization of technology and nature.

3.2.9 Economic factors

Actual capital costs for Units 1 and 2 and estimated·capital costs for Unit 3 are given in table 3.10. Average installed capital cost per kilowatt of capacity for the first two units is about US$825, including the $10 million cost of the waste disposal channel. Table 3.11 contains a summary of the cost of electricity generation at Ahuachapán over the history of the plant. The figures in table 3.11 do not include interest payments (which are made) or taxes (which are not paid). Even so, electricity produced from geothermal energy constitutes a very inexpensive source for El Salvador, as the round figures in table 3.12 for alternative sources show. The project has been financed through loans from the World Bank [Sheehan, 1977], bonds issued by CEL, and cash.

3.2.10 Operating experience

The operation of the Ahuachapán geothermal power plant has been highly successful. The plant is a vital link in the electricity supply system of El Salvador, which relies on three types of power stations: hydroelectric, thermal (fossil fuel), and geothermal. As can be seen from the figures in table 3.13, the Ahuachapán plant constitutes about 14% of the country's rated installed electric generating capacity. However, during the dry season when the actual capacity of the hydro plants falls to about 50% of their rated value, the geothermal units constitute nearly 20% of actual capacity. The electrical generation, annual capacity factors, and the percentage of total generation in El Salvador of Units 1 and 2 at Ahuachapán are given in table 3.14. Unit 1 began in June 1975 and Unit 2 in June 1976.

The geothermal plant has been essentially free of major breakdowns. This is reflected in an availability factor of 95% based on outages caused by breakdowns alone [availability = (total hours − outage hours caused by breakdowns)/total hours]. When scheduled maintenance is included with breakdown outage, the availability factor is 84%. For the sake of compari-

Table 3.10—*Capital costs for geothermal power units at Ahuachapán*

Item	Unit 1 (US$ 1975)	Unit 2 (US$ 1976)	Unit 3 (US$ est.) [a]
Turbogenerator, condenser, cooling water system, wellhead equipment, turbine controls	$3, 423, 740. 48	$5, 205, 803. 80	$8, 000, 000
Steam piping, including supports, insulation, and drains	536, 526. 00	991, 679. 20	1, 950, 000
Electrical equipment	790, 272. 18	613, 576. 60	700, 000
Auxiliary equipment	225, 484. 06	68, 313. 00	110, 000
Well drilling and piping	3, 200, 000. 00	3, 200, 000. 00	1, 000, 000
Other civil engineering, installation of equipment, studies, engineering, administration, etc.	[b] 23, 389, 967. 55	[b] 7, 952, 309. 00	8, 240, 000
Total	31, 565, 990. 27	18, 031, 681. 60	20, 000, 000

[a] Estimates shown were made in 1975; latest estimates (1979) for Unit 3 total about $35,000,000.
[b] Cost of waste water disposal channel ($10,000,000) has been allocated between Units 1 and 2; upon completion of Unit 3 it will be allocated among the three units.

Table 3.11—*Cost of electricity generation at Ahuachapán (exclusive of interest)*

Expenses	1975 [a]	1976 [b]	1977 [c]	1978 (est.)
Production	$82, 191	$398, 363	$584, 340	$613, 600
Depreciation	235, 619	591, 222	1, 350, 731	2, 040, 000
General	24, 313	114, 580	187, 650	192, 000
Total	342, 123	1, 104, 165	2, 122, 721	2, 845, 600
Generation (MW·h) [d]	72, 330. 6	279, 800	400, 051	418, 400
Specific cost (mill/kW·h) [e]	4. 73	3. 95	5. 31	6. 80
Specific cost (mill/kW·h) [f]	5. 14	4. 29	5. 77	7. 39

[a] Unit 1 for 7 months of the year.
[b] Unit 1 for full year; Unit 2 for 5 months.
[c] Both units for full year.
[d] Gross generation; net generation is 8% less.
[e] Based on gross generation.
[f] Based on net generation.

Table 3.12—*Cost of electricity from various sources in El Salvador*

Type of power plant	Cost of electricity (US mill/kW·h)
Hydroelectric	4
Geothermal	6
Thermal (bunker C oil)	25
Gas turbine (diesel)	50

Table 3.13—*Installed electric power capacity in El Salvador (as of mid-1978)*

Plant Type	Unit Name	Capacity	Total
Hydroelectric	Cerron Grande Nos. 1 and 2	135 MW	
	5th of November	82 MW	232 MW
	Guajoyo	15 MW	
Thermal	Acajutla	63 MW	
	Soyapango (gas turbine)	58. 6 MW	128. 2 MW
	Gas turbines	6. 6 MW	
Geothermal	Ahuachapán Units 1 and 2	60 MW	60 MW
1978 Total installed capacity			420. 2 MW

Table 3.14—*Electricity generation at Ahuachapán*

Year	Electrical generation	Capacity factor	Percentage of total generation
1975	72,331 MW·h	47%	11. 8%
1976	279,800 MW·h	67%	25. 4%
1977	400,051 MW·h	76%	32. 3%

son, table 3.15 provides 1977 figures for the hydroelectric plants, which are generically noted for high reliability.

A complete overhaul of each unit is carried out once every 2 years. It takes about 1 month to complete each inspection. Inspections are scheduled to coincide with the wet season so that sufficient hydroelectric capacity is available. Wellhead equipment is given a thorough inspection and cleaning at least once per year, and each well is visited on a rotating basis. It takes about 1 month to cover all the wells. Maintenance of the bore field is handled by the following teams: general maintenance by 1 engineer and 9 mechanics; measurements by 1 engineer and 7 persons; operations by 1 engineer and 6 persons. The geothermal plant is maintained by 5 engineers and 77 persons.

There have been relatively few problems with operation of the plant. The steam filters for the Unit 1 turbine failed and caused damage to the first- and second-stage buckets. The shell of one of the aftercoolers in the gas extraction system split along a weld that had been exposed to the direct impingement of the flow from the second-stage ejector. The power transformer for Unit 2 failed in June 1978. These have been the only major failures of the energy conversion equipment during the 3 years of operation.

In the wells and gathering system, premature failure of some of the rupture disks has occurred. These 305-mm-diameter (12-in), 3-mm-thick (⅛-in) aluminum disks have developed pits and have failed, requiring a shutdown of the well. The failures most likely are caused by a combination of stress cracking and fatigue. Pressure relief valves will be installed on Unit 3 and backfitted on Units 1 and 2 if these prove satisfactory.

Table 3.15—*Performance factors for hydroelectric plants in El Salvador (1977 data)*

Plant	Capacity factor	Availability factor
Cerron Grande..................	55%	96%
5th of November...............	43%	93%
Guajoyo.......................	65%	93%

In addition, one of the inlet elbows on a wellhead separator gave way. And finally, there was a collapse and eruption of well AH-24 in November 1976 and of well AH-20 in April 1978. The damage has been repaired and both wells are now in production [G. Cuéllar, personal communication].

During the 19 months from October 1976 to April 1978, little change in the total geofluid mass flow rate has been observed (see table 3.3). In fact, the original four wells feeding Unit 1 showed an increase of 7.8%, while the five wells serving Unit 2 dropped by 5%; overall, total flow increased by 1.8%. In that period a new well was brought on-stream (AH-24), but these percentages exclude the flow from this well. There has been no plugging of production or reinjection wells. Well AH-25 has not produced geofluid even though it is located within 200 m (650 ft) of four highly productive wells. Most likely a local zone of impermeable material surrounds this particular well. All other dry wells were at the periphery of the field and served to define the limits of the producing reservoir (see table 3.1 and figure 3.3).

Tests have shown that fluid reinjection plays an important role in maintaining reservoir pressure. While it does not seem necessary to reinject all the waste liquid, it is essential to control the amount being reinjected relative to the amount of fluid being withdrawn. At present about 63% of the waste liquid from the plant is reinjected; after Unit 3 comes into service, this percentage will fall to about 30%. Reinjection wells have been sited at the periphery of the field, downstream of the assumed northwestward recharge flow in the aquifer [Romagnoli et al., 1975]. In some instances, the reinjected fluid finds its way upward between the reinjection well casing and the bore hole (see figure 3.4), flowing into the reservoir instead of into the basement as intended. Reinjection may also have contributed to the lack of trouble with subsidence, although the formation is relatively hard and should not be subject to significant subsidence in any event.

3.3 Geothermal areas under exploration

To help keep pace with the 11% growth rate in electrical demand, El Salvador is exploring four additional geothermal areas. The actual total generation from all types of power plants and the maximum demand are shown in table 3.16 for 1975 and projected to 1980. To meet this expected

Table 3.16—*Actual and projected demands for electricity in El Salvador*

	1975 (actual)	1980 (projected)
Total electrical generation....	965.9 GW·h	1864.7 GW·h
Maximum demand..........	183.5 MW	351 MW

demand, CEL plans to have on-line in 1980 a total capacity of 455.2 MW with 95 MW being installed at Ahuachapán. Thus 21% of the country's installed electrical capacity in 1980 will be geothermal. The sites at which future geothermal power stations may be located are briefly described in the following sections.

3.3.1 Berlín

This field is located in eastern El Salvador, about 90 km (56 mi) from San Salvador, 18 km (11 mi) east of the Rio Lempa, and 6 km (4 mi) south of the Pan American Highway. Exploration took place at Berlín simultaneously with Ahuachapán in 1965 when two deep wells were drilled. The terrain is considerably more rugged than that at Ahuachapán, and this fact influenced the decision to proceed with Ahuachapán for the first geothermal power plant, although the results of the early exploratory studies at Berlín were encouraging. The first deep well was drilled to a depth of about 1800 m (5900 ft) and encountered a temperature of 271°C (520°F). The geofluid at Berlín contains roughly 10,000 ppm (1%) of total dissolved solids. This area is expected to be the site of the next (the fourth) geothermal unit in El Salvador by 1984/1985. Exploration is continuing at the site; ultimate capacity is estimated to be 100 MW(e).

3.3.2 Chinameca

This field is about 20 km (12 mi) east of Berlín and 17 km (11 mi) west of the city of San Miguel. Exploration is presently underway. A geothermal power unit is expected to be operating there by 1985 and may eventually support 100 MW(e).

3.3.3 San Vicente

The San Vicente geothermal field is located in east-central El Salvador, 50 km (31 mi) east of San Salvador and 40 km (25 mi) west-northwest of Berlín. Extensive exploration activity is taking place. The potential of this site is estimated at 100 MW(e).

3.3.4 Chipilapa

This area is about 5 km (3 mi) east of the Ahuachapán geothermal field and was the site of the first deep well drilled in El Salvador. Chipilapa may share Ahuachapán's geothermal field and is expected to develop about 50 MW(e) in the future.

REFERENCES [a]

CEL, 1976. "Annual Report 1976: Comisión Ejecutiva Hidroeléctrica del Río Lempa," San Salvador.

Cuéllar, G., 1975. "Behavior of Silica in Geothermal Waste Waters," *San Francisco, 1975*, Vol. 2, pp. 1337–1347.

Einarsson, S. S., Vides R., A., and Cuéllar, G., 1975. "Disposal of Geothermal Waste Water by Reinjection," *San Francisco, 1975*, Vol. 2, pp. 1349–1363.

Fuji, 1977. "Ahuachapán Unit No 3, CEL, El Salvador: Technical Particulars; Steam Flow Diagram, PL 311188; General Arrangement, PL 210640 and PL 210641; Geothermal Turbine Sectional View, ST 418390," Fuji Electric Company, Ltd., Tokyo.

MHI, 1977. "Geothermal Power Plant: Ahuachapán 30,000 kW×2 units in El Salvador, C. A.," Mitsubishi Heavy Industries, Ltd., Tokyo.

MHI, 1978: "List of Geothermal Power Plant," Mitsubishi Heavy Industries, Ltd., Tokyo.

Romagnoli, P., Cuéllar, G., Jimenez, M., and Ghezzi, G., 1975. "Hydrogeological Characteristics of the Geothermal Field of Ahuachapán, El Salvador," *San Francisco, 1975*, Vol. 1, pp. 563–574.

Sheehan, R. H., 1977. "Economic Aspects of Geothermal Energy," in *Minutes 6th CATMEC Meeting*, Brown Univ. Rep. No. CATMEC/5, DOE No. COO/4051-9, pp. 1–5 and Appendix A, Providence, R.I.

---

[a] See note on p. 24.

CHAPTER 4

ICELAND

4.1 Geological features of Icelandic geothermal regions

Iceland is perhaps better known for its direct use of geothermal energy in space-heating applications than for geothermal electric power generation. Roughly 65% of the population of this island heats its homes with geothermal hot water. Essentially all of the capital city of Reykjavik is heated by means of geothermal water at about 86°C (187°F). Plans call for the expanded use of geothermal hot water to provide up to 80% of the space-heating needs of the country within the next few years [Lindal, 1977]. There are only a few locations in Iceland where electric power is generated from geothermal resources. For the most part these lie inland and to the northeast, although exploration and the beginning of exploitation are now taking place on the Reykjanes Peninsula in southwestern Iceland.

Iceland is situated astride the Mid-Atlantic Ridge. The Icelandic graben which sweeps from the north to the southwest through the center of the island exhibits the great tension that exists in the crustal rift zone. Figure 4.1 shows this prominent geological feature along with the major cities and existing geothermal power plants. It has been observed that the rift zone is separating at a rate of about 2 cm/yr [Burke and Wilson, 1976]. A small geothermal plant at Svartsengi is at the southwestern extremity of the west branch of the rift valley. The geothermal power plants at Námafjall and Krafla are located within the rift zone at the northern end. The geological structure of these regions is highly unstable, creating serious problems related to well completions, reservoir engineering, and geothermal power production in general.

The Námafjall geothermal region has been the site of several projects making use of the thermal anomaly in the region. Wells had been drilled to permit the extraction of sulfur from the hydrogen sulfide that constitutes a portion of the geofluid. This mining operation gave rise to the name "Námafjall," which means "the mountain of the mines" [Ragnars et al., 1970]. The Námafjall thermal area covers about 400–500 ha (988–1235 acres) but is part of a much large thermal region, including Krafla, extending more than 5000 ha (12,350 acres). An abundance of surface thermal manifestations are found here, including boiling mud pools, steaming

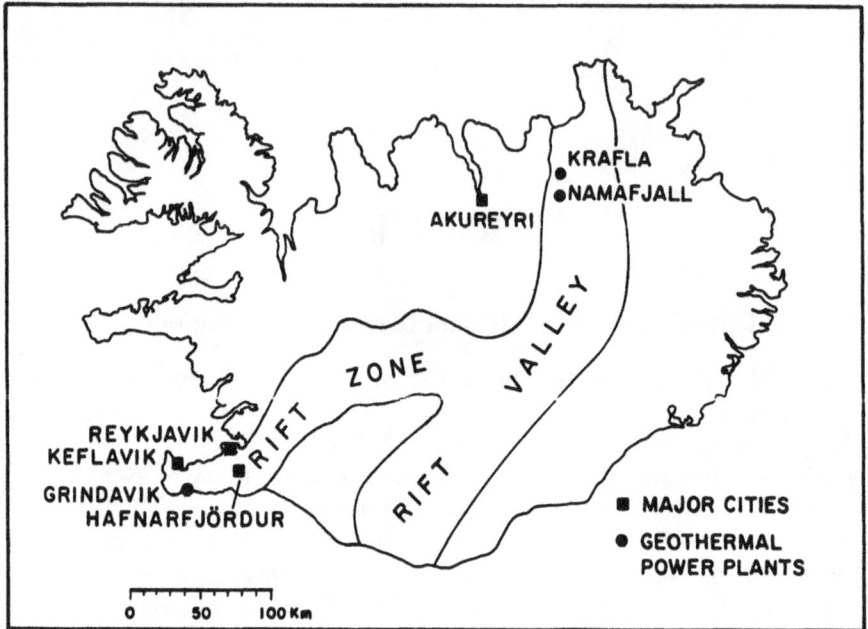

FIGURE 4.1—Map of Iceland showing rift zones, major cities, and sites of geothermal power plants.

ground, and fumaroles. The area is highly fractured with fissures and faults trending north-northeast/south-southwest. Rocks are of the silicic volcanic type and range in composition from basaltic andesites to rhyolites.

As can be seen from figure 4.1, Krafla lies in the same volcanic rift zone as does Námafjall. The Krafla area was most recently subjected to a series of strong seismic events. In July 1975 earthquake tremors were detected. Gradually these increased in strength, and on December 20, 1975, lava burst out in Leirhnjúkur, only 3 km (2 mi) from the site of the Krafla plant. Although the lava flow lasted only a few hours, steam continued to erupt until the end of the day. During this period 2000–4000 earth tremors were recorded each day.

During the first 3 months of 1976, there occurred seven earthquakes of magnitude greater than 4.0 on the Richter scale, with two of these exceeding magnitude 5.0. All of these were centered within a few kilometers of the plant site. By June 1976 most of the activity had ceased, but continuous vigilance is carried out by means of seismic monitors and field observations [Sólnes, 1976]. This recent activity has resulted in the formation of a large number of new fissures that traverse the geothermal fields in the Krafla area. These trend mostly north-south through the southern border of the caldera in which Krafla is located. A number of new hot springs and boiling mud pools were also created [Pálmason et al., 1975].

4.2 Námafjall

4.2.1 Well programs and gathering system

The geothermal resource at Námafjall is being employed to process diatomaceous earth, which is used as a filter aid. The diatomite plant at Mývatn is supplied with steam from wells at the Námafjall field [Birsic, 1977]. The geothermal power plant, also known as Kísilidjan, is relatively small and used to supply electricity for the processing plant. The arrangement of the wells relative to the power plant and adjacent diatomite plant is shown in figure 4.2. A number of shallow wells (not shown) were drilled

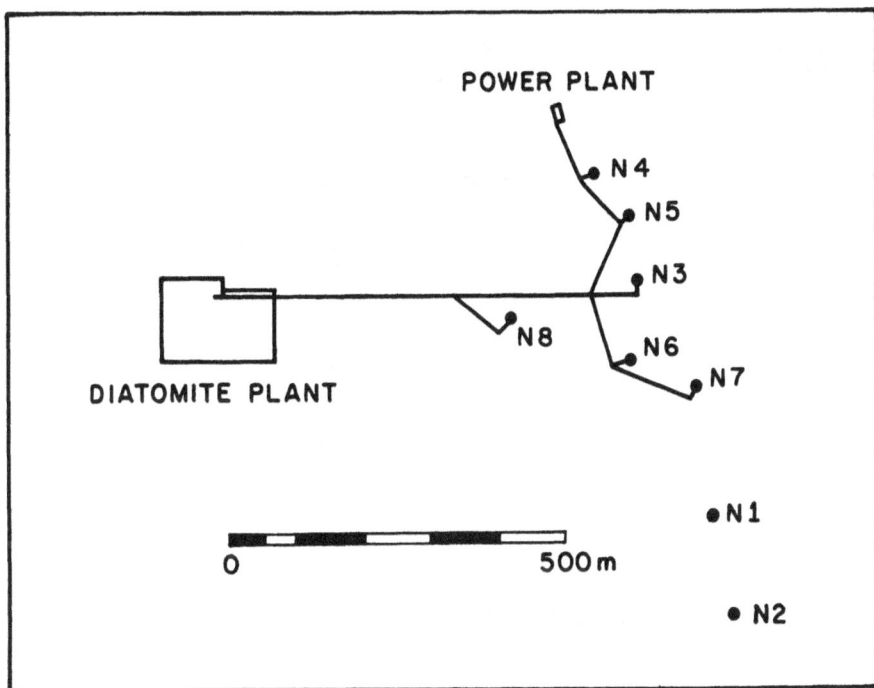

FIGURE 4.2—Arrangement of wells at Námafjall to serve diatomite processing plant and 3-MW noncondensing power unit [after Ragnars et al., 1970].

from 1947 to 1953 for sulfur mining. The present production wells were begun in 1963, with the first of these, N1, being completed in 1966, followed by N2 and N3 in 1968. The wells are located along two faults and are separated by about 90 m (295 ft).

Since the uppermost 180 m (590 ft) of the formation are permeable, loss of circulation is often encountered during the shallow drilling phase. Repeated cementing is required to prevent the wells from collapsing. The newest wells are cased as follows: (1) surface casing of 406 mm (16 in) to a depth of 30–40 m (100–130 ft); (2) anchor casing of 244 mm (9⅝ in) to 200 m (656 ft); (3) production casing of 194 mm (7⅝ in) to 600 m (1970 ft). If needed, 152-mm (6-in) slotted liners are installed below the production casing [Ragnars et al., 1970].

Wellhead equipment is shown schematically in figure 4.3. It was found necessary to install a U-pipe separator on the wellhead because of sand,

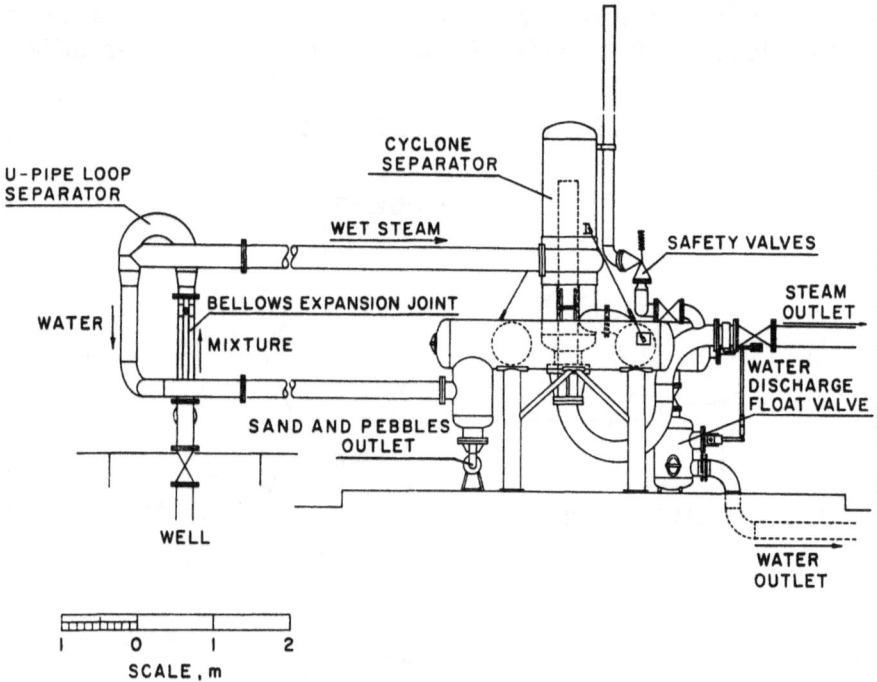

FIGURE 4.3—Typical wellhead equipment at Námafjall [Ragnars et al., 1970].

mud, pebbles, and other solid material ejected occasionally from the wells. A conventional cyclone separator is used to separate the vapor and liquid phases of the geofluid. Wellhead equipment is housed in a small shed to shelter it from severe Icelandic winters.

The steam pipelines are run 2 m (6.6 ft) above ground level to avoid burial in snow during winter and inundation during thaws. The pipes are 250 mm (10 in) in diameter and carry 50 mm (2 in) of glass wool insulation protected by a water proof sheathing of aluminum. Thermal expansion loops are provided along the lines to allow for temperature differences as large as 220°C (396°F), which amounts to a movement of about 0.26 m/100 m.

4.2.2   Geofluid characteristics

The wells produce a mixture of liquid and vapor plus an amount of non-condensable gases equal to about 1% (by weight) of the steam flow. The composition (by volume) of the noncondensable gases is roughly as follows: hydrogen sulfide, 52%; carbon dioxide, 32%; hydrogen, 12%; other gases such as nitrogen, methane, argon, etc., 4% [Bjornsson, 1968]. Down hole temperatures of nearly 300°C (572°F) have been observed; normal steam delivery temperature is about 183°C (361°F). Each well produces about 25 Mg/h (55,000 lbm/h) of separated steam at a pressure of 1078 kPa (156 lbf/in²). The total dissolved solids in the liquid (down-hole) is about 1000 ppm, nearly 60% of which is silica. At 25°C (77°F) the liquid has a pH≅7.

4.2.3   Energy conversion system

The power plant was of the noncondensing type with a nominal capacity of 3.0 MW. The original turbine-alternator was manufactured by British Thomson-Houston and had a rating of 2.5 MW. It was purchased second-hand by the Laxá Power Works to reduce construction time. The plant was, in fact, designed and built in 7 working months [Ragnars et al., 1970]. Recently the turbine had been upgraded to 3.0 MW.

The configuration of the power generation equipment is shown in figure 4.4. A final cyclone separator was required at the powerhouse to remove moisture that condenses out of the steam during transmission from the wells to the power plant. The exhaust stack/silencer stands beside the building, rising about 9 m (29 ft) above ground level. The powerhouse is quite compact, being 12.5 m wide, 8 m longer, and 7.5 m high (41 x 26 x 21 ft). Because of the climate, the transformer was located inside the building.

Since the plant was of the noncondensing type, it had a low resource utilization efficiency. On the assumption that 20% of the geofluid is vapor at the wellhead and that the appropriate ambient sink temperature for the region is 11°C (51.8°F), then the utilization efficiency is 14% and the plant consumed 82.5 kg of geofluid/kW·h (182 lbm/kW·h). Technical specifications for the Námafjall unit are contained in table 4.1 (see page 73).

4.2.4   Economic data and operating experience

Preliminary cost estimates made in 1967 showed that electricity could be generated at between 4.5 and 5.5 US mill/kW·h at a 91% capacity factor [Ragnars et al., 1970]. The capital costs have been computed [Leardini, 1974] as follows: power station, US$50/kW; wells and gathering systems, US$73/kW, for a total of US$123/kW. It should be noted, however, that the power station cost included a secondhand, recommissioned steam turbine. Had a new machine been purchased at that time (1968), the total capital cost for the plant would have been about US$147/kW.

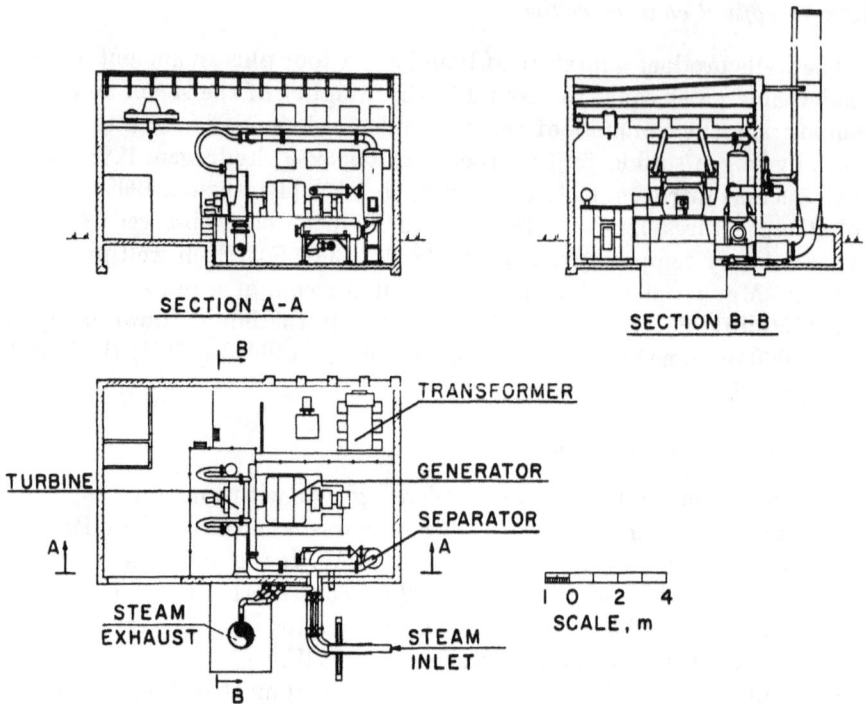

FIGURE 4.4—Layout of 3-MW geothermal power plant at Námafjall [Ragnars et al., 1970].

The power plant at Námafjall is no longer operational. Continuing volcanic activity in the region has damaged the boreholes, and there is only enough steam to supply the diatomite plant at Mývatn. There are no plans to set the power plant in operation again in the near future, and the turbine has been removed for use at another location.

4.3 Krafla

4.3.1 Planning

The geothermal power plant at Krafla is the first major power station of its type in Iceland. An advanced design, it uses a secondary flash process to generate additional steam for power generation from liquid that would otherwise be wasted. The project got underway officially in April 1974 when the Icelandic Parliament passed an act calling for construction of a geothermal power plant at Krafla. A number of agencies and organizations have been involved in the planning, design, construction, and operation of the plant. These include:

Krafla Geothermal Project Executive Committee—Responsibility for planning and construction of power plant, building and turbine.

National Energy Authority of Iceland—Well drilling, steam winning, design, and construction of steam gathering system and liquid
waste disposal system.

Nature Conservation Council—Consent for plant design, especially
waste water disposal as it impacts the environment, specifically Lake
Mývatn and River Laxá.

State Electricity Works—Electrical switchyard and transmission of
power away from the site.

A large number of subcontractors have been commissioned for the actual
operations, including the Rogers Engineering Company, Inc., of San Francisco who did the mechanical engineering for the power plant.

4.3.2 Well programs

A series of exploratory surveys were conducted from 1970 to 1973 in
the Krafla area. These included geological, geochemical, and geophysical
(electrical resistivity and magnetic) surveys. As a result of the magnetic
low (caused by hydrothermal alteration of magnetite) and the delineation
of the field by the resistivity survey, two exploratory wells were sited and
drilled to depths of about 1100 m (3600 ft).

One of these was quite hot, with a bottom-hole temperature of 298°C
(568°F), but was a poor producer (low permeability). It was estimated
that a temperature of 330°C (626°F) would be encountered at a depth of
2 km (6560 ft), the depth of the proposed production wells. The other
exploratory well was an excellent producer but yielded relatively cool
geofluid. The low temperatures were caused by an influx of cool water
from a shallow aquifer at a depth of about 340 m (1100 ft). Although a
high-temperature reservoir lay at deeper depths, the cold water from
the shallow reservoir resulted in bottom-hole temperatures of only 210°C
(410°F). In the case of production wells that intercept the shallow reservoir, the upper zone will be sealed off to prevent the degradation of fluid
temperature arising from mixing of the fluids from the hot and cold producing zones [J. T. Kuwada, personal communication].

A plan view of the power station site is given in figure 4.5 showing the
locations of the powerhouse, cooling tower, cooling pond, and proposed
wells. Well locations are tentative and subject to change. Also shown are
the sites of recent volcanic activity—Leirhnjúkur, mentioned earlier, and
Víti ("Hell"), a crater formed at the beginning of the "Fires of Mývatn"
in 1724. The proximity of these centers of volcanic action to the Krafla
bore field is evident. To date 12 wells have been completed and 2 more are
planned for the summer of 1979. Figure 4.6 shows a typical casing program for a production well. The wells are capped by a 254-mm (10-in)
gate valve rated at 6.2–10.3 MPa (900–1500 lbf/in²).

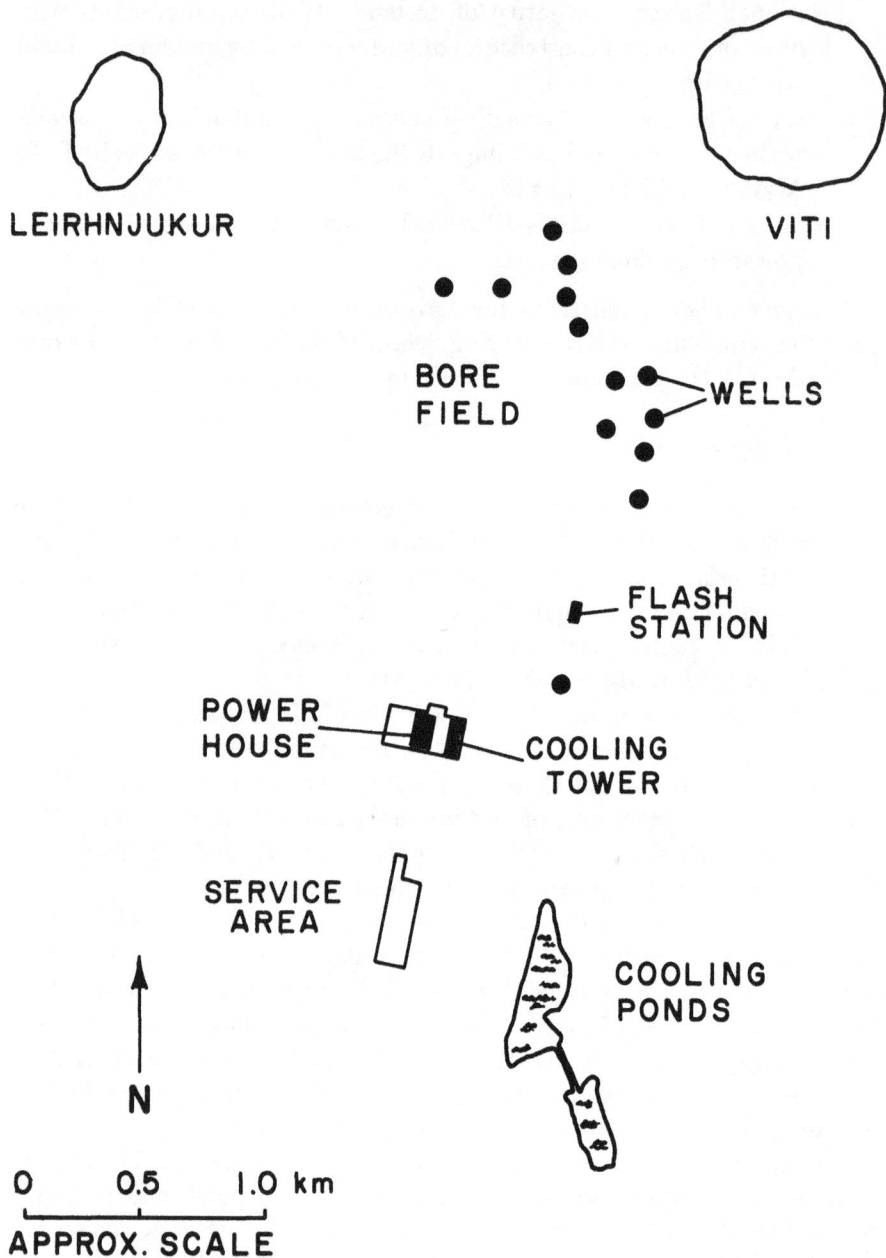

FIGURE 4.5—Arrangement of Krafla geothermal power plant and steam wells
(idealized) [after Sólnes, 1976].

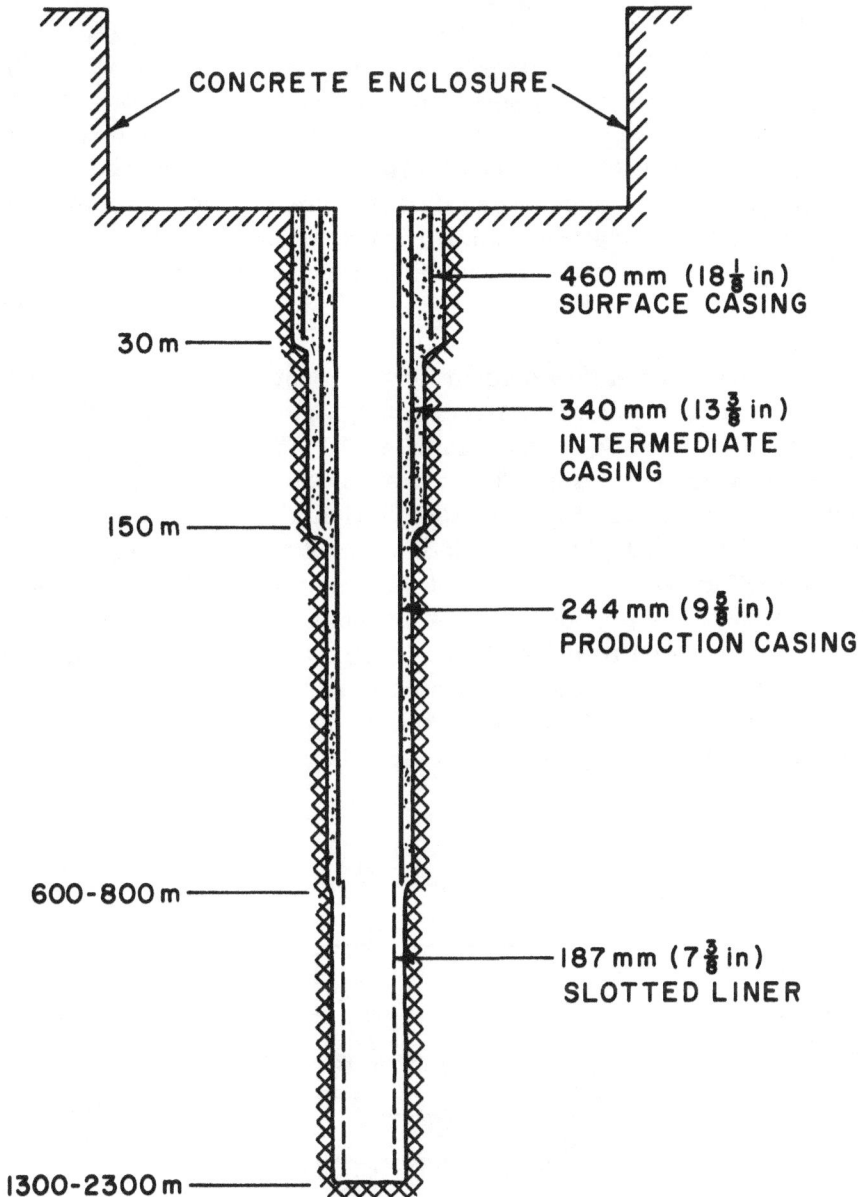

FIGURE 4.6—Typical casing program for production well at Krafla (not to scale) [after Sólnes, 1976].

4.3.3   Energy conversion system

The plant is of the separated-steam/single-flash (or "double-flash") type. Both high and low pressure steam are generated at a central station from which steam is piped to the powerhouse while the liquid is directed to a flash tank located 400–500 m (1312–1640 ft) from the plant. The turbine is a single-cylinder, double-flow, dual-pressure unit manufactured by Mitsubishi Heavy Industries, Ltd. It has five stages in each flow, with the secondary steam admitted to the machine through pass-in sections on each side where it mixes with the primary steam before expanding in the last stages.

A highly simplified flow diagram for the plant is shown in figure 4.7. Only one typical wellhead setup is depicted; there may be five or six wells required for each turbogenerator unit. Two 30-MW units are planned for Krafla, although there is insufficient steam available at present to supply fully even the one unit that is installed. Technical specifications are given in table 4.1 [MHI, 1978]; figure 4.8 is a photograph of the plant site.

The cooling pond can hold about 12,000 m³ (32×10^6 gal); a 4-day holdup time can be provided to allow for cooling of the waste water from about 90°C (194°F) to 10°–20°C (50–68°F). During this time, hydrogen sulfide will have escaped to the atmosphere and silica will have polymerized and settled out. The effluent will then be discharged into the Skarosels stream for eventual disposal in the Búrfell lava field [Sólnes, 1976].

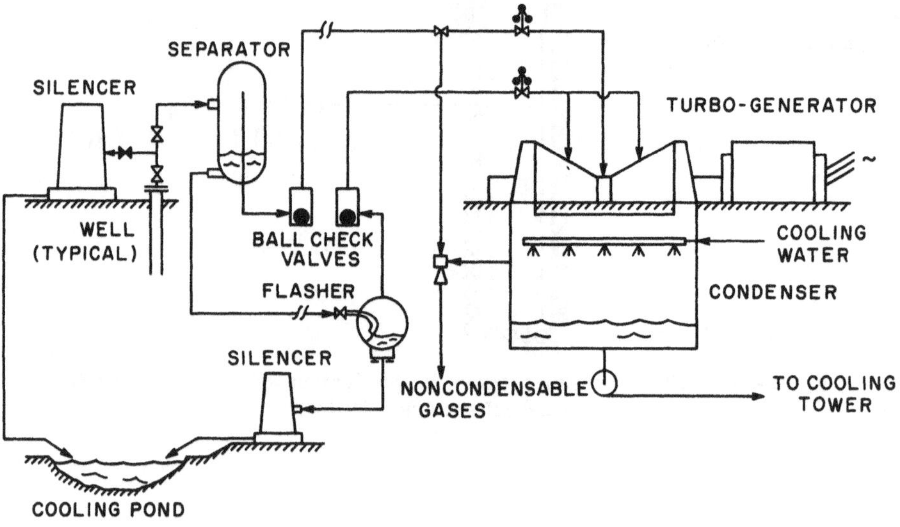

FIGURE 4.7—Simplified flow diagram for Krafla geothermal power station.

Table 4.1—*Technical specifications for Icelandic geothermal power stations*

	Námafjall 1969[b]	Krafla Unit 1 [a] 1977[b]	Svartsengi 1978[b]
Turbine data:			
Type	Single-cylinder, one Curtis stage, noncondensing	Single-cylinder, double-flow, dual-admission, impulse-reaction	AEG–KANIS type GT 63, geared
Rated capacity, MW (per unit)..	3.0	30.0	1.0
Maximum capacity, MW.......	3.4	35.0	1.0
Speed, rpm.	3000	3000	4479/1500
Main steam pressure, lbf/in²....	142.7	110.0	78.3
Main steam temperature, °F. .	354.5	334.4	311.0
Secondary steam pressure, lbf/in²	—	27.5	—
Secondary steam temperature, °F	—	244.4	—
Exhaust pressure, in Hg	31.4	3.5	50.2
Main steam flow rate, 10^3 lbm/h.	109	417	70.5
Secondary steam flow rate, 10^3 lbm/h	—	142	(c)
Condenser data:			
Type..	(None)	Low-level, direct-contact, tray type	(None)
Cooling water temperature, °F .	—	71.6	(c)
Outlet water temperature, °F...	—	115.2	(c)
Cooling water flow rate, 10^6 lbm/h	—	12.4	(c)

[a] Krafla Unit 2 is identical to Unit 1 and is under construction.
[b] Year of startup.
[c] Not available.

4.3.4 Economic data

The capital investment figures given in table 4.2 are projected values made in 1976; it is quite likely that the cost of the wells and steam transmission system will exceed the figures give in this table. In fact the latest estimates place the total cost at roughly $55 million. Cost per installed kilowatt is based on the full 60,000-kW station capacity.

4.3.5 Operating experience

At present about 7 MW is being produced from five wells that have proved troublesome. Although the geofluid is relatively clean (TDS≈

FIGURE 4.8—View of Krafla power plant and steam field looking east; flash station is at far left, high and low pressure steamlines approach powerhouse from left [photo courtesy of Mitsubishi Heavy Industries, Ltd.].

Table 4.2—*Estimated capital costs for Krafla geothermal power plant* [a]

Item	Capital Cost	Unit Cost
Powerhouse, energy conversion equipment, substation, staff housing, etc............................	$26, 000, 000	$433/kW
Production wells and complete steam gathering system..	10, 000, 000	167/kW
Transmission line from Krafla to Akureyri..........	3, 300, 000	55/kW
Total cost................................	39, 300, 000	655/kW

[a] Costs are given in US$.

1000 ppm with about 650 ppm silica), the wells have been subject to clogging. Two plugs seem to develop: a deep plug at about 1550 m (5085 ft) which consists of iron sulfide, and a calcium carbonate plug at about 700 m (2300 ft). In a short period of time, the 187-mm-diameter (7⅜-in) hole in the slotted liner is reduced to about 38 mm (1½ in). One of the two production zones lies between the two plugs, and it is possible that the calcium carbonate plug may be eliminated by blocking off the upper production zone [J. T. Kuwada, personal communication]. In any event, it now seems evident that the cause of poor production is the presence of these deposits in the bore holes rather than collapsing of the formation from earthquake activity as was earlier believed.

4.4 Svartsengi (Grindavik)

A 2×1-MW geothermal power unit is located in Svartsengi near Grindavik on the Reykjanes peninsula in southwestern Iceland [Gudmundsson, 1978]. It is part of the power plant supplying the Sudurnes District Heating that includes four 12.5 MW thermal units to serve the 12,000 people of the Reykjanes peninsula with 85°C (185°F) hot water. Bottom-hole temperature is 235°C (455°F), the temperature of the steam evaporator is 155°C (311°F). The plant is of the noncondensing type and makes use steam from 5 wells. It is expected that the capacity installed at the site will be increased as field development takes place and expansion of the plant can be justified. Table 4.1 contains what little information is known about the plant.

REFERENCES [a]

Birsic, R. J., 1977. "A Krafla Visit: Iceland's Major Project," *Geothermal Energy Magazine*, Vol. 5, No. 10, pp. 8–16.
Björnsson, S., 1968. "Aflmaeling á N–3 Námafjalli," Geothermal Division, National Energy Authority, Reykjavik. (In Icelandic.)

[a] See note on p. 24.

Burke, K. C. and Wilson, J. T., 1976. "Hot Spots on the Earth's Surface," in *Continents Adrift and Continents Aground*, W. H. Freeman and Co., San Francisco, pp. 58–69.

Dewey, J. F., 1976. "Plate Tectonics," in *Continents Adrift and Continents Aground*, W. H. Freeman and Co., San Francisco, pp. 34–45.

Gudmundsson, J. S., 1978. Personal communication to D. J. Ryley, Geothermal Division, National Energy Authority of Iceland, Reykjavik.

Leardini, T., 1974. "Geothermal Power," *Phil. Trans. R. Soc. Lond. A.*, Vol. 276, pp. 101–120.

Lindal, B., 1977. "Geothermal Energy for Space and Process Heating," in *Energy Technology Handbook*, D. M. Considine, ed., McGraw-Hill, New York, pp. 7.43–7.58.

MHI, 1978. "List of Geothermal Power Plant," Mitsubishi Heavy Industries, Ltd., Tokyo.

Pálmason, G , Ragnars, K., and Zoëga, J., 1975. "Geothermal Energy Developments in Iceland, 1970–1974," *San Francisco, 1975*, Vol. 1, pp. 213–217.

Ragnars, K., Saemundsson, K., Benediktsson, S., and Einarsson, S. S., 1970. "Development of the Námafjall Area, Northern Iceland," *Pisa, 1970*, Vol. 2, pp. 925–935.

Sólnes, J., 1976. "Krofluvirkjun-Krafla Geothermal Power Plant," Krafla Geothermal Project Executive Committee, Akureyri, Iceland.

Wilson, J. T., 1976. "Continental Drift," in *Continents Adrift and Continents Aground*, W. H. Freeman and Co., San Francisco, pp. 19–33.

CHAPTER 5

ITALY

5.1 Historical use of geothermal energy in Italy

Documentation shows that the natural steam fields in Tuscany were recognized as early as the third century and that the commercial potential of these mineral-laden waters led to wars between the Tuscan republics during the Middle Ages [ENEL, 1970]. It was not until 1904, however, that the power of natural steam was first harnessed to produce electricity, the accomplishment being credited to Prince Piero Ginori Conti.

Conti's original system used a reciprocating engine that received steam separated from the geothermal fluid. The engine was of the noncondensing type, exhausted to the atmosphere, and generated about 15 kW of electric power. The output from the DC generator provided lighting for the boric acid factory at Larderello in the boraciferous region of Italy. This primitive engine was replaced by a turboalternator of 250-kW capacity in 1913, thus marking the beginning of the production of electricity from geothermal sources on a commercial scale [Conti, 1924]. Since that time endogenous fluid has been tapped at two other sites, Monte Amiata and Travale, and the total installed geothermal electric generating capacity in Italy has grown to 420,000 kW.

In the following sections we will describe some of the geologic features of the main geothermal regions currently under exploitation—Larderello, Monte Amiata, and Travale—as well as the technical details related to the gathering and distribution of the geothermal fluid, the energy conversion systems and associated auxiliaries, and the economic and operating experiences of the plants.

5.2 Boraciferous region (Larderello)

5.2.1 Geology and exploration techniques

The Larderello region in general structural terms corresponds to a tectonic high located between the Era Graben to the north and northwest and the positive feature of the crystalline basement evident in outcroppings to the south and southeast [ENEL, 1970]. The presence of a deep magmatic intrusion at about 6–8 km (4–5 mi) is inferred from the huge gravity deficit. The structural outline of the region is caused by the apennine and plutonian tectonics in conjunction with the magmatic intrusion. The up-heaval of Pliocenic coastal deposits to about 600 m (1970 ft) gives evidence of the plutonian tectonic.

The high heat flow in the region is generated by the gross interaction between the African and Eurasian tectonic plates and several smaller plates that are in contact in the area. The Larderello geothermal field is part of an arc of high heat flow extending along the west coast of the Italian peninsula from Tuscany to Sicily [Mongelli and Loddo, 1975].

Figure 5.1 shows a geologic map of the boraciferous region in both plan and cross-sectional views [ENEL, 1970]. The major outcrops may be grouped into three main complexes: (1) the upper complex (denoted by regions "2" in the figures), which comprises shales, limestones, and sand-stones ("argille scagliose") and constitutes, for the most part, the imper-meable cap rock for the underlying reservoir; (2) the main permeable complex (denoted by "4"), which forms the circulation region for the endogenous fluid and comprises the highly pervious Tuscan formations ranging from radiolarites to evaporites; and (3) the basement complex (denoted by "5"), consisting of phyllitic-quartzitic formations, which is highly impervious where phyllites predominate but which can be highly pervious where intercalations of quartzites or crystalline limestones occur [ENEL, 1970]. The outcroppings of the main permeable complex in the southern region contribute importantly to the recharge of the aquifer through the absorption of rainfall.

During the exploration phase, geologic, geochemical, and geophysical methods have been employed, of which, geochemical techniques have proven highly effective. Normal analytical techniques have been used along with specialized ones to determine the isotopic relationships of certain elements in the geothermal fluid and the rocks including oxygen, hydro-gen, and carbon. Of the geophysical methods, the Schlumberger quadri-pole technique has been used extensively because of its relatively low cost and high efficiency. This method is favored since the reservoir is usually located at depths less than 1000 m (3280 ft) and is characterized by a distinct resistivity high ($>100 \ \Omega \cdot m$) relative to the overlying cap ($\sim$2–40 $\Omega \cdot m$).

Heat flow, thermal gradients, and thermal conductivity measurements have also been employed as prospecting tools. The area is characterized by exceptionally high thermal gradients, on the order of 30°C/100 m (16°F/100 ft) and in some places as high as 100°C/100 m (55°F/100 ft). The accepted normal gradient is about 3°C/100 m (1.6°F/100 ft). The geo-thermal field at Larderello is believed to cover about 25,000 ha (62,000 acres) [Koenig, 1973], although the drilled area extends over only about 18,500 ha (45,700 acres) [Ceron et al., 1975; Ellis and Mahon, 1977].

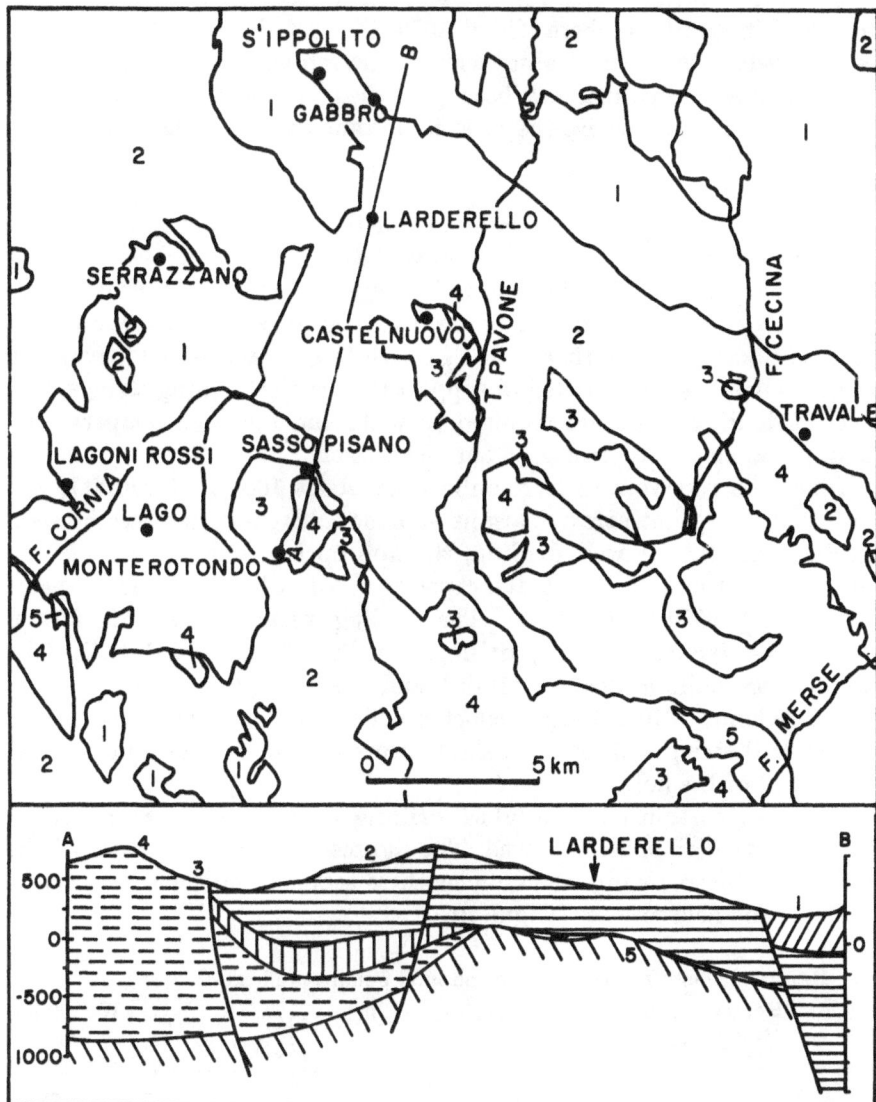

FIGURE 5.1—Geologic map and cross-section of the boraciferous region (Larderello), Tuscany, Italy. Geologic formations: 1. Clays, sandstones, conglomerates, etc.; 2. Shales, marls, limestones, etc.; 3. "Macigno," "polychrome shales"; 4. Tuscan formations; 5. Phyllites, quartzites [after ENEL, 1970].

5.2.2 Well programs and gathering system

There are roughly 190 producing wells in the Larderello region out of a total of 511 drilled [Overton and Hanold, 1977]. Average depth of all wells is 656 m (2152 ft); wells drilled since 1969 average 1129 m (3704 ft) in depth [Ceron et al., 1975]. The techniques used in the drilling, casing, and cementing of the wells in the Italian geothermal fields have been reported in detail elsewhere [Cigni, 1970; Cigni et al., 1975]. Furthermore, an extensive discussion dealing with the design and construction of geothermal steam pipelines has been published [Pollastri, 1970]. We will therefore only summarize the most important operations as described by these authors.

5.2.2.1 *Drilling of wells.* Geothermal drilling operations are in some ways similar to those employed in oil well drilling, the notable exceptions being that geothermal wells tend to be larger in diameter, formations are of higher temperature, and flow velocities tend to be larger. These factors lead to the conclusions that drill rigs with larger capacities (for a given depth capability) are needed to support the heavier drilling strings, that special drilling muds are required to withstand the high temperatures, and special materials are needed to resist erosion.

Since the reservoir at Larderello is at about 1000 m (3280 ft), and taking into account safety margins and the inherent differences between geothermal and oil well drilling, the appropriate size rig will be one of 1600–1800 m (5250–5900 ft) capacity. Some of the rigs are capable of reaching 3000 m (9800 ft). The drilling string consists of 127-mm (5-in) and 168-mm (6⅝-in) drill pipe with 203-mm (8-in) O.D. drill collars for drilling 406-mm (16-in) and 508-mm (20-in) holes, respectively [Cigni, 1970]. Large-diameter collars are preferred because of added drill string stability and the improvement in maintaining vertical alignment of the hole.

The power required to conduct drilling operations varies according to the particular phase involved. The figures below are estimates of the maximum power requirements under three sets of conditions for various machine and field services [Cigni, 1970]:

Regular drilling (rotary machine and mud pump at full rate):

Rotary machine.................	75 kW (100 hp)
Mud pump...........	520 kW (700 hp)
Field services..	35 kW (50 hp)
	630 kW (850 hp)

Round trip for pipe extraction from maximum depth or casing operations:

Draw-works.....	330 kW (440 hp)
Field services.......................................	35 kW (50 hp)
	365 kW (490 hp)

Emergency hoisting (during fishing operations):

Draw-works...............................	300–330 kW (400–440 hp)
Mud pump (half load).	185–260 kW (250–350 hp)
Field services...............................	35 kW (50 hp)
	520–625 kW (700–840 hp)

The accepted procedure for drilling wells in the geothermal fields of Italy is to use drilling muds while passing through cap rock [Celati et al., 1975]. The drilling mud must be carefully selected since the high temperatures encountered stimulate chemical reactions in the mud, altering the fluid viscosity, its free-water content, and other properties [ENEL, 1970]. The presence of clays, anhydrites, and gypsum lenses in the formation makes it difficult to maintain mud properties during drilling. In addition, the permeable layers that are encountered lead to frequent loss of circulation. The muds used are dispersed ferro-chrome-lignosulfonate-treated types which exhibit good dispersion characteristics owing to their protective effect on clay particles. Furthermore, in high concentrations they inhibit exchange reactions [Cigni et al., 1975]. Drilling operations in the permeable reservoir and in the basement rock are usually carried out with fresh, cold water without a return because of the high loss of circulation.

5.2.2.2 *Well casings and cementing operations.* A pair of typical well profiles are shown in figure 5.2; a standard casing program is shown on the left and a casing program for an exploratory or a relatively deep well is shown on the right. The standard well produces natural steam through a 311-mm (12¼-in) open hole and a 400-mm (13⅜-in) casing that is cemented within a 406-mm (16-in) hole. The deeper well has a 216-mm (8½-in) open hole throughout the permeable zone with a 244-mm (9⅝-in) production casing. In this case the 400-mm (13⅜-in) casing serves as an intermediate casing for safety purposes. The casings are J-55 API heavy-wall pipe to withstand the corrosive nature of the geothermal fluid and the severe temperature cycling to which the wells may be subjected.

Cementation of the casings to the formation or to other casings is a critically important operation. A proper cementation job must result in complete and uniform filling of the well casing annulus in order to withstand the various strains undergone by a casing, to guarantee good bonds between the cement and the casing and between the cement and the formation to strengthen and protect the entire casing column, and finally, to allow for well checks during drilling to assure proper production during later stages. The art of casing and cementing geothermal wells requires attention to many factors, such as correct choice of cement and additions, exact centering and placement of the casing, and proper flushing of the annulus prior to cementation to avoid mud contamination [Cigni et al., 1975].

The cement used for wells in the boraciferous region consists of a mixture of portland 425 cement and a fine-grained silica flour in 60/40 proportions. In laboratory tests under a pressure of 9.8 MPa (1420 lbf/in²) and a temperature of 150°C (320°F) this cement showed a compressive strength of 34 MPa (4950 lbf/in²) after a 28-day curing period [Cigni et al., 1975].

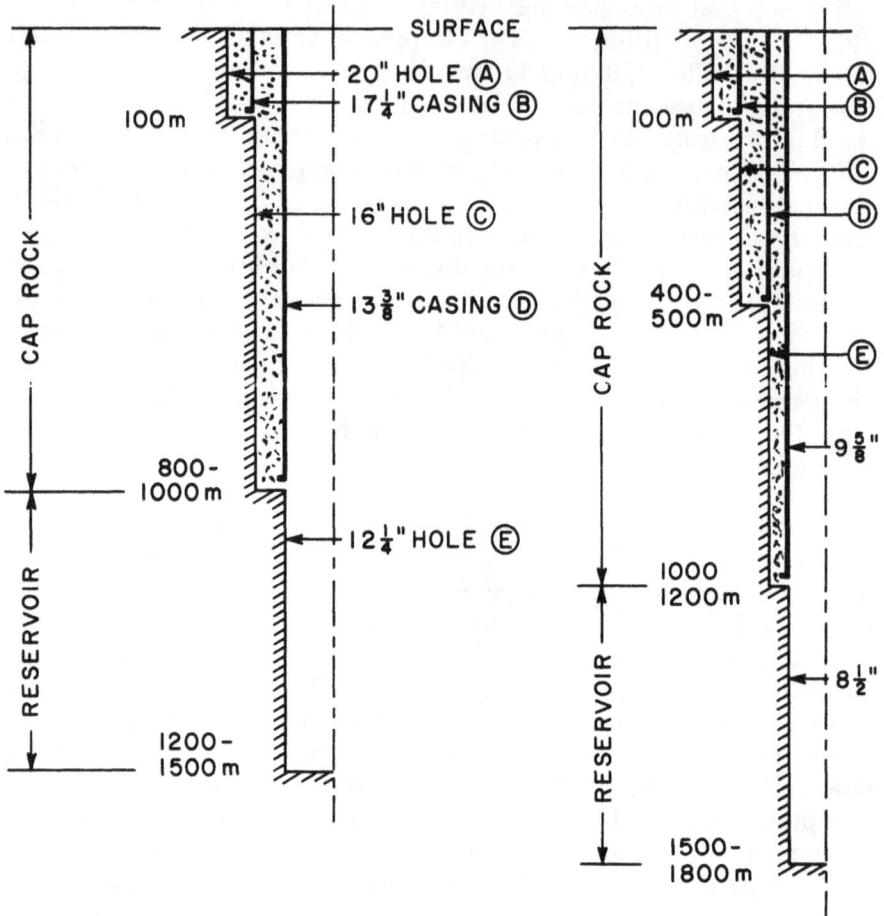

FIGURE 5.2—Typical well profiles in Italian geothermal fields: production well (left) and exploratory or deep well (right) [after Cigni et al., 1975].

5.2.2.3 *Design of wellhead equipment.* The design of wellhead equipment for geothermal applications differs considerably from that used for oil wells. The most important unique characteristics of geothermal wells are [Cigni, 1970]:

High fluid velocities
Relatively low well closing and operating pressures
High mud and steam temperatures
Strong corrosive nature of fluid
Connection to large-diameter surface pipelines.

Four wellhead arrangements are shown in figure 5.3; each is used during various stages of the drilling operation, as described below.

Arrangement I

Arrangement II

Arrangement III

Arrangement IV

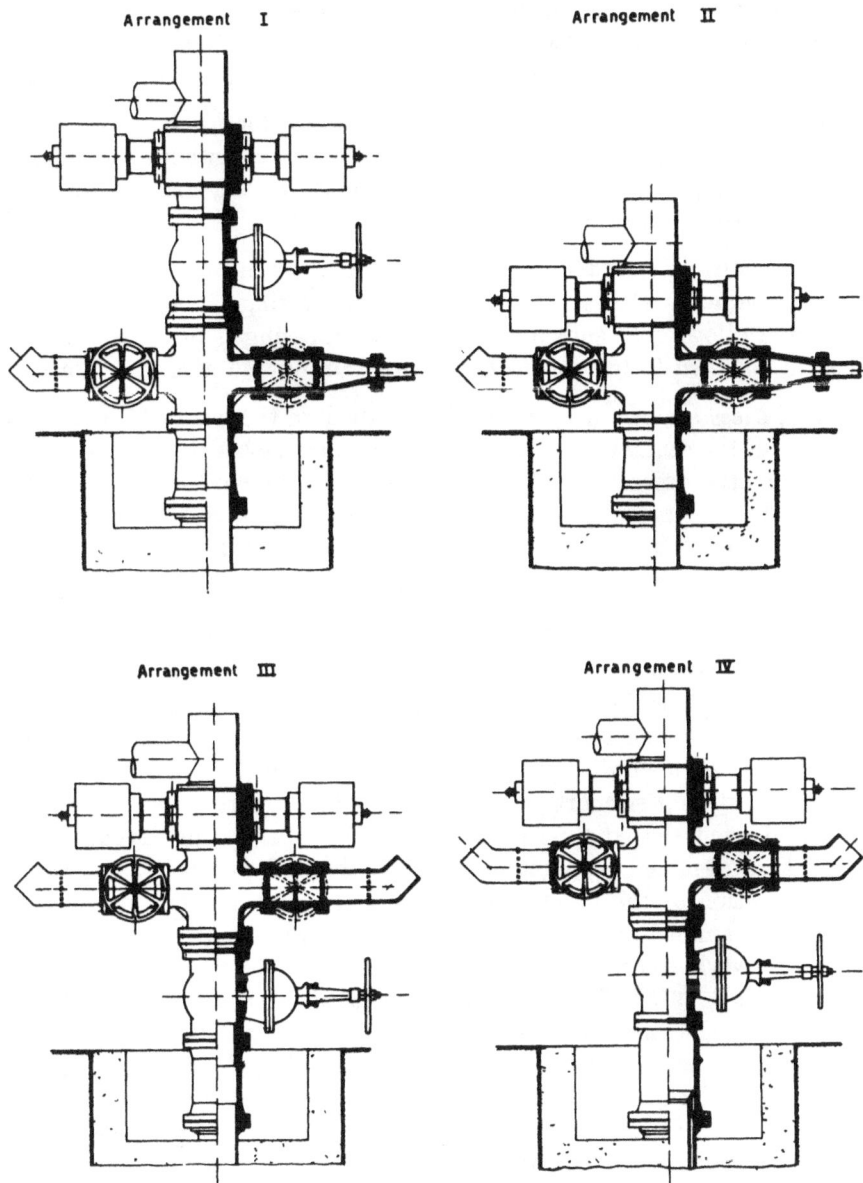

FIGURE 5.3—Typical wellheads for various phases of drilling operations at Italian geothermal fields [Cigni, 1970].

Arrangement I: used during drilling through the cap rock; wellhead mounted on the surface casing with a guiding bore of 311 mm (12¼ in). Side valves are 254-mm (10-in); mud filling is accomplished through a side connection.

Arrangement II: used during widening of the hole in cap rock from 311 mm (12¼ in) to 406 mm (16 in) ; wellhead mounted on surface casing; central valve eliminated since blowouts are not expected in this phase.

Arrangement III: used during drilling in the production zone; wellhead mounted on production casing. Since flow of steam is expected in this phase, all equipment is in place to handle the situation, starting with the 356-mm (14-in) central valve.

Arrangement IV: used when production casing is in place but only partially cemented; wellhead mounted on surface casing. A device that supports and allows for centering the production casing is provided. Equipment is installed to handle the expected steam flow in case of a blowout.

In all cases a blowout preventer is installed on the wellhead. Designed to close off the central bore in case of emergencies, even when the drilling string or other pieces of equipment are below the surface, it works by means of a mechanically or hydraulically actuated valve fitted with jaws to clamp around anything positioned in the well. High-temperature gaskets are required for steam-well drilling [Cigni, 1970].

5.2.2.4 *Steam pipelines.* Steam is conveyed from the individual wellheads to the power stations through an interconnected network of pipes. The photograph in figure 5.4 shows a typical wellhead connection; it is characterized by the sweep of a large-radius expansion bend from the wellhead to an anchor. Types of supports used include fixed moorings (anchors), sliding supports, turning type, and supporting trestles of the fixed and fixed slotted type [Pollastri, 1970]. The zig-zag layout of the network is designed to accommodate the thermal expansion associated with operation of the plants. Expansion caused by temperature fluctuations from ambient values to 260°C (500°F) can be absorbed.

The network consists of more than 118 km (73 mi) of steam pipes [Ceron et al., 1975]. The pipes are fabricated from weldable steel and have a wall thickness of 6–8 mm (0.24–0.31 in) and diameters of 250, 350, 450, 650, and 810 mm (10, 14, 18, 26, and 32 in). Asbestos fiber is used for insulation, in thicknesses of 30, 60, 90, and 120 mm (1.2, 2.4, 3.5, and 4.7 in) [DiMario, 1961]. Pipe and insulation are protected within an aluminum-plate jacket or painted with a coating of bituminous material [ENEL, 1970].

5.2.3 Geofluid characteristics

The geothermal fluid produced at the wells in Larderello consists of steam (dry, saturated or slightly superheated) and a mixture of noncondensable gases. The amount of noncondensables is relatively high, ranging from 1% to 20% by weight of the total fluid flow, on average, with some new wells showing even higher percentages. As a rule, the gas content of the steam at Larderello has remained roughly constant dur-

FIGURE 5.4—Typical steam pipeline connection at wellhead [Pollastri. 1970].

ing the period of exploitation, attributable to the fact that the natural surface thermal manifestations that have existed for centuries have prevented a buildup of large amounts of gas. The reservoir is thus viewed as being in a steady state as regards the evolution of noncondensable gases. One well, drilled a few kilometers east of Larderello and believed isolated from the main thermal area (and thus not benefitting from the purging action in the main field), produced geofluid containing 98% gas and only 2% steam. The composition of the gas was essentially identical to that found in the fluid from the wells of the main field [ENEL, 1970].

The noncondensable gas consists of carbon dioxide, for the most part, with small amounts of hydrogen sulfide, hydrogen, methane, and nitrogen. Recent reports [ENEL, 1970] put the CO_2 percentage at about 4.8% (by weight of geofluid), H_2S at 0.5%, and all others at less than 1%. Table 5.1 lists the composition of the noncondensables for specific areas within the boraciferous region.

Table 5.1—*Composition of noncondensable gases found in geofluid produced in the boraciferous region of Italy* [a]

Area	Percentage by volume of total noncondensables				
	CO_2	H_2S	H_2	CH_4	N_2
Castelnuovo.............	95. 98	1. 75	1. 08	0. 73	0. 46
Lago....................	89. 48	3. 02	4. 50	1. 95	1. 05
Lagoni Rossi.............	88. 60	4. 00	5. 17	1. 33	0. 90
Larderello...............	93. 82	2. 56	1. 87	1. 10	0. 65
Monterotondo............	89. 30	2. 20	3. 57	3. 74	1. 19
Sasso Pisano.............	91. 77	2. 77	2. 56	2. 14	0. 76
Serrazzano...............	91. 32	3. 03	3. 53	1. 50	0. 67

[a] Source: Pollastri, 1970.

The highest reservoir temperature encountered so far has been 300°C (570°F) [Overton and Hanold, 1977]; the maximum pressure is 3.1 MPa (450 lbf/in²). Steam is produced at temperatures ranging from 140° to 220°C (285° to 430° F) and at pressures from 200 to 700 kPa (29 to 102 lbf/in²) [Ellis and Mahon, 1977]. Average fluid flow rate per producing well is about 17 Mg/h (37,500 lbm/h) [Ceron et al., 1975], although maximum flow rates may range from 50 to 100 Mg/h (110,000 to 220,000 lbm/h) and in some cases may even exceed 300 Mg/h (660,000 lbm/h) [ENEL, 1970]. Flow rate varies considerably from well to well and depends strongly on the age of the well, particularly in the early stages of production. Figure 5.5 shows the production history of two wells in the Larderello field; each exhibits an approach to a steady-state flow rate after an initial transient period during which flow rate decreases by a significant amount. Figure 5.6 gives well productivity curves as a function of well-

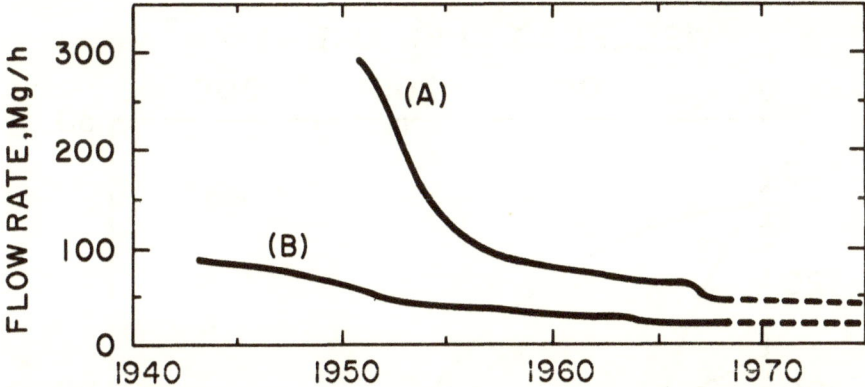

FIGURE 5.5—Production history of wells No. 85 (A) and Fabriani (B) at Larderello;
wellhead pressures roughly 0.4 MPa (57 lbf/in²) [after ENEL, 1970].

head pressure for three wells, one each from Sasso Pisano, Lagoni Rossi,
and Gabbro. All three curves exhibit the expected behavior for gas flow
and may be represented analytically by an equation of the form:

$$(p/p_o)^n + (m/m_o)^n = 1, \tag{5.1}$$

where the exponent n lies in the range $1.5 < n < 1.85$. The subscript o denotes
the maximum value of either the pressure or the flow rate.

5.2.4   Energy conversion systems

Power is produced at the present time in Larderello by means of two
types of energy conversion systems: direct-intake, noncondensing, impulse
or impulse-reaction turbines ("Cycle 1"), or direct-intake, condensing,
impulse-reaction turbines ("Cycle 3"). Prior to 1968 another type was in
operation, one using pure steam generated from geothermal steam in heat
exchangers and expanded in impulse or impulse-reaction, condensing turbines ("Cycle 2"). These three schemes are shown schematically in figure
5.7.

Cycle 1 plants are installed at locations with high noncondensable gas
content in the geothermal steam or at sites insufficiently developed to
justify construction of steam lines to join the field to the main network.
Such plants are extremely simple, highly reliable, easily assembled or dis-
assembled, and offer low costs because they may be remote controlled from
a nearby power station.

Cycle 2 plants were used when it was desirable and economic to extract
chemicals such as boric acid and ammonia from the geothermal fluid, while
at the same time avoiding materials corrosion problems in the turbine and
taking advantage of the improved power output associated with condens-

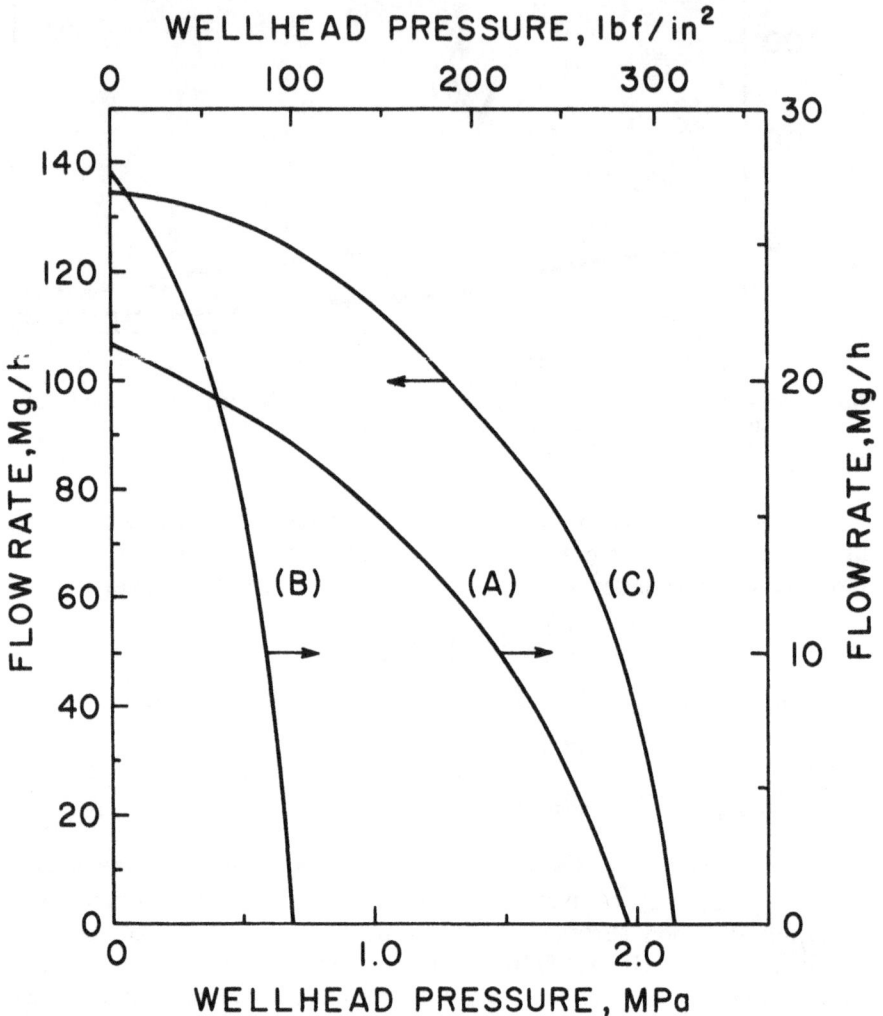

FIGURE 5.6—Flow rate versus wellhead pressure for wells: (A) St. Silvestro (Sasso
Pisano); (B) Scarzai 3 (Lagoni Rossi); and (C) Gabbro 9 [after Pollastri, 1970].

ing operation. Considerable difficulty was encountered in the operation
of the heat exchangers, however, because the water tubes that formed the
boiler section were subject to deposits of iron sulfide or breakage, depending on whether iron or aluminum was used for the tube material [Hahn,
1924]. Since chemicals are no longer extracted from the fluids and the
problems of corrosion of turbine blades can be avoided, this energy conversion scheme has been eliminated.

Cycle 3 plants form the mainstay of the Italian geothermal system. The
effects of impurities or corrosive substances in the steam can be reduced by

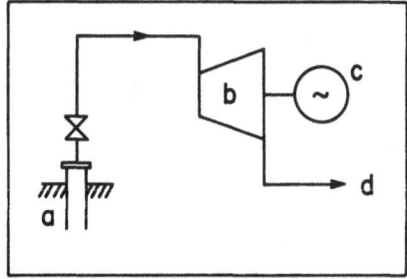

(a)

FIGURE 5.7—Energy conversion schemes at Larderello. (a) Direct-intake, noncondensing "Cycle 1" plant: a=steam well, b=turbine, c=generator, d=exhaust to atmosphere. (b) Pure-steam, condensing "Cycle 2" plant: a=steam well, b=heat exchanger, c=turbine, d=generator, e=degassing plant, f=condenser, g=liquid discharge, h=to and from cooling tower. (c) Direct-intake, condensing "Cycle 3" plant: a=steam well, b=water injection (scrubber), c=axial separator, d=turbine, e=generator, f=condenser, g=gas compressor, h=to and from cooling tower.

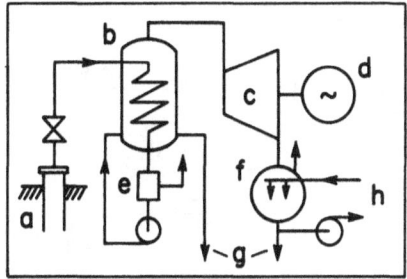

(b)

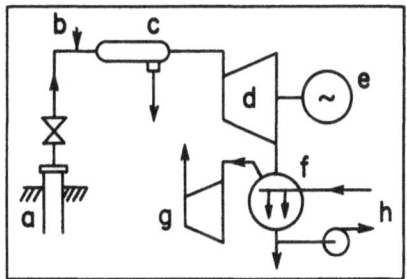

(c)

scrubbers located upstream of the turbine inlet. Pure water or alkaline solutions may be injected to wash the steam; axial separators then remove the injected liquid prior to admission into the turbine. The large amount of noncondensable gases in the steam requires the use of high-capacity turbocompressors to remove the gases from the condensers. The schematic layout diagram of figure 5.8 shows a typical arrangement for a Cycle 3 power unit. Of particular interest are the three stages of intercooling used with the gas compressor, the first stage of which is integral with the condenser.

A typical flow diagram for a 14.8-MW (gross)/13.4-MW (net) power unit is given in figure 5.9. Inlet steam is 185°C (365°F) and 443 kPa (64.3

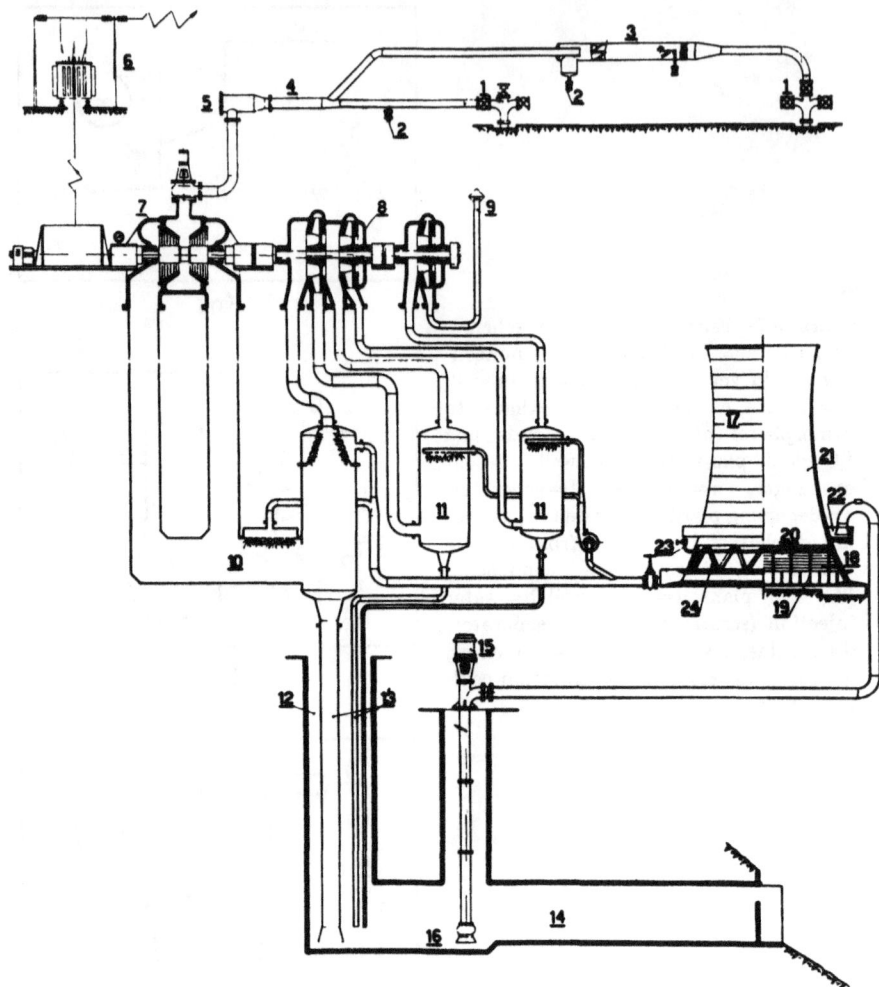

FIGURE 5.8—Typical arrangement for Cycle 3 power unit at Larderello [Allegrini and
Benvenuti, 1970].

lbf/in²) with about 4% (by weight) of noncondensable gases. The geothermal resource utilization efficiency, based on the available work of the
geofluid relative to the design wet-bulb temperature of 19.4°C (67°F), is
about 52%. However, none of the actual units analyzed and described below
have efficiencies as high as this; highest actual efficiency is 47.4% for the
two units located at the Sasso 2 geothermal field.

5.2.4.1 *Condensing units.* Condensing units in the boraciferous region are
in operation at the following sites: Larderello, Gabbro, Castelnuovo, Serrazzano, Lago, Sasso Pisano, and Monterotondo. There are 27 such units

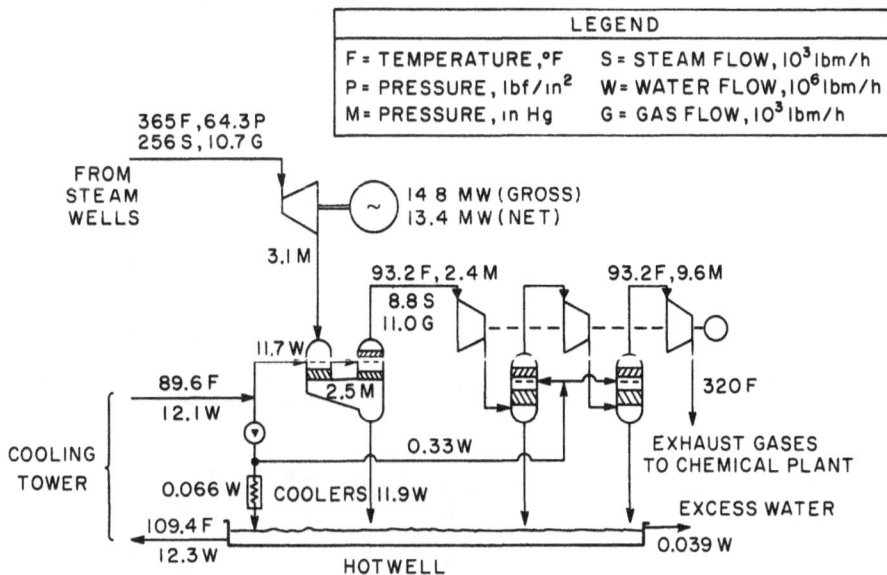

FIGURE 5.9—Typical flow diagram for Cycle 3 power unit with 14.8-MW installed capacity [Dal Secco, 1970].

installed, with a combined capacity of 362.7 MW. These range in size from a 2-MW unit at Castelnuovo to 26-MW units, three of which are located at Larderello 3 and one at Castelnuovo.

All plants use natural-draft cooling towers designed to handle flow rates of 9, 12, 15, and 18×10^3 m³/h (20, 26, 33, and 40×10^6 lbm/h) of water depending on the unit rating. The water is cooled through a range of 41°–31°C (105.8°–87.8°F), typically, in an environment at 25°C (77°F) and a relative humidity of 60%, i.e., at a wet-bulb temperature of 19.4°C (67°F). The choice of natural-draft over forced-draft towers was dictated by lower operating costs and the need for reliable, continuous operation. Cooling water is circulated by means of 2-speed helico-centrifugal pumps with vertical axes and adjustable blades, and with a maximum flow capacity of 9000 m³/h (20,000 lbm/h) [DiMario, 1961]. The following sections offer more details on the condensing power units in the boraciferous region. A listing of technical particulars is given in table 5.2.

Larderello 2. The indirect-steam system (Cycle 2) that generated clean steam by means of heat exchangers has been abandoned in favor of direct-intake, condensing plants. Cycle 2 plants allowed for the recovery of various chemicals from the geothermal fluid (boric acid, ammonia, carbon dioxide, hydrogen sulfide), but were costly both in capital and operating expenses. Furthermore, they required about 14.5–16.0 kg of steam per kW · h of electricity produced at the busbar. With the advent of turbine

Table 5.2—*Power system specifications for condensing units in the boraciferous region of Italy*

	(A) *1938	(B) *1952–54	(C) *1969	(D) *1967	(E) *1967	(F) *1967	(G) (a)	(H) *1960	(L) *1960	(J) (a)	(K) (a)
Turbine data:											
Type.	(b)	(b)	(c)	(b)	(b)	(b)	(c)	(d)	(b)	(c)	(b)
Installed capacity, MW	e 69	f 120	15	26	11	2	k 47	6.5	h 27	j 15.7	12.5
Speed, rev/min	3000	3000	3000	3000	3000	3000	3000	3000	3000	3000	3000
Steam inlet pressure, lbf/in²	59.7	62.6	103.8	61.1	27.0	15.6	69.7	29.9	76.8	71.1	64.0
Steam inlet temperature, °F	385	387	433	370	345	302	385	289	352	365	370
% (wt.) noncondensable gases	7.0	6.8	6.7	14.3	3.8	2.4	3.8	1.8	2.2	3.0	1.7
Exhaust pressure, in Hg	3.0	3.5	3.0	2.6	2.6	2.6	2.9	2.0	2.0	3.0	(a)
Steam flow rate 10³ lbm/h	j 899	j 1480	238	375	127	61.7	j 633	154	j 558	j 357	269
Condenser data:											
Type.	All units have low-level, direct-contact, barometric condensers										
Cooling water temperature, °F	87.8	82.9	87.8	73.4	73.4	73.4	8.42	78.8	78.8	(a)	(a)
Outlet water temperature, °F	105.8	106.3	105.8	98.6	98.6	98.6	104.0	95.0	95.0	(a)	(a)
Cooling water flow rate, 10⁶ lbm/h	(a)	(a)	17.8	(a)	(a)	(a)	(a)	(a)	(a)	(a)	(a)

		A	B	C	D	E	F	G	H	I	J	K
Gas extractor data:												
Type		All units have multistage centrifugal turbocompressors with interstage coolers										
Gas capacity, 10^3 ft³/min		ʲ~310	ʲ~330	196	182	(ᵃ)	(ᵃ)	134	134	ⱼ134	(ᵃ)	(ᵃ)
Power consumption, kW		ʲ~4625	ʲ~5580	1625	2270	(ᵃ)	(ᵃ)	1760	1760	ⱼ1760	(ᵃ)	(ᵃ)
Heat rejection system data:												
Type		All units have natural-draft water cooling towers										
No. of towers		3	4	1	1	(ᵃ)	(ᵃ)	1	(ᵃ)	(ᵃ)	(ᵃ)	(ᵃ)
Design wet-bulb temp., °F		67.0	58.4	67.0	63.3	63.3	63.3	58.4	58.4	58.4	58.4	(ᵃ)
Water pump power, kW		(ᵃ)	855	(ᵃ)	750	(ᵃ)	(ᵃ)	560	58.4	365	(ᵃ)	(ᵃ)

(ᵃ) Not available.
(ᵇ) Single-cylinder, double-flow.
(ᶜ) Tandem-compound, single-flow (HP) and double-flow (LP)
(ᵈ) Single-cylinder, single-flow.
(ᵉ) Four 14.5-MW units; one 11-MW unit.
(ᶠ) Three 26-MW units; one 24-MW unit; two 9-MW units.
(ᵍ) One 15-MW unit; two 12.5-MW units; two 3.5-MW units
(ʰ) One 14.5-MW unit; one 12.5-MW unit.
(ⁱ) One 12.5-MW unit; one 3.2-MW unit.
(ʲ) Total for all units.
(ᵏ) For the 14.5-MW unit only.

*Year of startup.
(A) Larderello 2.
(B) Larderello 3.
(C) Gabbro.
(D) Castelnuovo-V.C.
(E) Castelnuovo-V.C.
(F) Castelnuovo-V.C.
(G) Serrazzano.
(H) Lago 2.
(I) Lago 2.
(J) Sasso Pisano.
(K) Monterotondo.

materials that effectively resist corrosion by the geofluid and the decreasing interest in chemical recovery, the simpler and more efficient direct scheme, Cycle 3, has been fully adopted [Cataldi et al., 1970].

Five units are installed at Larderello 2: four 14.5-MW units and one 11-MW unit. Whereas total installed capacity is 69 MW, the actual rating is only 37.3 MW owing to insufficient steam supply [Ceron et al., 1975]. Overall geothermal resource utilization efficiency, based on actual power output and available work of the geothermal steam at the plant relative to the design wet-bulb temperature of 19.4°C (67°F), is 43%.

Larderello 3. This complex, the largest of the geothermal units in Italy, is located adjacent to Larderallo 2. The powerhouse contains six individual units: three 26-MW units, one 24-MW unit, and two 9-MW units. A view of the turbine hall is shown in figure 5.10. The net capacity of the Larderello 3 plant is only 65.4 MW as compared with the installed capacity of 120 MW. Net geothermal utilization efficiency is 44.3%.

Gabbro. The Gabbro plant consists of a single 15-MW (gross)/11.8-MW (net) unit. The turbine is of the tandem-compound design with a single-flow, high-pressure cylinder followed by a separate double-flow, low pressure section. The HP cylinder exhausts at essentially atmospheric pressure and may be uncoupled from the subatmospheric section for noncondensing operation during periods of shutdown of the LP section. Although the unit has been designed to accommodate gas content as high as 8% (by weight) of steam [Corti et al., 1970], most recent reports show only 6.7% gas concentration [Ceron et al., 1975]. The unit has a 46% geothermal utilization efficiency. This plant is equipped with a closed-circuit television system for remote-controlled operation from the Larderello 2 power station located about 5 km (3 mi) away [ENEL, 1970].

Castelnuovo-Val Cecina. Originally Castelnuovo-V.C. employed three 11-MW Cycle 2 units and one 2-MW unit for auxiliary services. The geothermal fluid temperature was about 195°C (383°F) and the gas content was 10% by weight. Specific steam consumption was relatively high at 17 kg/W·h (37.5 lbm/kW·h) [Corti et al., 1970]. A conversion resulted in the installation of four units in 1967 with a total installed capacity of 50 MW: one 26-MW unit, two 11-MW units, and one 2-MW unit. All are condensing units of the Cycle 3 type. The largest unit is supplied with the highest pressure steam available from the field, about 420 kPa (61.1 lbf/in²); the mid-sized units receive steam at about 186 kPa (27 lbf/in²); the smallest unit is fed by very low pressure jets about 108 kPa (15.6 lbf/in²). The four units have a combined geothermal utilization efficiency of 36% and consume, on the average, 14 kg (30.9 lbm) per net kW·h of electricity. The main unit has been designed to handle up to 12% (by weight) of noncondensable gases and to have a specific fluid consumption

FIGURE 5.10—Larderello 3 turbine hall. Turbogenerator units in foreground and at left background; gas extractors and turbocompressors at right and in background [ENEL, 1970].

of less than 10 kg/kW· h (22 lbm/kW·h). Figure 5.11 is a photograph of the turbine, generator, and gas compressor arrangement for the 26-MW unit.

Serrazzano. There are five condensing units, all of the Cycle 3 variety, in operation at Serrazzano: one 15-MW unit, and two units each of 12.5-MW and 3.5-MW. The turbocompressors for the two 12.5-MW units are shown in figure 5.12.

Lago 2. Total installed capacity at Lago 2 is 33.5 MW, comprising units of 14.5, 12.5, and 6.5 MW. The three units have a combined geothermal utilization efficiency of 46% and a specific steam consumption of 10.1 kg/kW· h (22.3 lbm/kW· h). The arrangement of the turbogenerator and gas compressors for the larger units at Lago 2 is shown in figure 5.13.

Sasso Pisano (Sasso 2). Although total installed capacity at Sasso Pisano is 15.7 MW, it was operated at 17.3 MW in 1974 [Ceron et al., 1975]. There are two units in operation, one of 12.5 MW and one of 3.2 MW. Combined specific steam consumption is one of the best for the Italian geothermal plants: 9.4 kg/kW · h (20.7 lbm/kW · h); utilization efficiency is 47.4% This plant is shown in figure 5.14.

Monterotondo. One 12.5-MW unit is in operation at Monterotondo. The percentage of noncondensable gases in the steam has fallen steadily from a value of 2.5% (by weight) in 1960 [Cataldi et al., 1970], to 2.2% in 1969, and to 1.7% in 1974 [Ceron et al., 1975]. This is the lowest gas content reported for any geothermal plant in Italy. Specific steam consumption is 10 kg/kW· h (22 lbm/kW· h) and geothermal utilization efficiency is 45%. Figure 5.15 shows the turbine hall at Monterotondo with the 12.5-MW unit in the right background.

5.2.4.2 *Noncondensing units.* Noncondensing units are installed at six sites in the boraciferous region: Sant'Ippolito-Vallonsordo, Lagoni Rossi 1 and 2, Sasso 1, Capriola, and Molinetto. All operate on Cycle 1 and exhaust to the atmosphere at a pressure of 102.9 kPa (30.4 in Hg), slightly above atmospheric pressure. Noncondensable gas content ranges from 2.7% (by weight) at Sasso to 4.0% at Capriola. The units are relatively small in capacity, with the 900-kW unit at Sant'Ippolito-Vallonsordo being the smallest geothermal unit installed in Italy. The largest unit of this group is at Sasso 1 and has a rated capacity of 7 MW.

All of these small plants are remote controlled from a larger plant. The control equipment used has been standardized and includes the following capabilities [Corti et al., 1970]:

From pilot to satellite plant:

> Six controls: opening and closing of generator breaker, increase and decrease of governor setting, excitation increase and reduction.
> One control for emergency plant shutdown.

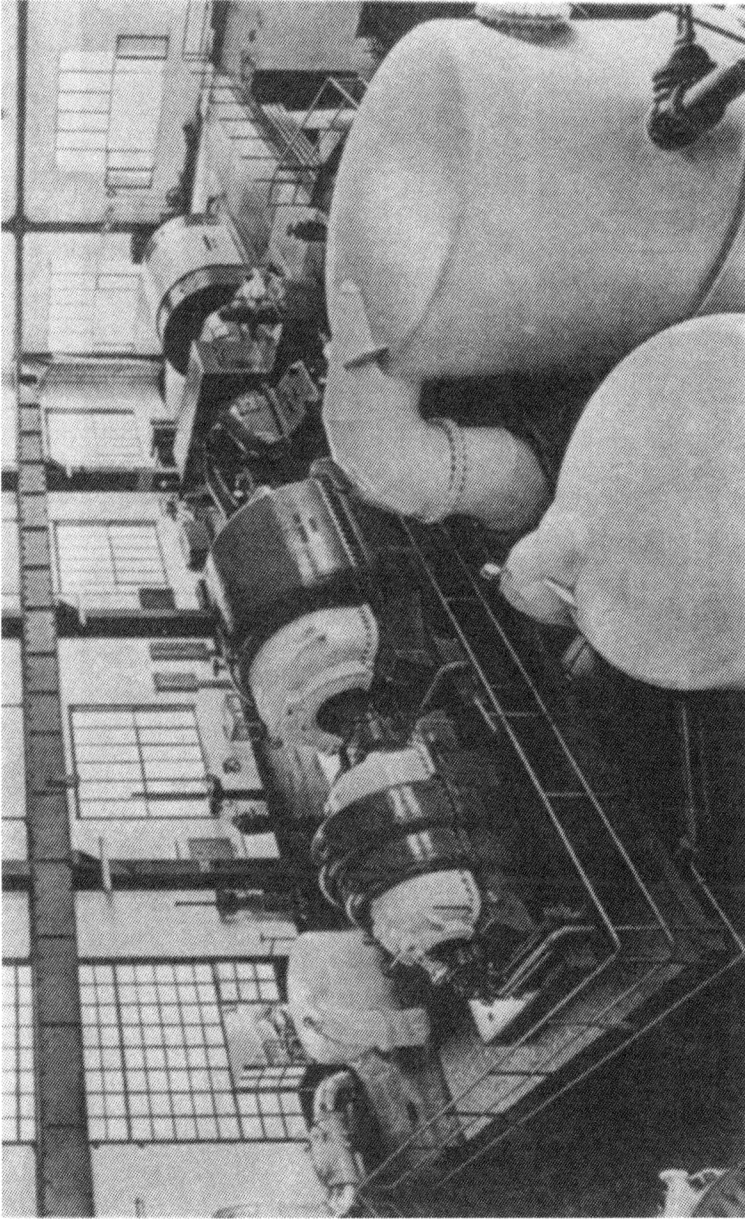

FIGURE 5.11—Castelnuovo-V.C. 26-MW turbogenerator unit. Gas compressors are at left end of the shaft; double-flow turbine and generator are at right end. The two large vessels in foreground and the one to left of the compressors are gas intercoolers [Villa, 1975].

FIGURE 5.12—Serrazzano 12.5-MW units. Turbocompressors for two 12.5-MW units are in foreground; one turbogenerator unit is at rear [Dal Secco, 1970].

FIGURE 5.13—Lago 2 turbine hall. Two units are arranged back-to-back; turbocompressors are in center foreground; one of the turbine-generator sets is at right rear [Dal Secco, 1970].

FIGURE 5.14—Sasso Pisano (Sasso 2) power station. A total of 15.7 MW is installed in this plant. Turbogenerators are mounted on elevated pedestals in the machine room [Saporiti, 1961a].

FIGURE 5.15—Monterotondo turbine hall. The 12.5-MW unit in background is the main unit operating at this site; unit at left is rated at 700 kW and provides power for compressor operation [Saporiti, 1961a].

From satellite to pilot plant:

Five signals: unit shutdown, unit in operation, switch opened, switch closed, alarm.

Eight measurements, six of which are contemporary, including: (with breaker open) generator voltage, generator frequency, excitation current, synchronism, line voltage, and line frequency; (with breaker closed) generator voltage, generator frequency, excitation current, generator current, output power, and line voltage.

The remote control operation of these plants is characterized by extreme simplicity and high reliability.

The following sections describe each of these units in more detail. Technical particulars may be found in table 5.3.

Sant'Ippolito-Vallonsordo. This unit is designed for continuous operation under a wide range of inlet steam conditions: temperatures from 185° to 250°C (365° to 482°F), pressures from 441 to 686 kPa (64 to 99.5 lbf/in^2), and gas content from 4% to 7% (by weight). The original steam flow contained about 20% noncondensables [Cataldi et al., 1970] but the most recent reports indicate gas content of only 3.3% [Ceron et al., 1975]. The unit consumes 26.7 kg of steam per net kW·h (58.9 lbm/kW·h) and has a geothermal resource utilization efficiency of only 15.3% because a significant fraction of the available work is lost when the steam is exhausted at atmospheric pressure. In the photograph of the turbine-generator assembly shown in figure 5.16, the compactness and simplicity of the unit are evident.

Lagoni Rossi 1 and 2. These two units of 3.5 and 3.0 MW, respectively, are characterized by a high degree of flexibility in meeting variable steam conditions. The turbines are capable of performing efficiently with steam conditions ranging in temperature from 230° to 235°C (446° to 455°F), in pressure from 686 to 1079 kPa (99.5 to 156 lbf/in^2), and in gas content from 4% to 20% by weight). The flexibility in performance is due to three factors: (1) A wide degree of control is available in the first, impulse stage; (2) A significant part of the reaction blading is mounted on removable inner shells; (3) The control valve is equipped with interchangeable internal elements [Saporiti, 1961b]. In this way extreme conditions (say, low pressures and temperatures—and corresponding high steam flow rates) can be accommodated by increasing the effective active nozzles in the impulse stage while reducing the reaction blading and increasing the internal steam passages in the control valve. The other extreme can be handled by reversing the effects. Such design flexibility allows good energy conversion efficiency to be maintained over wide variations in steam conditions with only relatively simple modifications to the apparatus. For example, Lagoni Rossi 1 has a specific steam consumption of 18.2 kg/kW · h (40.1

Table 5.3—*Turbine specifications for noncondensing units in the boraciferous region of Italy*

	Sant'Ippolito-Vallonsordo *1963	Lagoni Rossi 1 *1961	Lagoni Rossi 2 *1969	Sasso 1 *1969	Capriola *1969	Molinetto (a)
Turbine type	(b)	(c)	(c)	(c)	(c)	(c)
Installed capacity, MW	0.9	3.5	3.0	7.0	3.0	3.5
Speed, rev/min	3000	3000	3000	3000	3000	3000
Steam inlet pressure, lbf/in²	109.5	75.4	65.4	71.1	56.9	72.5
Steam inlet temperature, °F	419	313	356	369	379	370
% (wt) noncondensable gases	3.3	3.2	3.8	2.7	4.0	3.3
Exhaust pressure, in Hg	30.4	30.4	30.4	30.4	30.4	30.4
Steam flow rate, 10³ lbm/h	52.9	88.2	121	117	112	39.7

*Year of start up.
a Not available.
b Single-cylinder, single-flow, impulse blading.
c Single-cylinder, single-flow, impulse-reaction blading.

FIGURE 5.16—Sant'Ippolito-Vallonsordo 900-kW unit. Turbine-generator assembly for this noncondensing plant, smallest of the geothermal power units operating in Italy [ENEL, 1970].

lbm/kW · h) and a utilization efficiency of 25.2%, which is quite good for a noncondensing unit.

Sasso 1, Capriola, and Molinetto. These three units have a combined installed capacity of 13.5 MW, although net capacity is only 4.8 MW. Gas content averages 3.3% (by weight); steam inlet temperature averages $189° \pm 3°C$ ($372° \pm 5°F$). Geothermal utilization efficiency for the three units taken together is 17.7% and combined specific steam consumption is 25 kg/kW · h (55 lbm/kW · h).

5.2.5 Construction materials

Materials used in the fabrication of steam wells and transmission pipelines have been described in sections 5.2.2.2 and 5.2.2.4, respectively. This section will focus on materials used for the energy conversion equipment. In order to reduce corrosion by the geothermal fluid, sodium carbonate is added to the water used for scrubbing the steam and to the condensate. This neutralizes the pH and minimizes corrosion of iron and steel parts [DiMario, 1961]. Erosion of turbine blades can be reduced and their life extended by the use of 13% chrome/0.1% carbon alloy steel. The jet condenser is made of cast iron covered on the inside by lead-plated sheet. Since direct impingement can wear away the lead plating, areas subjected to jets are further protected with sheets of AISI 316 stainless steel (0.1% C/16%–18% Cr/10%–14% Ni/2%–3% Mo/2% Mn) [Ciapica, 1970; Ricci and Viviani, 1970].

The multistage, centrifugal compressors together with the associated intercoolers are key elements in the Italian geothermal plants because of the high concentration of noncondensable gases in the natural steam. In general, austenitic and ferritic stainless steels and lead-clad carbon steel are used for these systems. The body casings for both the high- and low-pressure casings are of meehanite cast iron. The LP body rotor is of AISI 431 alloy steel (0.2% C/15%–17% Cr/1.25%–2.50% Ni); there are six impellers of the same material. The HP unit has a shaft of carbon steel with AISI 431 sleeves and seven impellers in two stages. Journal and thrust bearings are of carbon steel with linings of white metal [Dal Secco, 1970].

Gas coolers are fabricated from carbon steel welded plates and are lined with lead, ebonite, or AISI 316 stainless steel on the inside. Cooler dimensions are large to keep the fluid velocities as low as possible in order to minimize erosion. The pipes conveying the hot compressed gases to the coolers are made of AISI 403 alloy steel (0.15% C/11.5%–13% Cr); the pipes carrying cool gases are of carbon steel with lead lining [Dal Secco, 1970]. Layers of sandstone rings are suspended in AISI 316 wire hampers downstream of the gas coolers to control the humidity and thus reduce the deposition of solid particles, a process enhanced by the presence of liquid droplets [Ricci and Viviani, 1970].

Pumpes for circulating water are equipped with impellers of 13% chrome/0.1% carbon alloy steel. Pipes carrying water and exposed to the air are made of either stainless steel or iron with stainlss steel liners. The hyperbolic-shaped, natural-draft cooling towers are of reinforced concrete.

5.2.6 Effluent and emissions handling systems

No controls or abatement systems are used for the liquid and gaseous effluents from the plants. In the case of the gaseous emissions, hydrogen sulfide (H_2S) is potentially the most serious because of its corrosive effects and its toxicity. The data in tables 5.1–5.3 may be used to estimate total H_2S emissions in the boraciferous region caused by geothermal power production. The condensing units produce about 3100 kg/h (6800 lbm/h) or roughly 14,000 g/MW·h (31 lbm/MW·h); the noncondensing units emit nearly 190 kg/h (419 lbm/h) or 19,000 g/MW·h. The weighted average for both types of plant in the region is 14,300 g/MW·h (32 lbm/MW·h). This may be compared with the maximum of 200 g/MW·h (0.44 lbm/MW·h) suggested by the Environmental Protection Agency for geothermal plants in the United States.

Liquid discharge from the power stations equipped with condensing units may be estimated using the flow diagram shown in figure 5.9. About 15% of the entering steam flow (by weight) is produced as excess liquid at the cold well of the cooling tower. Assuming that this percentage holds on the average for all plants, the total liquid effluent from the condensing units would amount to about 360,000 kg/h (794,000 lbm/h or 95,000 gal/min). The noncondensing or exhausing-to-atmosphere units discharge all of the steam flow directly to the surroundings; the total amounts to 241,000 kg/h (530,000 lbm/h), as can be seen from the data in table 5.3.

Roughly one-fifth of the waste liquid from the plants in the Larderello region is returned to reservoir by reinjection through wells at the periphery of the field. Generally this practice has been successful, although interference with a production well has been noted recently [Defferding and Walter, 1978]. About 80% of the waste liquid is disposed of by surface discharge into streams in spite of the relatively high concentration of boron in the waste. Eventually a greater percentage of the liquid waste may have to be reinjected.

5.2.7 Economic factors

The most recent cost figures for Italian geothermal power plants were reported in 1974 [Leardini]. At that time the actual capital costs for constructing a plant consisting of two 26-MW condensing steam turbine units in the Larderello region were US$170/kW. The costs for one 15-MW condensing unit in the same area were US$226/kW. For noncondensing units, a 15-MW unit cost US$95/kW and a 4-MW unit cost US$105/kW. The figures for the condensing units were roughly comparable to the costs for fossil power plants at that time. The Italian geothermal plants incurred higher capital costs than dry steam geothermal plants in other countries because of the large gas compressors required to expel the large amounts of noncondensable gases. About 20% of capital costs are allocated to gas compressors.

Leardini [1974] also quoted typical operating costs for several types of plant. The largest condensing units (26 MW) had operating costs of 2 US mill/kW·h of net electricity generated. Noncondensing units are simpler to operate and maintain, their operating costs being 1.6 US mill/kW·h. Remote controlled stations are even less expensive to run; condensing units had operating costs of 0.8 mill/kW · h and noncondensing units were reported at 0.6 mill/kW · h.

5.2.8 Operating experience

In general the operation of Italian geothermal power plants has been highly successful. One measure of this is plant availability, which is consistently above 90% and often exceeds 95% [Cataldi et al., 1970; Ricci and Viviani, 1970]. Most of the problems are related to the corrosive effects of the natural steam. In plants of the Cycle 1 type (exhausting-to-atmosphere), scale forms on turbine blades, leading to clogging of the passageways. The presence of chlorine ions causes the damage, particularly at locations where the superheated steam becomes saturated or wet steam. Cycle 2 (indirect, pure steam) plants were relatively free of problems except for clogging of boiler tubes. Although plants of this type are now obsolete in Italy, when they were in operation turbine blades could be counted on to last 100,000 hours or more. Plants of the Cycle 3 (condensing) variety experience attacks by corrosion and clogging in all elements of the plant including the turbine, condenser, water circulating pumps, gas compressors, and interstage gas coolers. These effects can be mitigated to some extent by proper choice of materials and anticipatory design features (see section 5.2.5). For example, it has been found that when the turbine blading is treated with an oil mixture bath during overhaul, scale buildup is reduced and its removal is facilitated. The results shown in figure 5.17 reveal that the oil treatment can reduce power loss by about 40% after about 2 year of operation [Ricci and Viviani, 1970].

A program of preventive maintenance is employed by monitoring of a number of key indicators for signs of deterioration or malfunction. These include:

Pressure drop across the steam filter
Steam pressure at turbine inlet
Vacuum pressure in the condenser
Temperature of cooling water into the condenser
Suction pressure at gas extraction point
Temperature of cold water into gas coolers

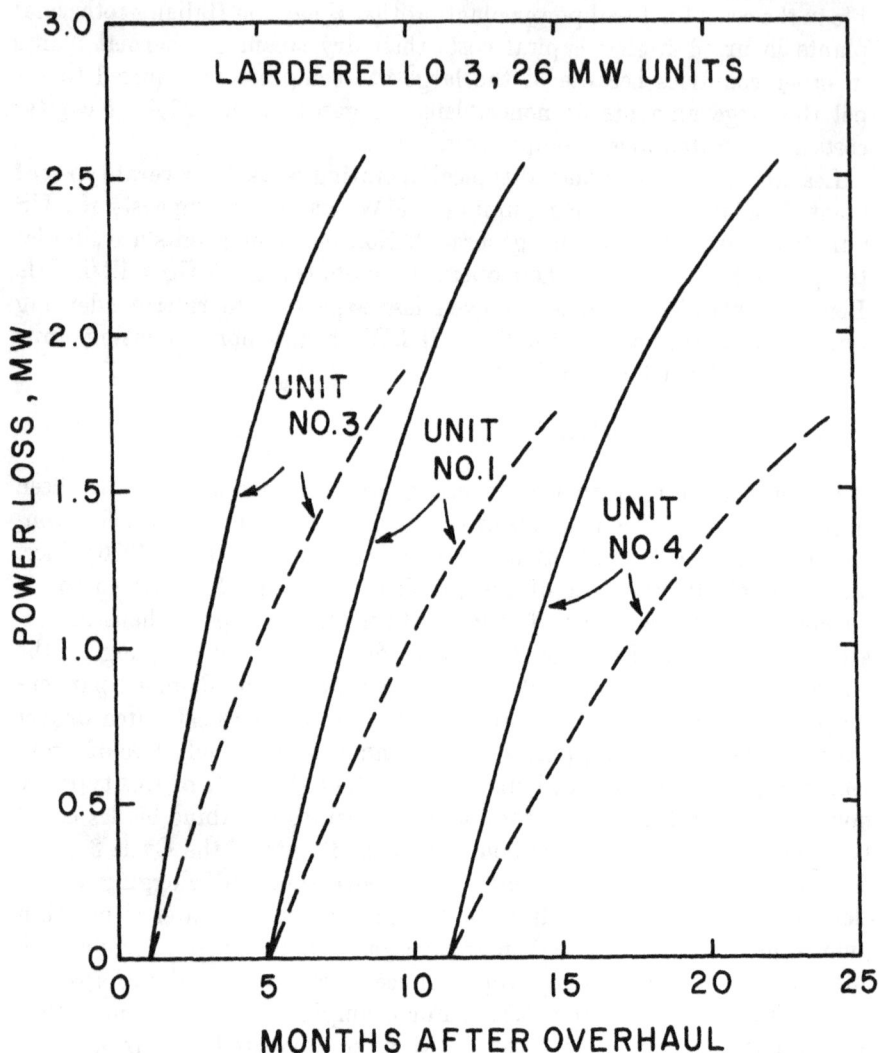

FIGURE 5.17—Power loss as a function of time after overhaul for three 26-MW units at Larderello 3. Solid lines—without oil washing; broken lines—with oil washing. Steam temperatures: 190°C (374°F) for Units 1 and 3, 201°C (394°F) for Unit 4 [after Ricci and Viviani, 1970].

Temperature of hot water out of gas coolers
Temperature of noncondensable gases at exhaust
Temperature of drain water
Temperature of lubricating oil
Pressure of lubricating oil
Composition of lubricating oil
Temperature of bearings
Vibration.

To continue operation as smoothly as possible, a large store of spare parts is maintained at each plant. In particular, one spare bladed rotor shaft is kept on hand for each two units at a given plant. Some sacrifice in machine efficiency is made when a spare rotor is installed because clearances are generally larger than for a custom-matched rotor. However, repairs on the original shaft are made as soon as possible to permit reinstallation and restoration of design performance [Ricci and Viviani, 1970]. The maintenance procedures described above apply to all geothermal power stations in Italy, those in the Monte Amiata and Travale regions as well as in the Larderello or boraciferous area.

5.3 Monte Amiata

5.3.1 Geology

A schematic geologic map of the Monte Amiata region is shown in figure 5.18 [ENEL, 1970]. The area is located about 70 km (44 mi) southeast of Larderello. Although the geology of the site is similar to that of Larderello, there are some noteworthy differences. The outcroppings consist of essentially three major complexes: (1) the volcanic complex (denoted by regions "2" in the figure), which includes the volcanites of Radicofani and Monte Amiata and constitutes a fairly pervious layer; (2) the upper complex (denoted by "4" and "5"), consisting of shales, marls, limestones, calcarenites, sandstones, etc.; and (3) the main pervious complex (denoted by "6") of Tuscan formations, from maiolica to evaporites, which constitutes the main geothermal reservoir [ENEL, 1970].

As can be seen from figure 5.18, the area is marked by magmatic extrusions, a feature absent at Larderello. The products of this volcanic activity are acid in nature and cover about 10,000 ha (24,700 acres) in the vicinity of Monte Amiata. Unlike the case at Larderello, there are relatively few outcroppings of the main aquifer complex. The main source of recharge fluid for the reservoir is the pervious volcanic formation linked to the aquifer by fractures, extrusion chimneys, and volcano-tectonic faults. The two areas of the field where power plants are located, Bagnore and Piancastagnaio, are characterized by extremely high thermal gradients of about 50°C/100 m (27°F/100 ft), nearly 17 times the normal gradient. The gradient exceeds 10°C/100 m (5.5°F/100 ft) over a wide area of 40,000 ha (100,000 acres) [ENEL, 1970].

5.3.2 Geofluid characteristics

The wells in this region produce dry, slightly superheated steam as at Larderello, but at generally lower temperatures. Steam temperature in the Bagnore area averages about 138°C (280°F), at Piancastagnaio 180°C (361°F). There has been a considerable decline in the shut-in pressure of the wells since the field was first exploited nearly 20 years ago. At that

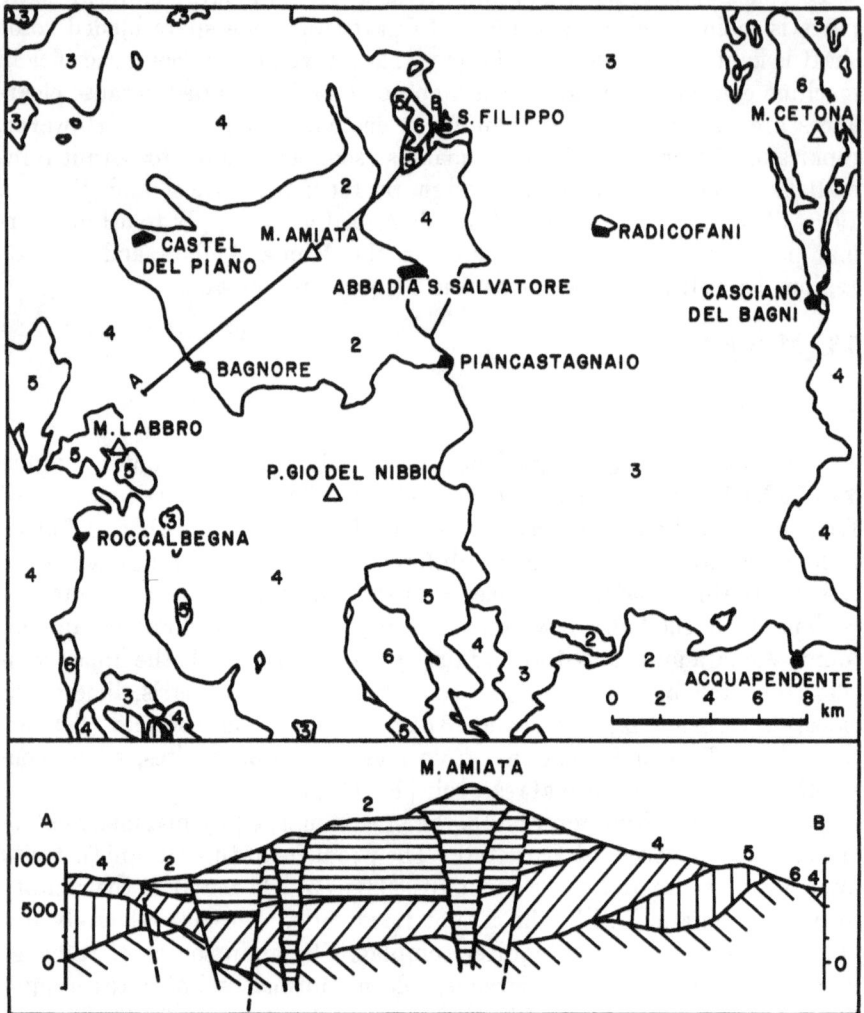

FIGURE 5.18—Geologic map and cross-section of the Monte Amiata region, Tuscany,
Italy. Geological formations: 1. Travertines; 2. Volcanites; 3. Clays, sands, conglomerates; 4. Shales, marls, limestones, etc.; 5. Calcarenites; 6. Tuscan formations
[after ENEL, 1970].

time the closed-in wells showed pressures of 2157 kPa (313 lbf/in²) and
4118 kPa (597 lbf/in²) at Bagnore and Piancastagnaio, respectively.
These values have now fallen to 588 kPa (85 lbf/in²) and 1961 kPa (284
lbf/in²), respectively. Wellhead operating pressures at the two sites are
about 309 kPa (45 lbf/in²) and 804 kPa (117 lbf/in²).

The amount of noncondensable gas in the geothermal steam is significantly larger than in the boraciferous region. At the time the field was
being developed, gas content exceeded 90% (by weight) of the natural

vapors. The earliest power plants encountered "steam" that contained between 30% and 80% (by weight) of noncondensables. This percentage has declined during exploitation and now ranges from 7% to 20%. On average, the noncondensable gas contains (by volume) 95% carbon dioxide, 0.4% hydrogen sulfide, 0.4% hydrogen, 3.5% methane, and 0.7% nitrogen [ENEL, 1970].

5.3.3 Energy conversion systems

The only geothermal power stations in the Monte Amiata region are of the noncondensing or Cycle 1 type. In the late 1960's four units were in operation, two at Bagnore and one each at Piancastagnaio and Senna. The last of these was a 3.5-MW unit, very similar to the 3.5-MW unit installed at Lagoni Rossi 1 (see section 5.2.4.2). It was remote controlled from Piancastagnaio but has since been shut down. Technical particulars for the remaining three units are listed in table 5.4.

Table 5.4—*Turbine specifications for geothermal units in the Monte Amiata and Travale regions of Italy*

	Bagnore 1 *1959	Bagnore 2 *1960	Piancastagnaio *1969	Travale *1973
Turbine type	(a)	(a)	(a)	(a)
Rated capacity, MW.	3. 5	3. 5	15. 0	15. 0
Speed, rev/min.	3000	3000	3000	3000
Steam inlet pressure, lbf/in². . .	42. 7	46. 9	116. 6	159. 3
Steam inlet temperature, °F	275	286	361	414
% (wt) noncondensable gases. .	8. 5	7. 2	21. 1	10. 6
Exhaust pressure, in Hg.	30. 4	30. 4	31. 3	31. 3
Steam flow rate, 10³ lbm/h.	97. 0	110. 0	483. 0	419. 0

* Year of startup.
(a) Single-cylinder, single-flow, impulse-type, noncondensing.

Bagnore 1 and 2. During the 15-year lifetime of the plants at Bagnore there has been a 40% decline in operating pressure which has resulted in a significant loss of net electrical capacity, as may be seen from the values shown in table 5.5. The thermodynamic performance of these units is rather poor: 31.3 kg of geothermal fluid are required to generate each kW·h of electricity (69 lbm/kW·h). The geothermal utilization efficiency for the two units combined is only 16%.

Piancastagnaio. The 15-MW unit at Piancastagnaio began operating in 1969 and is one of the two largest plants of the noncondensing type in existence. It receives steam from a single well, P.C./8, which produced 300 Mg/h (660,000 lbm/h) at a wellhead pressure of 980 kPa (142 lbf/in²)

Table 5.5—*Production history of the noncondensing units at Bagnore, Monte Amiata*

	Bagnore 1	Bagnore 2
Installed capacity, MW..	3. 5	3. 5
1960 [a]		
Pressure, lbf/in²............	71. 1	7. 11
Temperature, °F	311	320
Gas content, % (wt).......	30	80
Net capacity, MW......	3. 70	0. 96
1969 [a]		
Pressure, lbf/in²......... ...	45. 5	50. 0
Temperature, °F......... .	277	286
Gas content, % (wt) . .	S	S
Net capacity, MW...........	1. 50	2. 00
1974 [b]		
Pressure, lbf/in².	42. 7	46. 9
Temperature, °F............	275	286
Gas content, % (wt).........	8. 5	7. 2
Net capacity, MW...........	1. 20	1. 80

[a] Source: Cataldi et al., 1970.
[b] Source: Ceron et al., 1975.

in 1965. Since that time, the output has fallen 29%; in 1974 the well produced 219 Mg/h (483,000 lbm/h) at 804 kPa (117 lbf/in²) [Corti et al., 1970; Ceron et al., 19775].

The powerhouse, shown in figure 5.19, is designed to allow for easy disassembly in the event that the steam flow becomes too low to justify operation. In such a case the entire plant may be readily relocated at another site. Figure 5.20 shows the turbogenerator unit. The turbine is encased in a single cylinder, employs combined impulse-reaction blading, and has an adjustable intake to allow efficient operation over a wide range of steam pressure from 490 to 1079 kPa (71 to 156 lbf/in²). The turbine is, in fact, identical to the high-pressure turbine installed in the condensing plant at Gabbro described in section 5.2.4.1. A subatmospheric turbine may be attached to the existing machine should steam conditions at Piancastagnaio warrant conversion to a condensing unit. Gas content would have to drop to about 4% to justify such a change.

The turbine incorporates the same features described in section 5.2.4.2 for the Lagoni Rossi 1 and 2 units. The cutaway view of the turbine in figure 5.21 reveals three sets of blading, each of which has stationary blades mounted on removable inner shells. The first two sets of blades are essentially impulse-type, while the last set is impulse-reaction. The adjustable inlet nozzle can also be seen. This plant operates at a geothermal utilization efficiency of 24% and consumes 17.7 kg/kW·h (39 lbm/kW·h) of net electricity generated.

FIGURE 5.19—Piancastagnaio powerhouse, Monte Amiata. Steam line from well P.C./8 may be seen at right. Silencer and exhaust stack is visible on left side of the building, which is designed for portability in the event the field should become nonproductive [ENEL, 1970].

FIGURE 5.20—Piancastagnaio turbogenerator. This exhausting-to-atmosphere unit is rated at 15 MW and designed to accommodate a wide range of inlet steam pressures [Ciapica, 1970]

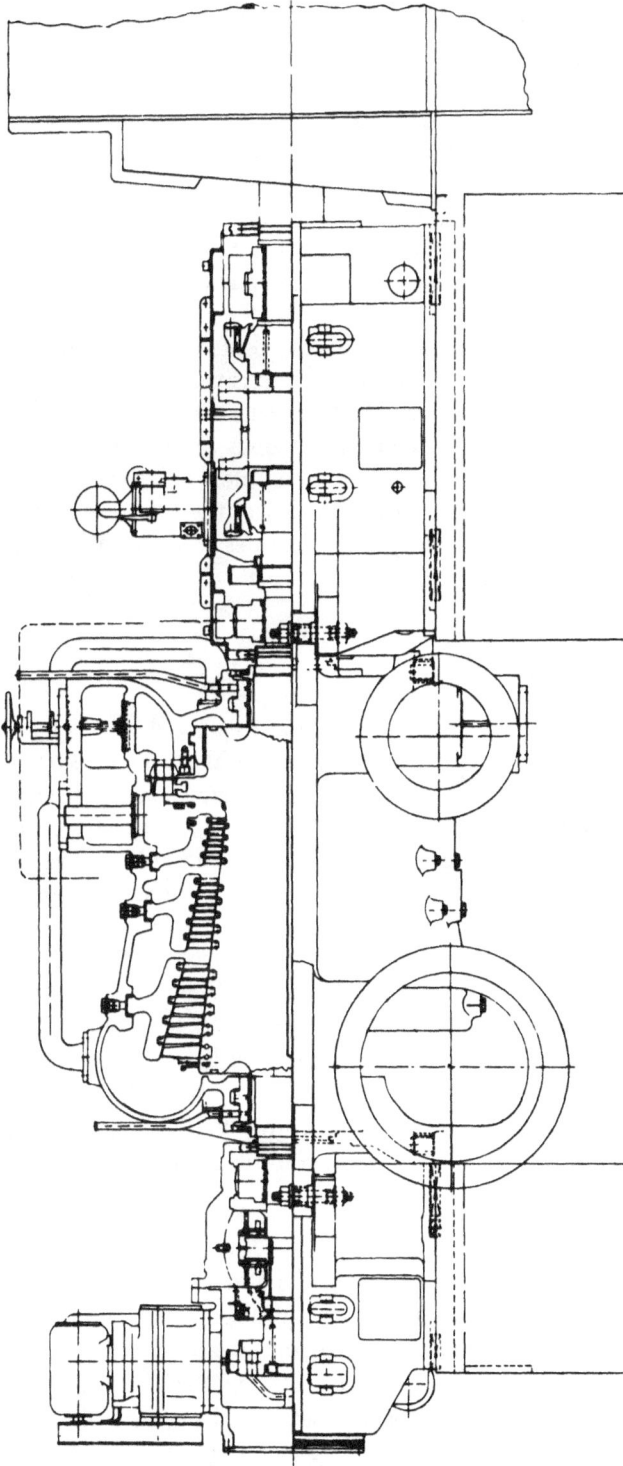

FIGURE 5.21—Piancastagnaio noncondensing turbine. Removable sections of blading and adjustable inlet nozzles permit high efficiency over wide operating conditions [Ciapica, 1970].

5.4 Travale

5.4.1 Geology

The Travale geothermal field is located on the southwest edge of the Era Graben, a northwest-southeast trending feature. The geology of the site is similar to that of the boraciferous region 10–15 km (6–9 mi) to the west-northwest. In fact, the nature of the boundary between the hydrological systems of Larderello and Travale is not well known, even though both regions have been the subject of many surveys [Petracco and Squarci, 1975].

The hydrogeological complexes may be described as follows: (1) surface layer (recent alluvial, coastal deposits, travertines, and magmatic rocks); (2) upper complex (Pliocene and Miocene marine, lagoonal, and lacustrine deposits, flysch-facies formations), which forms the impervious cap rock for the reservoir; and (3) lower complex (Mesozoic carbonate formations), which is fractured and constitutes the permeable reservoir. Where the lower complex outcrops, it provides a reasonably absorptive area allowing for recharge and pressure control in the reservoir [Burgassi et al., 1975].

5.4.2 Well programs

Locations of the wells in the Travale area are shown in figure 5.22. Eight wells were drilled prior to 1969 ("old" wells); as of 1975 six additional wells had been drilled ("new" wells). Information on the new wells is given in table 5.6. In the vicinity of well T22, the main permeable zone is believed to trend northeast-southwest, perpendicular to the direction of the Era Graben [Cormy and Musé, 1975]. This seems to account for the lack of production from new wells R1, R2, and R3. Well R4, which lies 0.8 km (2600 ft) northeast of well T22, is a good producer. The wide discrepancy in the geofluid characteristics of wells T22 and R4 is related to the fact that these two wells lie on opposite sides of a step-fault that forms one of the boundaries of the Era Graben. As can be seen from table 5.6, the production zone is down-shifted by about 680 m (2230 ft) from well T22 to well R4 because of the step-fault feature. Well C1 is a replacement for well R3, which was not completed because of technical difficulties.

5.4.3 Energy conversion system

One power plant is operating at the Travale field, a noncondensing unit (Cycle 1) of 15 MW nominal capacity. The plant is essentially identical in design to the one at Piancastagnaio, in the Monte Amiata region, described in section 5.3.3. The unit was installed in 1973 and utilizes the geofluid from well T22. Technical particulars are listed in table 5.4. This plant is reported to have the best operating efficiency of any exhausting-to-atmos-

phere geothermal plant is Italy. Specific steam consumption is 13.5 kg/
kW·h (29.8 lbm/kW·h) [Ceron et al., 1975], and the geothermal energy
net utilization efficiency is 29%.

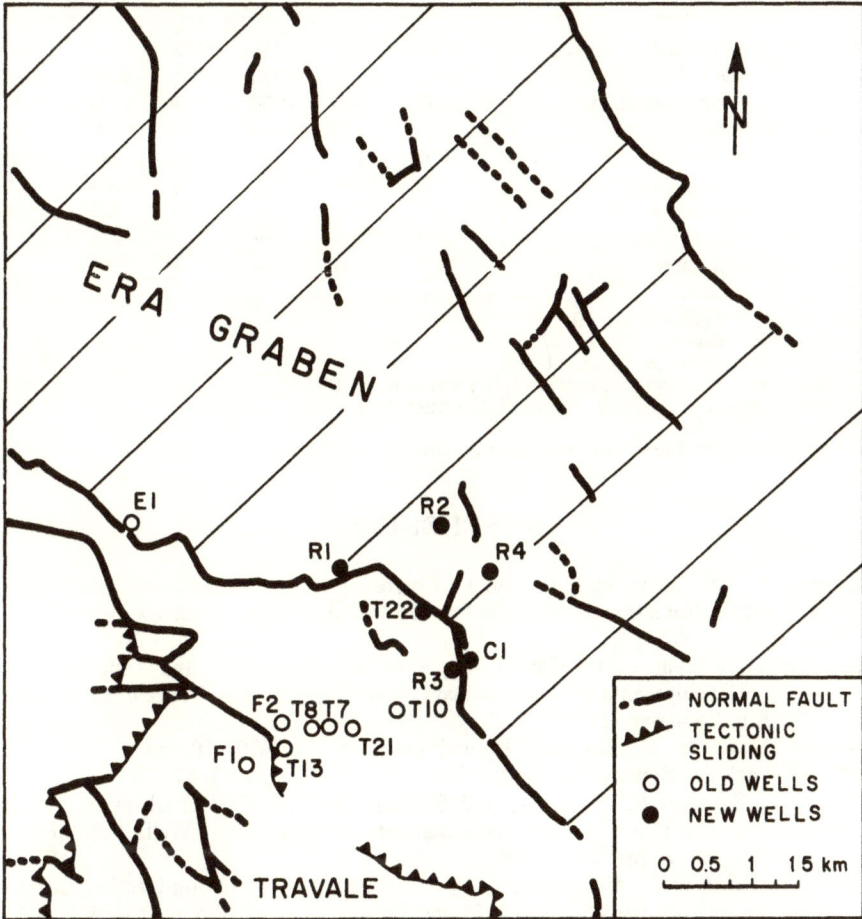

FIGURE 5.22—Well locations at Travale, Italy [after Burgassi et al., 1975].

Table 5.6—*Information on new wells drilled at Travale* [a]

	Well					
	[b]T22	R1	R2	R3	[b]R4	C1
Wellhead elevation, ft [c].	1247	1371	1214	1378	1116	1394
Total well depth, ft.	2267	4922	6040	3340	4498	6004
Production casing depth, ft	2087	3763	4856	2359	4429	2290
Casing diameter, in.	13⅜	9⅝	9⅝	13⅜	9⅝	13⅜
Reservoir temperature, °F.	507	486	516	[d]428	387	>446
Delivery pressure, lbf/in².	356	—	—	([e])	84	([f])
Delivery temperature, °F	462	—	—	—	331	([f])
Total flow rate, 10³ lbm/h	388	[g] dry	dry	—	201	([f])
Gas content, % (wt)	10. 3	—	—	—	66	([f])

[a] Source: Burgassi et al., 1975.
[b] Measured during March 1975.
[c] Relative to sea level (s.l. = 0 ft).
[d] Nonequilibrium value; extrapolated value is ≈ 482°F.
[e] Not completed because of technical difficulties.
[f] Not available.
[g] Originally produced a steam-gas mixture intermittently.

REFERENCES [a]

Allegrini, G. and Benvenuti, G., 1970. "Corrosion Characteristics and Geothermal Power Plant Protection (Collateral Processes of Abrasion, Erosion and Scaling)," *Pisa, 1970*, pp. 865–881.

Burgassi, P., Stefani, G. C., Cataldi, R., Rossi, A., Squarci, P., and Taffi, L., 1975. "Recent Developments of Geothermal Exploration in the Travale-Radicondoli Area," *San Francisco, 1975*, Vol. 3, pp. 1571–1581.

Cataldi, R., Ceron, P., DiMario, P., and Leardini, T., 1970. "Progress Report on Geothermal Development in Italy," *Pisa, 1970*, Vol. 2, pp. 77–87.

Celati, R., Squarci, P., Taffi, L., and Stefani, G. C., 1975. "Analysis of Water Levels and Reservoir Pressure measurements in Geothermal Wells," *San Francisco, 1975*, Vol. 3, pp. 1583–1590.

Ceron, P., DiMario, P., and Leardini, T., 1975. "Progress Report on Geothermal Development in Italy from 1969 to 1974 and Future Prospects," *San Francisco, 1975*, Vol. 1, pp. 59–66.

Ciapica, I., 1970. "Present Development of Turbines for Geothermal Application," *Pisa, 1970*, Vol. 2, pp. 834–838.

Cigni, U., 1970. "Machinery and Equipment for Harnessing of Endogenous Fluid," *Pisa, 1970*, Vol. 2, pp. 704–713.

Cigni, U., Fabbri, F., and Giovannoni, A., 1975. "Advancement in Cementation Techniques in the Italian Geothermal Wells," *San Francisco, 1975*, Vol. 2, pp. 1471–1481.

Conti, Prince P. G., 1924. "The Larderello Natural Steam Power Plant," *Proc. First World Power Conf.*, London.

Cormy, G. and Musé, L., 1975. "Utilization of MT-5-EX in Geothermal Exploration," *San Francisco, 1975*, Vol. 2, pp. 929–935.

Corti, R., DiMario, P., and Mondolfi, F., 1970. "New Trends in the Planning and Design of Geothermal Power Plants," *Pisa, 1970* Vol. 2, pp. 768–779.

[a] See note on p. 24.

Dal Secco, A., 1970. "Turbocompressors for Geothermal Plants," *Pisa, 1970,* Vol 2, pp. 819–833.

Defferding, L. J. and Walter, R. A., 1978. "Disposal of Liquid Effluents from Geothermal Installations," *Geothermal Resources Council Trans.,* Vol. 2, Sec. 1, pp. 141–143.

DiMario, P., 1961. "Remarks on the Operation of the Geothermal Power Stations at Larderello and on the Transportation of Geothermal Fluid," *Rome, 1961,* pp. 334–353.

Ellis, A. J. and Mahon, W. A. J., 1977. *Chemistry and Geothermal Systems,* Academic Press, New York.

ENEL, 1970, *Larderello and Monte Amiata: Electric Power by Endogenous Steam.* Ente Nazionale per L'Energia Elettrica, Compartimento di Firenze, Direzione Studi e Richerche, Roma. (In English.)

Hahn, E., 1924. "Some Unusual Steam Plants in Tuscany," *Power,* Vol 57, pp. 882–885.

Koenig, J. B., 1973. "Worldwide Status of Geothermal Resources Development," in *Geothermal Energy,* P. Kruger and C. Otte, eds., Stanford Univ. Press, Stanford, CA, pp. 15–58.

Leardini, T., 1974. "Geothermal Power," *Phil. Trans. Royal Soc. Lond. A,* Vol. 276, pp. 507–526.

Mongelli, F. and Loddo, M., 1975. "Regional Heat Flow and Geothermal Fields in Italy," *San Francisco, 1975,* Vol. 1, pp. 495–498.

Overton, H. L. and Hanold, R. J., 1977. "Geothermal Reservoir Categorization and Stimulation Study," Los Alamos Scientific Laboratory, Int. Rep. LA–6889–MS, Los Alamos, NM.

Petracco, C. and Squarci, P., 1975. "Hydrological Balance of Larderello Geothermal Region," *San Francisco, 1975,* Vol. 1, pp. 521–530.

Pollastri, G., 1970. "Design and Construction of Steam Pipelines," *Pisa, 1970,* Vol. 2, pp. 780–811.

Ricci, G. and Viviani, G., 1970. "Maintenance Operations in Geothermal Power Plants," *Pisa, 1970,* Vol. 2, pp. 839–847.

Saporiti, A., 1961a. "Progress Realized in Installations with Endogenous Steam Condensing Turbine-Generator Units," *Rome, 1961,* pp. 380–394.

Saporiti, A., 1961b. "Progress Realized in Installations with Endogenous Steam Turbine-Generator Units without Condenser," *Rome, 1961,* pp. 397–409.

Villa, F. P., 1975. "Geothermal Plants in Italy: Their Evolution and Problems," *San Francisco, 1975,* Vol. 3, pp. 2055–2064.

CHAPTER 6

JAPAN

6.1 Overview

Japan is the only country which has operated geothermal plants of the dry steam, single-flash, double-flash, and binary type. Although only 165 MW are installed at this time, an ambitious and aggressive development program is underway, aimed at putting 48,000 MW on-line by the year 2000 from all geothermal sources, including tapping volcanic magma and hot dry rocks.

The full range of geothermal activities in Japan is directed by the government's Ministry of International Trade and Industry (MITI) through the Sunshine Project. Problems relating to fundamental research and development are handled through the Geothermal Energy Research and Development Company, Ltd. (GERD), an organization that enjoys the participation of 30 institutions, mainly from industry. The Agency of Industrial Science and Technology (AIST) also oversees certain research projects, in particular those of the Geological Survey of Japan (GSJ). Projects related more to development than research are also administered by MITI but are channeled through the Japan Geothermal Energy Development Center (JGEC), where funding is shared between government and industry on a 90/10 ratio. Thus a concerted and effective effort is underway in Japan to develop the geothermal resources of that country through a close partnership between government and industry.

Exploitation of geothermal energy for electric power has, nevertheless, been slow in Japan because nearly all of the outstanding geothermal prospects are located in national parks, which are enthusiastically protected for their natural beauty. The construction and operation of geothermal power plants thus are subject to rigid and stringent controls.

A summary of the geothermal plants in Japan is given in table 6.1, where information is provided for plants in operation, in testing, under construction, and in planning. Each will be described in the following sections.

6.2 Matsukawa

The Matsukawa geothermal power plant was completed in 1966 by the Japan Metals and Chemicals Company, Ltd. Located in Iwate Prefecture

Table 6.1—*Geothermal power plant development in Japan*

Plant	Type	Capacity, MW	Location		Status
			Island	Prefecture	
Matsukawa	Dry steam	20	Honshu	Iwate	Operational since 1966
Otake	Single-flash	10	Kyushu	Oita	Operational since 1967
Onuma	Single-flash	10	Honshu	Akita	Operational since 1973
Onikobe	Single-flash	25	Honshu	Miyagi	Operational since 1975
Hatchobaru	Double-flash	50	Kyushu	Oita	Operational since 1977
Kakkonda	Single-flash	50	Honshu	Iwate	Operational since 1978
Otake	Binary	1	Kyushu	Oita	Tested 1977–1979
Mori (Nigorikawa)	Binary	1	Hokkaido	Hokkaido	Tested 1977–1979
Mori	Single-flash	55	Hokkaido	Hokkaido	Under construction
(To be named)	Double-flash	55	Kyushu	Kumamoto	In planning

in the northern part of Honshu Island, it is one of the few sites in the world where essentially dry, natural geothermal steam is used to produce electric power. Matsukawa was the first geothermal power plant installed in Japan. The turbine-generator is rated at 20 MW and was supplied by the Tokyo Shibaura Electric Company (Toshiba). The power produced is supplied to the network of the Tohoku Electric Power Company, Ltd. A site photograph appears in figure 6.1.

6.2.1   Geology and exploration

The Matsukawa Valley is a Quaternary composite volcano with a caldera. The southern margin of the caldera shows thermal outcrops and steam vents (fumaroles). The volcano sits on the andesite lava of yet another volcano, which in turn rests on a Tertiary basement. Surface manifestations result from fractures in the caldera and a deep-seated magmatic intrusion. The hydrothermal reservoir consists of porous beds in the upper portion of the Tertiary, with the Quaternary andesite lavas forming an overlayer [Saito, 1961].

In 1953 area residents began to drill wells in the hope of discovering a source of hot water to supply a health spa. Instead, many of the wells produced steam at depths of 160–300 m (525–984 ft) [Saito, 1961]. Since that time, numerous methods have been applied to discern the geologic nature of the field, including detailed geologic mappings, geophysical prospecting by means of seismic and electric resistivity methods, and geophysical well loggings [Saito, 1970; H. Nakamura et al., 1970].

Before the power station was constructed, tests were run for 18 months on a 450-kW back-pressure turbine to assess the corrosion effects on various materials from exposure to geothermal steam and its condensate. Table 6.2 lists the characteristics of the geothermal steam and its condensate. Fourteen materials were tested; table 6.3 shows the types of materials used for various components and the observed corrosion rate after about 3 years of operation.

6.2.2   Steam gathering system

Five wells supply geothermal steam to the plant. Four of these produce superheated steam at a running pressure of 441.2 kPa (64.0 lbf/in²) and at temperatures of 153, 170, 183, and 190°C (307, 338, 361, and 374°F). Recording instruments at each wellhead keep track of pressure, temperature, flow rate, and degree of penetration (because of fine rock dust). One of the original wells was found to have a scale buildup that caused a steady, linear decrease in flow rate. Following a workover of the well, the flow increased somewhat, but not to the full initial value.

Until 1973 all drilling was carried out with mud. With the failure of the No. 3 well at Matsukawa owing to erosion of the casing, air drilling was initiated for the new No. 3 well. Mud was used to a depth of 815 m (2673

FIGURE 6.1—Matsukawa geothermal power plant. The 45-meter-high natural-draft cooling tower dominates the scene at the Matsukawa plant [GERD, 1975].

Table 6.2—*Characteristics of geothermal steam and condensate at Matsukawa* [a]

Noncondensable gas content, percent by volume of steam	0. 2%–0. 6%
Composition of noncondensable gases (volume percentage):	
Carbon dioxide, CO_2	79. 3%–85. 2%
Hydrogen sulfide, H_2S	12. 9%–17. 7%
Methane, CH_4	1. 15%
Hydrogen, H_2	0. 28%
Composition of condensate (ppm):	
Sulfate, SO_4	1708. 1
Amorphous silicon dioxide, $SiO_2 \cdot H_2O$	795. 5
Sodium, Na	280
Potassium, K	180
Hydrogen sulfide, H_2S	10–42. 4
Chloride, Cl	9. 2
pH of condensate	4. 35–4. 85

[a] Source: Kawasaki, 1977.

Table 6.3—*Materials and corrosion test data: Matsukawa* [a]

Component	Material name (ASTM)	Corrosion rate, μm/yr
Turbine casing	Carbon steel plate (A283–6rD) . .	636
Impingement shield . . ' . .	Stainless steel plate (410)	21. 3
Rotor	1% Cr/1.25% Mo/0.25%V, forged steel .	623
Buckets	Low carbon, 12% Cr steel (410)	21. 3
Nozzle diaphragms	Carbon steel plate (A283–6rD) . .	636
Nozzle partitions	12% Cr Al alloy steel	49. 4
Labyrinth packing strips	15% Cr/1.75% Mo steel	—
Valve bodies	Carbon steel plate	636
Valve seats	Stellite welding	—
Bearing babbits	White metal (D–23–6)	—
Oil cooler tubes	Aluminum	—
Tube sheets	Naval brass sheets and plates (B–111) . .	[b] 4. 92
Water boxes	Cast iron (48–35)	—
Condenser shell and tail pipe . .	Epoxy-coated carbon steel . . .	[b] 2. 84
Condenser tray	Stainless steel (304)	[b] 21 2
Gas ejector	Stainless steel (304)	[b] 21. 2

[a] Sources: Akiba, 1970; Mori, 1970.   [b] In condensate.

ft), while air was used from there down to 1170 m (3838 ft). The new well produces 70 Mg/h (0.154×10^6 lbm/h) [Iga and Baba, 1974]. Wells range in depth from 800 to 1200 m (2625 to 3940 ft). At Matsukawa, and typically elsewhere in Japan, geothermal wells are completed as follows: a surface casing of 457-mm diameter (18-in) to a depth of about 30 m (98 ft), a 340-mm (13⅜-in) conductor casing to 250 m (820 ft), a 244-mm (9⅝-in) production casing to about 500 m (1640 ft), or to the depth of the reservoir, plus an additional 700–800 m (2300–2625 ft) of 219-mm

(8⅝-in) slotted liner in production zones subject to sluff-in. Straight cement is used with no additives. Drilling is done by rigs with a 2-km depth capability; drilling mud is used throughout.

The total length of steam pipe is 2.2 km (7218 ft). Pipelines are inspected periodically with thickness gauges at points of high erosion such as bends. Since air tends to enter the pipe during shutdown, the steam is kept flowing even when the plant is down for inspection or maintenance to prevent oxidation. Pipes or valves exposed to the atmosphere tend to corrode rapidly. The main problem with valves is that they stick. Gate and butterfly valves are used, but it is difficult to secure proper seating. Spring-loaded valves are used for safety valves, but these also tend to stick and require periodic rapping to maintain them in working condition.

6.2.3   Energy conversion system

A schematic heat balance diagram is shown in figure 6.2 and technical specifications of the main mechanical components are listed in table 6.4. The turbine is coupled to an air-cooled generator with a rating of 23,500 kVA, a 0.85 power factor, 11 kV, 3000 rpm, and with a static excitation system. The 4-stage impulse turbine uses blades of 12% chrome with a carbon steel casing. During its early operation, the unit was damaged by mud and dust deposits. In January 1967, after about 50 days of operation, the unit was forced out of service by abnormal vibrations. Inspection showed that 82 buckets of the fourth stage were damaged, the shroud and bucket tennons of the third stage were worn out, and the entire third- and fourth-stage blades and nozzles were embedded in mud. This mud was formed when the moisture present in the last two turbine stages came in contact with an excessive amount of dust expelled from one of the wells. The well was identified and disconnected.

The blade failures were attributed to a change in the natural frequency of vibration, which is normally far in excess of the nozzle-passing frequency. The encrustation of mud and scale caused the natural frequency to equal the nozzle-passing frequency, leading to vibration fatigue. The fatigue strength of the blades, furthermore, had been reduced by the corrosive atmosphere. Scale was observed in a number of locations:

Inside steam receiver at branch point of pipeline
Drain separator
Main parts of main stop valve and control valve
Steam inlet port of turbine casing
Drain pipe from nozzle diaphragm
Steam path through nozzle
Outside perimeter of diaphragm
Effective part of buckets
Inner surface of shroud
Gland packing and wheel of rotor.

LEGEND

S = Mg/h, STEAM (———)   P = abs PRESSURE, KPa

SA = Mg/h, STEAM+AIR (——)   H = ENTHALPY, KJ/Kg

W = Mg/h, WATER (-----)   G = Mg/h, GASES (=====)

FROM COOLING WATER TANK
25 0 °C
4847 4 W

TO TANK FOR HOT WATER PUMP

TO HOT WATER TANK
122 4 W
30 0 °C

TURBINE OIL COOLER
32 4 W

0 06 W

AFTER-CONDENSER

91.3 P
97 0 °C

3 48 G

175 W

2nd STAGE STEAM EJECTOR

0 08 G

0 63 SA
97 6 °C

6 87 S

INTER-CONDENSER

42.1 P
77.2 °C

230 W

1st STAGE STEAM EJECTOR

6 43 S

4320 W

12.80 G

12.7 P
50.5 °C

20 MW

GENERATOR

0 13 G

BAROMETRIC CONDENSER

2343.3 H

TURBINE

193 1 S

0.40 S

EJECTOR FOR STEAM SCALING

0.43 S
96 °C

0 20 S

MAIN STEAM
207 0 S
441.2 P
147. 2 °C
2723. 6 H

HOT WELL

4928 65 W
47 0 °C

TO HOT WATER TANK

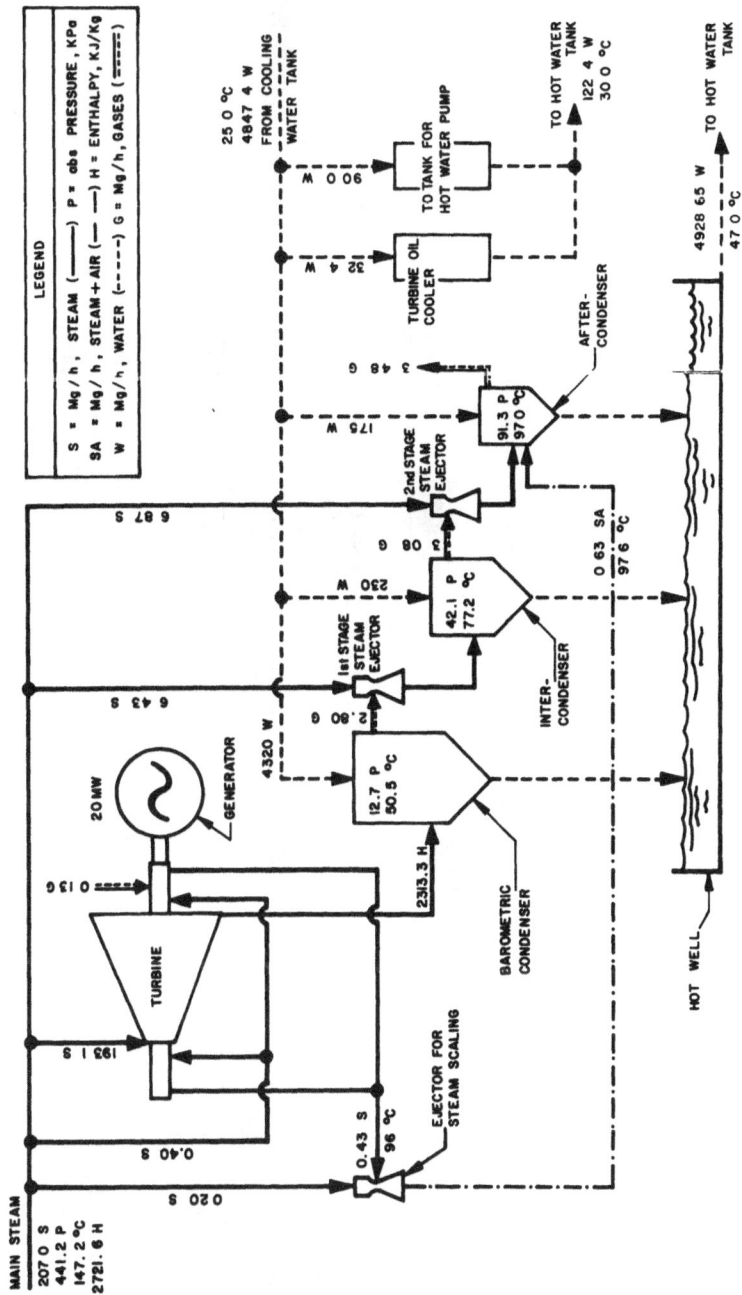

FIGURE 6.2.—Heat balance diagram of Matsukawa geothermal power plant [A-iba, 1970].

Table 6.4—*Technical specifications for Japanese geothermal power plants*

	Matsukawa *1966	Otake *1967	Onuma *1973	Onikobe *1975	Hatchobaru *1977	Kakkonda *1977
Turbine data:						
Type [a]	SCSF4I	SCSF4I	SCSF4I	SCSF5PI	SCDF5PI	SCDF4IR
Rated capacity, MW	20	10	10	25	50	50
Maximum capacity, MW	22	13	12.5	25	55	50
Speed, rev/min	3000	3600	3000	3000	3600	3000
Main steam pressure, lbf/in²	64	35.6	35.6	64	68.9	64
Secondary steam pressure, lbf/in²	—	—	—	—	16.5	—
Main steam temperature, °F	297	261	261	297	327.6	297
Secondary steam temperature, °F	—	—	—	—	215.6	—
Exhaust pressure, in Hg	4.0	3.2	3.2	3.1	2.9	4.0
Main steam flow, 10³ lbm/h	456	249	236	503	847	877
Secondary steam flow, 10³ lbm/h	—	—	—	—	256	—
Last-stage blade height, in	23.0	16.5	19.7	24.7	25.0	23.0
Condenser data:						
Type [b]	DCEB	DCEB	DCEB	DCEB	DCLL	DCLL
Cooling water temperature, °F	77.0	78.8	73.4	78.8	78.8	(c)
Outlet water temperature, °F	116.6	106.5	110.1	107.8	110.3	(c)
Cooling water flow rate, 10⁶ lbm/h	9.52	8.60	6.28	15.60	27.10	(c)

Gas extractor data:

Type [d]	SJE	MDRVP	MDRVP	SJE/VP	SJE/2RB	SJE
Suction pressure, in Hg	3.74	2.66	2.66	3.00	2.75	4.00
Gas capacity, ft³/min	10,600	2,719	618	(°)	9,594	(°)
Steam consumption, 10^3 lbm/h	(°)	—	—	33	11.5	(°)
Power consumption, kW	—	53 (steady)	30 (steady)	—	—	—

Cooling tower data:

Type [e]	ND	CXMD	CXMD	CTMD	CTMD	CXMD
Number of cells	(Tower)	3	3	5	4	—
Design wet-bulb temperature, °F	(°)	62.6	57.2	62.6	62.6	(°)
Fan motor power, kW	—	66.0	86.0	(°)	213	(°)

*Year of startup

[a] SCSF4I=single-cylinder, single-flow, 4-stage impulse; SCSF5IR =single-cylinder, single-flow, 5-stage impulse reaction; SCDF5PI =single-cylinder, double-flow, 5-stage pass-in; SCDF4IR=single-cylinder, double-flow, 4-stage impulse-reaction.

[b] DCEB=direct-contact, external, barometric type; DCLL= direct-contact, low-level type.

[c] Not available.

[d] SJE=steam jet ejector; MDRVP=motor driven reciprocating vacuum pump; SJE/VP=steam jet ejector with vacuum pump; SJE/2RB=steam jet ejector with 2-stage radial blower.

[e] ND=natural draft; CXMD=crossflow, mechanically-induced draft; CTMD=counterflow, mechanically-induced draft.

Composition of the scale from parts of the turbine is given in table 6.5. These elements formed such compounds as SiO_2, $CaSO_4$, $NaSO_4$, Al_2O_3, MgO, $FeSO_4$, and FeS. The decrease in SO_4, Na, and K is attributed to their solubility in the moisture of the last stages of the turbine. The insoluble SiO_2 is believed to be flushed out mechanically by the condensed steam.

Table 6.5—*Chemical composition of scale after 50 days of operation at Matsukawa* [a]

Element	Control valve (percent)	First-stage buckets (percent)	Second-stage buckets (percent)	Third-stage buckets (percent)
SiO_2.	37	24	11	4
SO_4.	44	50	24	18
Fe.	1. 3	10	51	56
Al	0. 9	0. 5	0. 2	0. 2
Ca.	1. 0	0. 8	0. 3	0. 04
Mg.	0. 3	0. 2	0. 1	(Trace)
S.	(Trace)	0. 4	2. 3	4. 3
Na.	11. 5	9. 3	2. 3	0. 6
K.	4. 3	2. 8	0. 6	(Trace)

[a] Source: Akiba, 1970.

The condenser is of the direct-contact, barometric type and is designed to condense 194 Mg/h (0.428×10^6 lbm/h) at a vacuum of 12.7 kPa (3.74 in Hg), absolute pressure. The condenser, located outside the powerhouse, is made of carbon steel plate lined with stainless steel and coated with epoxy paint on the inside surfaces. The shell is 4.7 m (15.4 ft) in diameter and is 10.3 m (33.8 ft) in height. The tail pipes of the inter- and after-condensers for the gas extraction system are now made of stainless steel to protect them against corrosion. The original carbon steel pipes failed from corrosion, leading to loss of vacuum in the condenser.

The water cooling tower is of the natural-draft type, the only one of its kind in use in Japanese geothermal power stations. It has a ground-level diameter of 45 m (147.6 ft), a height of 44.6 m (146.3 ft), and is sheathed with 10-cm-thick (3.94-in) vacuum concrete panels set to steel pipe shells. The tower is fitted with 1.6 m (5.25 ft) of plastic packing. Water is distributed by means of steel pipes with plastic spray nozzles. Make-up water is obtained from the Matsukawa River. Some trouble has been experienced at times when the screens are unable to trap all the dust and sediment. Such contaminants can foul and clog the spray nozzles, thereby lowering the efficiency of the cooling tower. This clogging tends to occur at flood time. Since the ambient temperature during the winter can be quite low, ranging from $-25°$ to $-5°C$ ($-13°$ to $+23°F$), the draft ports have to be covered to control the flow rate of air through the tower to prevent freezing.

The main step-up transformer is rated at 23,500 kVA, 11 kV/155 kV, 3-phase, 50 Hz; the plant auxiliaries are fed from a 2500 kVA transformer, 11 kV/3.5 kV, 3-phase. Electrical power from the plant is fed to the grid of the Tohoku Electric Power Company by means of a 20-km (12.4-mi), 155-kV transmission line. Because of the extremely corrosive atmosphere in the vicinity of the powerhouse, the substation is located about 600 m (0.4 mi) from the plant in an area where the concentration of H_2S is extremely low. Main electrical equipment is sealed in oil-filled containers; terminals are coated with conductive grease; bare wires are made of aluminum; and supporters are coated with antirust paint. It is difficult to protect meters and relays from the corrosive action of the atmosphere with antirust paint because these coatings tend to cause sticking. Since the trip time for a relay is crucial for the safety of the plant, movement tests of all relays are conducted more frequently than in a conventional power station. The concentration of H_2S in the turbine room is between 0.06 and 0.40 ppm [Sato, 1970].

6.2.4 Environmental effects and co-utilization

Water is taken from the Matsukawa River to provide for cooling water make-up and tower blowdown. The Akagawa River, adjacent to the plant, receives the excess condensate. There are no controls on the emission of hydrogen sulfide. The main steam contains between 0.03% and 0.11% (by volume) H_2S. The 2-stage steam jet ejectors remove the H_2S and the other noncondensable gases and discharge them to the atmosphere through the cooling tower. The concentration of H_2S in the atmosphere is described [Sato, 1970] as exerting "no remarkable influence" on the power plant. The specific emissions level of H_2S from the plant ranges from 5050 to 20,800 g/MW · h, a moderate-to-high level of emissions.

Since 1971 the waste heat from the power plant has been used for space-heating at nearby villages [Mori, 1975]. Recipients of this thermal energy include Hachimantai Hot Springs in Matsuo Village, Hachimantai Heights (a recreational center for employees of small companies run by the government's Employment Promotion Corporation), a regional forestry office (working site), and three inns at Matsukawa Hot Springs. The supplier of the heated water is the Hachimantai Hot Springs Development Company, Ltd., which comprises Matsuo Village, Japan Metals and Chemicals Company, Ltd., and the Northern Iwate Bus Company.

6.2.5 Economic factors

Construction cost of the Matsukawa geothermal power plant was one and a half to two times the figure for a thermal power station per kW installed. Higher costs resulted from the added expenses of building in a very rugged region and the distances between geothermal wells. The figures in table 6.6 show construction costs in 1968 yen [S. Nakamura, 1970]. The cost of electricity for one of the early years of operation turned out to be 2.63 yen/kW·h (~ 7 mill/kW·h), roughly equivalent to the cost of electricity from fossil-fuel power plants at that time. Since fossil-fuel costs have increased by a factor of three or four since then, the present cost of geothermal electricity from the Matsukawa plant is clearly considerably less than that from a fossil-fuel plant.

Table 6.6—*Construction costs of Matsukawa plant*
(*in millions of yen, 1968*)

Geological surveys and exploration	160
Steam wells	439
Steam-gathering system	193
Buildings	216
Cooling tower	357
Power generation equipment	407
Substation	170
Incidental equipment	49
Interest and other related costs	66
Total	ª 2057

(~ US $6 million, 1968)

ª Note: The total cost is about $285/kW in 1968 US dollars.

6.2.6 Operating experience

The Matsukawa plant has been running in a highly reliable fashion since it was commissioned in 1966. A summary of its performance for the period 1967–1974 is given in table 6.7. Recent availability factors have

Table 6.7—*Performance of Matsukawa dry-steam 20-MW geothermal power plant* ª

Year	Electricity Generation, MW·h Gross	Net	Peak load MW	Net Capacity factor,[b] (percent)	Availability factor, (percent)
1967	67,439	62,853	13.0	49.0	78.4
1968	106,673	101,873	20.5	58.0	80.7
1969	142,354	136,660	20.6	78.0	96.2
1970	162,518	155,530	21.0	88.8	95.5
1971	153,108	146,371	21.1	83.5	93.1
1972	176,463	169,228	22.5	ᶜ 96.3	96.3
1973	183,680	176,884	22.5	ᶜ 101.0	96.7
1974	185.910	178.660	22.5	ᶜ 102.0	95.5

ª Source: JGEDC, 1978.
[b] Based on nominal capacity of 20 MW.
ᶜ When based on 22.5 MW, these three values become 85.6, 89.8, and 90.7%, respectively.

been greater than 95%, and net capacity factors are consistently above 85%, based on the rated capacity of 20 MW. In recent years the plant has operated at peak loads of 22.5 MW at times, leading to capacity factors in excess of 100%.

6.3 Otake

The Otake geothermal power plant was the second to be commissioned in Japan and came on-line in August 1967. It was the second central station in the world to use a liquid-dominated hydrothermal reservoir, the Wairakei plant in New Zealand having been the first. Otake is rated at 10 MW, with a maximum capacity of 13 MW. The Kyushu Electric Power Company, Inc., owns and operates the plant, which is shown in figure 6.3.

6.3.1 Geology and exploration

The highly volcanic region near Mount Aso on the island of Kyushu, Oita Prefecture, is the site of the Otake power station. Mount Aso is an active volcano of the caldera type. The Otake field is about 20 km (12 mi) southwest of the popular hot-springs resort city of Beppu. Exploration at the Otake geothermal area began as early as 1927, when a 94-m (308-ft) test well was drilled by the Tokyo Electric Power Company. No temperature data were taken because the well blew out. Beginning in 1951 geologic, geophysical, and geochemical surveys got underway. These included electrical resistivity (1951, 1963–1965), magnetic (1963–1965), gravity (1963–1966), radioactivity (1951), and thermal gradient (1951). From 1953 to 1966, the Kyushu Electric Power Co. drilled six exploratory wells ranging in depth from 300 ot 1030 m (984 to 3380 ft). Although five of these wells blew out, maximum temperatures of from 148° to 205°C (298° to 401°F) were observed.

The early test wells showed that the reservoir was liquid dominated, yielding a mixture of steam and liquid. Exploitation of such fields, in contrast to dry steam fields, was unknown in the early 1950's. Thus the Otake field lay unused until the experience at Wairakei, New Zealand, showed that such fields could be tapped to produce electricity. As a result of the exploration conducted in the early 1960's, it was concluded that a geothermal power plant of at least 10-MW capacity could be built at Otake, given the available supply of steam.

6.3.2 Well programs and gathering system

The plant originally received steam from five relatively shallow production wells, 350–600 m (1150–1970 ft) deep. However, three of these have ceased production, leaving only two wells to supply the plant. Recently a program was undertaken to increase the number of wells, particularly reinjection wells. Table 6.8 gives a summary of the drilling experience at

FIGURE 6.3—Otake geothermal power plant. Situated on filled land, the Otake power station is located in rugged terrain near the active volcano Mount Aso [MHI, 1976a].

Table 6.8—*Drilling experience at Otake*

Well No.	Drilling period	Size,mm	Depth, m	Comments
OT-1....	Feb. 1953–Jun. 1953	63. 5	300	Successful; stopped prod. Dec. 1953
OT-2 ...	Mar. 1953–Jun. 1953	142. 9	300	Successful; stopped prod. Dec. 1953
OT-3	Dec. 1953–Mar. 1955	142. 9	900	Successful; stopped prod. Mar. 1957
OT-5 ...	Jul. 1954–Dec. 1955	193. 7	450	Successful; stopped prod. Apr. 1965
O-6.....	Dec. 1963–Feb. 1964	193. 7	500	Successful; stopped prod. Dec. 1967
O-7 ...	Mar. 1964–May 1964	193. 7	350	Successful, stopped prod. Oct. 1975 [a]
O-8.....	Oct. 1964–Feb. 1965	142. 9	497	Successful; stopped prod. Dec. 1977
O-9. ...	Jan. 1965–Mar. 1965	193. 7	550	Successful; still in use
O-10....	Jan. 1966–Apr. 1966	193. 7	600	Successful; still in use
OS-1....	Oct. 1971–Mar. 1972	193. 7	1550	Failure; low temperature [b]
OS-2.. .	Dec. 1972–Mar. 1974	193. 7	1300	Failure; dry well
OS-3....	Jun. 1978–Dec. 1978	193. 7	~700	
OR-1...	Nov. 1971–Feb. 1972	193. 7	329	Successful; still in use
OR-3	Nov. 1974–Feb. 1975	193. 7	600	Successful; still in use
OR-4.	Apr. 1976–Sep. 1976	269. 9	600	Failure
OR-5.	Nov. 1976–Jun. 1977	269. 9	530	Successful; still in use
OR-6.	Oct. 1977–Feb. 1978	269. 9	501	Successful; still in use
OR-7.	Mar. 1978–Mar. 1978	269. 9	351	Successful; still in use
OR-8..	Apr. 1978–Jun. 1978	269. 9	470	Successful; still in use
OR-9...	Dec. 1978– —	269. 9	~700	Under construction

[a] Converted to a reinjection well, O–7R; currently in use.
[b] Converted to a reinjection well, OR–2, plugged at 500 m; currently in use.

Otake. Through the end of 1978 there were two production wells (O-9 and O-10) and eight reinjection wells (O-7R, OR-1, OR-2, OR-3, and OR-5 through OR-8) in operation. Figure 6.4 shows the profiles of the two currently producing wells. A typical wellhead arrangement is shown in figure 6.5; a schematic of typical steam separator equipment in figure 6.6. The arrangement of steam pipelines is given in the plan view in figure 6.7 for the original production wells, O-6 through O-10. The gathering system covers an area of about 0.38 ha (0.94 acres), and comprises three main steam lines with a total length of roughly 900 m (2950 ft) and a maximum diameter of 700 mm (27.5 in).

The characteristics of the geothermal fluid produced by wells O-9 and O-10 are listed in table 6.9. The amount of noncondensable gases, particularly hydrogen sulfide, is relatively low. The bulk of the noncondensable gases is carbon dioxide [Kawasaki, 1977].

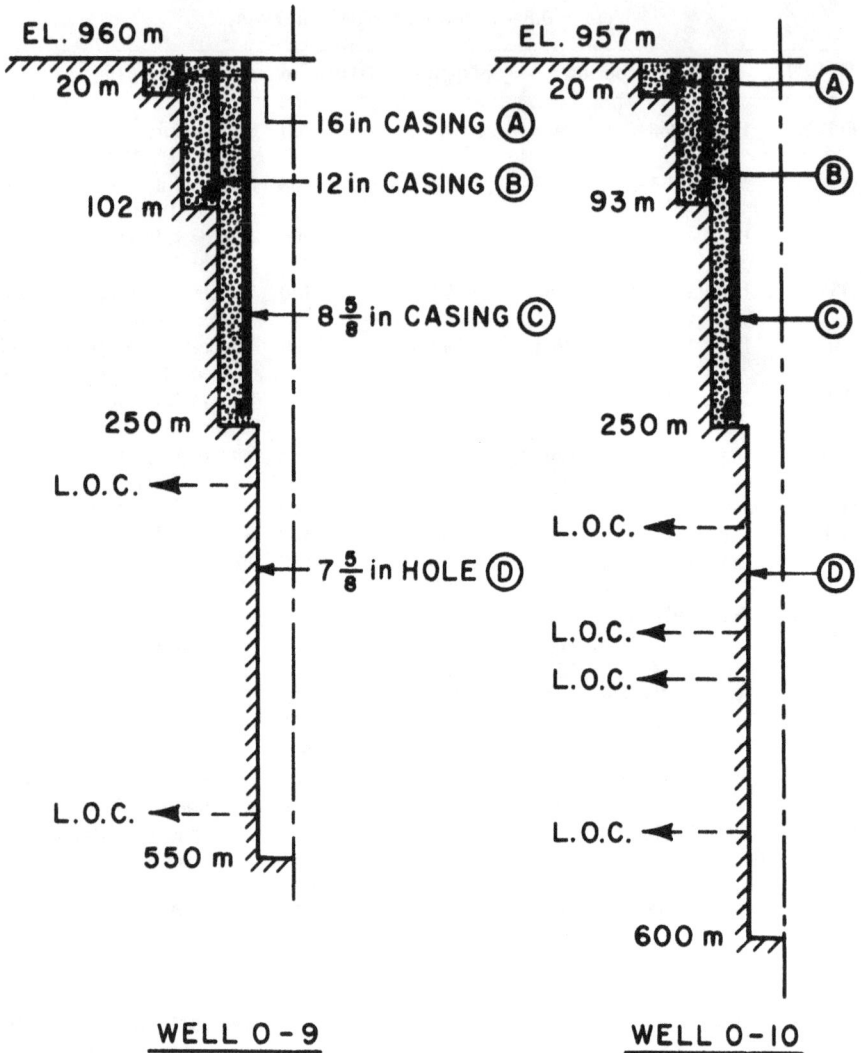

FIGURE 6.4—Well casings for Otake wells O-9 and O-10.

6.3.3 Energy conversion system

A thorough investigation of site and reservoir characteristics was conducted for about 18 months prior to the decision to construct a power plant.
The well flows were traced for 1 year. The optimum turbine inlet pressure
selected was 245.1 kPa (35.6 lbf/in²) in order to yield maximum power
output while guarding against future power dropoff in case of scale buildup on the turbine nozzles. Since scaling of nozzles increases working pressures, the pressure chosen is somewhat less than the exact value corresponding to the power peak. To produce the desired turbine inlet pressure,

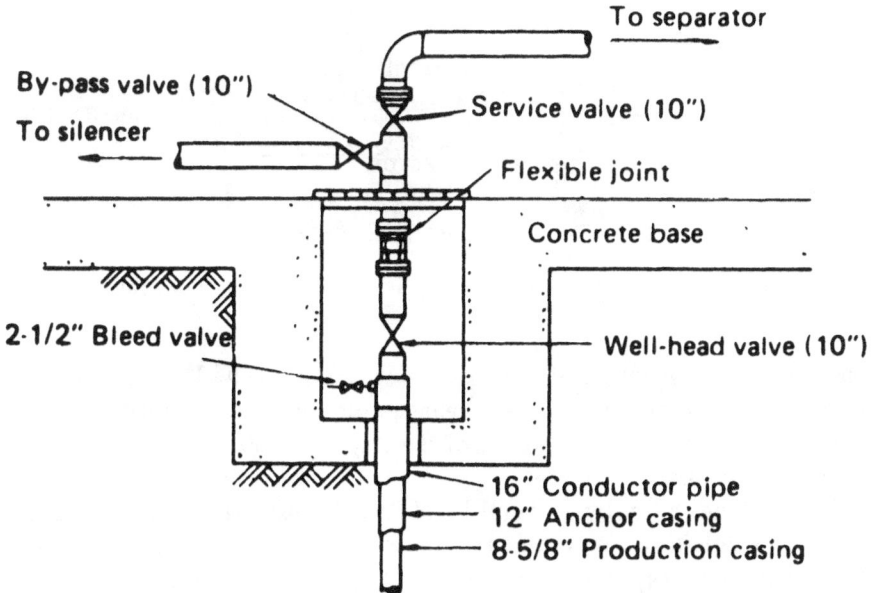

FIGURE 6.5—Typical wellhead apparatus at Otake [MHI, 1976b].

Table 6.9—*Characteristics of the geofluid from Otake wells No. O-9 and O-10*

	Well O-9	Well O-10
Shut-in pressure, kPa...	882	745
Separator pressure, kPa........	245	245
Flow rate of steam, Mg/h.	29	30
Flow rate of liquid, Mg/h........	61	135
Total flow rate, Mg/h...........................	90	165
Wellhead quality, %............................ ..	32	18
Noncondensable gas content [a].	0. 228	0. 05
Composition of NC gases [b]:		
Carbon dioxide..........	96. 0	93. 0
Hydrogen sulfide...	1. 4	1. 0
Others; O_2, N_2, etc......	2. 6	6. 0
Analysis of hot water:		
pH............................	6. 7	8. 0
Total dissolved solids, ppm.....	3810	4030
Chloride, ppm...............................	1630	1720
Silica, ppm	668	612
Calcium, ppm................................	29	—
Magnesium, ppm..........	17	—
Sodium, ppm.	940	1060
Potassium, ppm.........	110	140
Sulfate, ppm................................	145	—

[a] As a percentage by volume of steam.
[b] As a percentage by volume of total noncondensables.

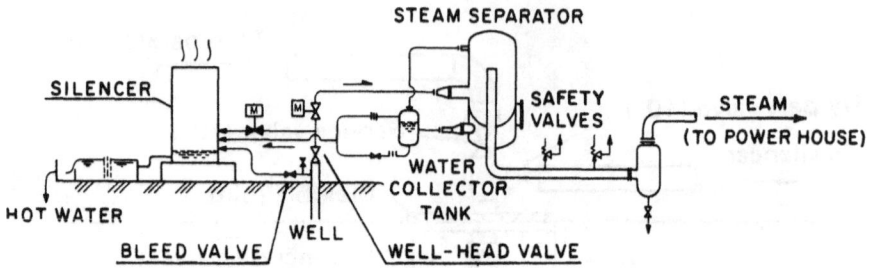

FIGURE 6.6—Schematic of typical steam separator equipment at Otake [MHI, 1976b].

the pressure at the outlet of the separator must be 303.9 kPa (44.1 lbf/in²) [Usui and Aikawa, 1970]. The schematic for the power plant is given in figure 6.8. Technical specifications of the major mechanical components are listed in table 6.4.

The turbine, built by Mitsubishi Heavy Industries, Ltd., is a single-cylinder, single-flow, four-stage impulse machine. It is rated at 10 MW (13 MW maximum capacity), turns at 3600 rpm, and accepts essentially saturated steam at 127°C (261°F) and 245 kPa (35.6 lbf/in²). To cope with the high moisture content of the steam, the turbine's last stage is fitted

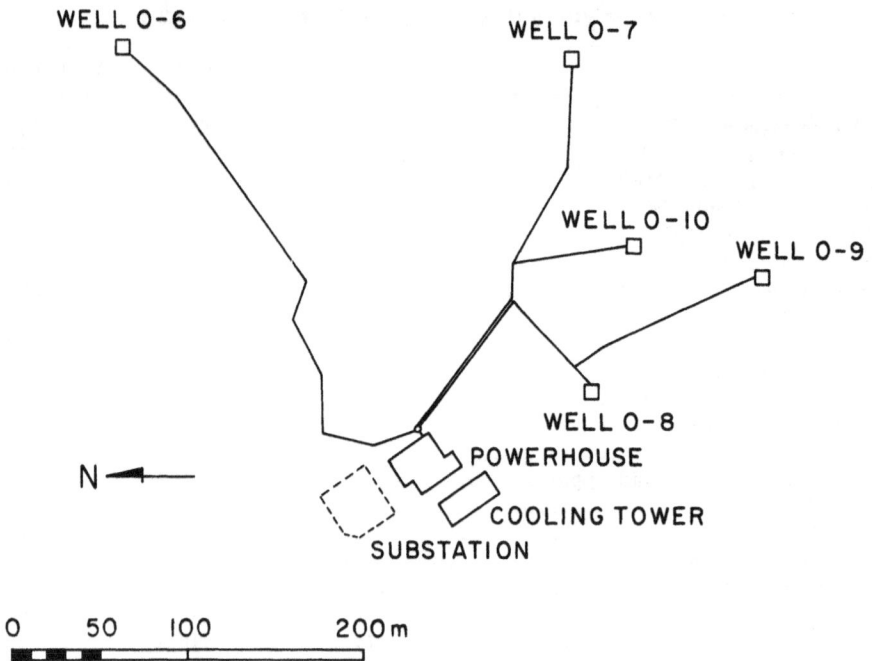

FIGURE 6.7—Layout of production wells, powerhouse, cooling tower, and substation at Otake [after Kubota and Aosaki, 1975].

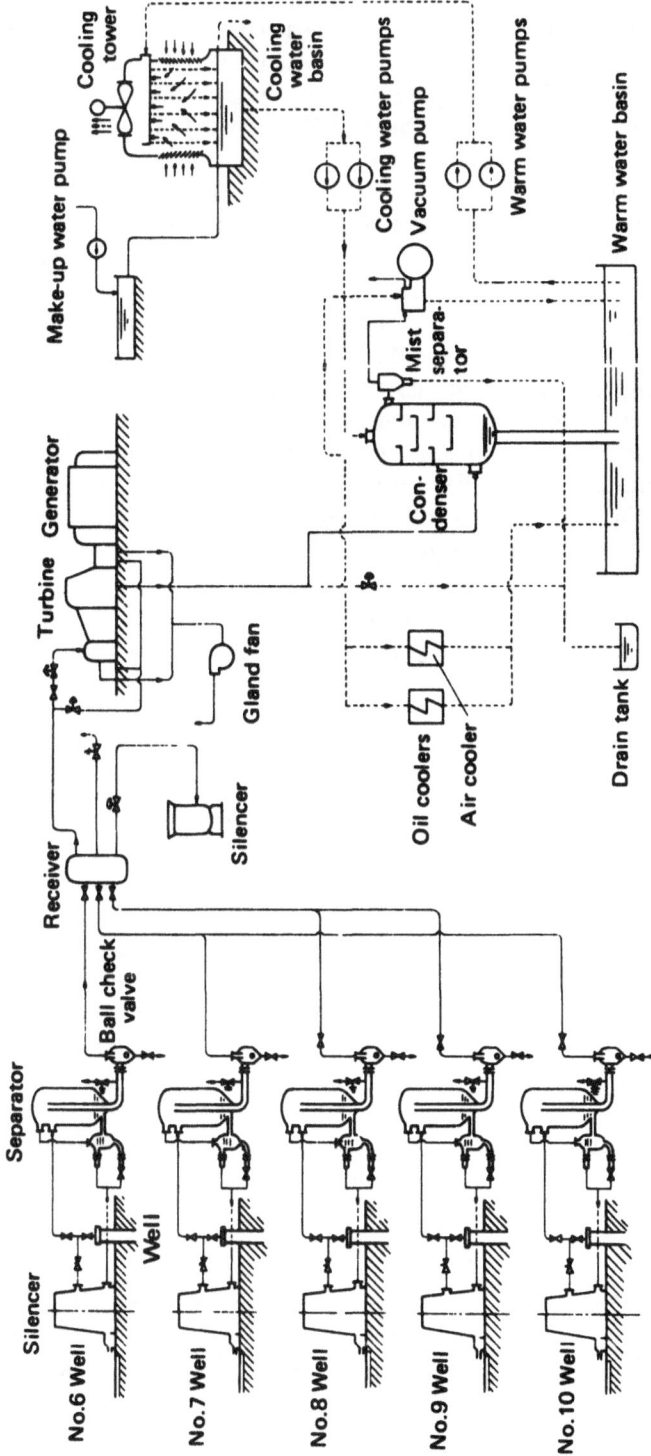

FIGURE 6.8—Schematic diagram of Otake power plant [MHI, 1976b].

with a recessed stellite strip and each stage has a drain catcher. Liquid droplets are induced to travel to the periphery by centrifugal force enhanced by a higher-than-normal velocity ratio (wheel velocity to steam velocity), leading to a higher whirl component for the steam. No erosion of turbine blades was discerned at the first 2-year overhaul after 16,830 hours of operation.

Two main-steam stop valves of the swing-check type are located on each side of the turbine to allow a steam-free test under full-load conditions. In series with the main-steam stop valves are regulating butterfly valves. Since the valves in geothermal plants are subject to sticking because of the deposition of scale and since base-load operation means full, rated flow almost all the time, the regulating valves were designed to move over a large range of angle openings in the vicinity of full steam flow. Thus small departures from full-load operation result in reasonably large angular movement of the regulating valves, and this movement tends to keep the valves moving freely. Figure 6.9 shows the steam flow regulating valve characteristic curve. It is believed that this steam turbine was the first to be controlled with a butterfly-type regulating valve from zero speed to full load [Usui and Aikawa, 1970].

The generator was built by the Tokyo Shibaura Electric Company, Ltd. (Toshiba), and was used previously at another Kyushu Electric power

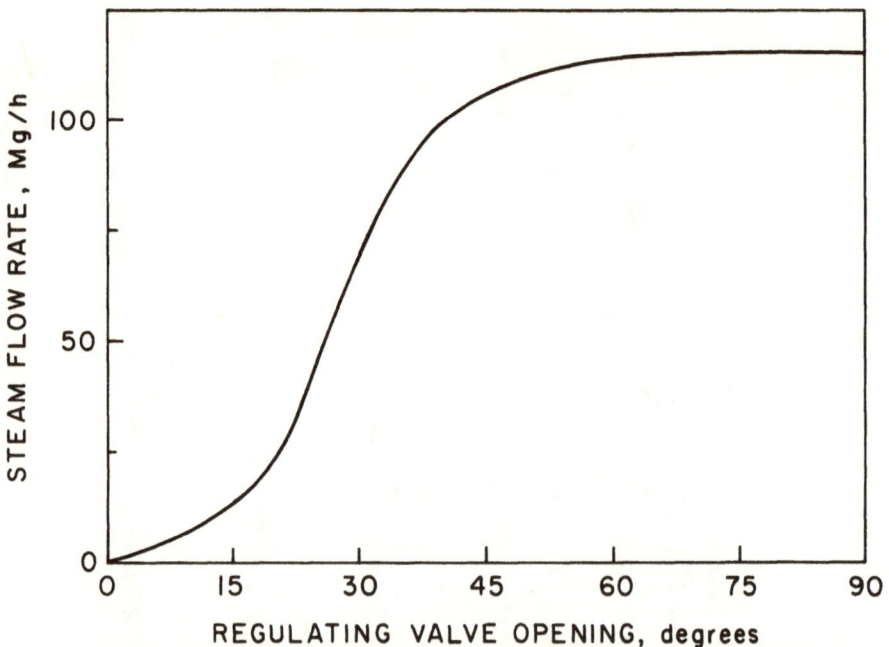

FIGURE 6.9—Butterfly regulating valve flow characteristic at Otake [Usui and Aikawa, 1970].

station. It is of the horizontal, cylindrical, rotating-field type, air-cooled, rated at 15,000 kVA, 6.6 kV, and has a power factor of 0.9.

The condenser, of the direct-contact, barometric type, is situated outside the powerhouse about 10 m (33 ft) above the hot well. The pressure is 9.8 kPa (2.90 in Hg), the cooling water enters at 26°C (78.8°F) and leaves at 41.4°C (106.5°F), and the flow rate of the cooling water is 3900 Mg/h. These are four spray trays; the top two serve as gas coolers. The basins of the top three trays are made of stainless steel because they are exposed to the most corrosive fluids and cannot be coated with epoxy paint owing to the presence of small spray holes.

The gas extractor consists of a set of three motor-driven reciprocating vacuum pumps, each rated at 110 kW. A steam jet gas extractor was considered and rejected because the low pressure of the geothermal steam would have required about 6 Mg/h (13,000 1bm/h), an amount of steam equivalent to 600 kW of electrical output from the plant. Despite the additional maintenance, the reciprocating pump was selected over other types (axial blower, radial blower, Nash pump, water ejector, Roots blower, and screw pump) because it was the most effective in terms of operating range, efficiency, number required, and cost. The limitation in capacity of reciprocating pumps was not a handicap since the amount of noncondensable gases in the case of Otake is small.

The plant uses a 3-cell, crossflow, mechanically-induced-draft cooling tower. Its relatively low profile gave this type an advantage over the taller natural-draft type because the plant is located in scenic Aso National Park. The tower receives 4233 Mg/h (9.33×10⁶ lbm/h) at 41.4°C (106.5°F) and cools it to 26°C (78.8°F) at a design wet-bulb temperature of 17°C (62.6°F). Wet-bulb temperature at Otake reaches a maximum of 22°C (71.6°F) during the summer; the yearly average is 13°C (55.4°F). The cells are constructed of ferro-cement and have vinyl chloride elements. Each cell is fitted with a 6-bladed propeller of 5.7 m (18.7 ft) diameter to extract the cooling air. The total air flow is 49,000 m³/min (1.75×10⁶ ft³/min); the three fans consume about 200 kW under normal operating conditions.

6.3.4 Construction materials

Table 6.10 lists the main elements of the plant that are exposed to the geothermal fluid and their construction materials. The selection of materials was made after 1 year of testing 20 types of material and paint for anticorrosiveness, stress corrosion, and scale deposition.

6.3.5 Environmental effects and co-utilization

Owing to its location in beautiful Aso National Park, serious attention was paid to harmonizing the plant with its surroundings. Only 40 ha (99 acres) are owned by the Kyushu Electric Power Company at the site, and most of this area consists of rugged terrain. The power plant occupies about 0.86 ha (2.1 acres), which includes 0.48 ha (1.2 acres) for the power station and 0.38 ha (0.94 acres) for the wells and pipeline system. The powerhouse itself takes up 393 m² (4230 ft²).

Table 6.10—*Construction materials for Otake power plant* [a]

Material		
Type	ASTM Specification	Component
Rolled steel for general structures	A6-68a	Silencers; steam pipes [>400-mm (>15.75-in) dia.]; receiver turbine exhaust casing; exhaust duct [b]; condenser (exc. spray basins); mist separators for gas extractors [b]; cooling tower spray basins [b]
Rolled steel for welded structures	A242-68	Shell of steam separators; shell of ball-check valves; tanks for separated hot water; drain tank
Gray iron castings. . .	A48-64, No. 30B or 35B..	Turbine inlet casing; turbine diaphragms; gas extractors (without feather valves); warm and cold water pump casings; main water pipes
Carbon steel castings..	A27-65 N-2... . .	Sluice valves and safety valves for steam line (with stainless steel parts)
Carbon steel pipes.....	A120-68a.....	Steam pipes [< 400-mm (<15.75-in) dia.]; hot water pipes
Carbon steel pipes....	A53-68 Gr. A	Gas extractor and discharge pipes;[b] barometic pipe for jet condenser [b]
Low Cr Mo steel.....	. ...	Turbine rotor
Stainless steel.........	403 or 304.....	Turbine nozzles, blades, gland packing fins; strainer for main steam; balls of ball-check valves; expansion joints; elements of steam separators and steam receiver; feather (plate) valves of gas extractor; condenser spray basins; shafts of warm water pumps and cooling water pumps; cooling water pipes for auxiliary
Stainless steel castings	A296-68, CA40....	Impellers of warm water pumps and cooling water pumps
Anticorrosion aluminum alloy	B211, alloy 5052... ...	Fan impellers of cooling tower

[a] Source· Usui and Aikawa, 1970.
[b] Painted with protective epoxy coating.

Reinjection wells are used for the disposal of the waste liquid from the wellhead separators and the excess steam condensate from the cold well of the mechanically-induced-draft cooling tower. Originally there was no reinjection done at the Otake plant; the first reinjection well was completed in February 1972. Prior to that time the liquid was held in a settling pond, allowed to cool, and then discharged into the Kusu River. Because the geothermal fluid contains arsenic, this practice was abandoned. The reinjection wells are located along or near the river, but monitoring shows that arsenic is not now entering the waterway.

Reinjection is carried out under atmospheric pressure to guard against any chance of inducing earthquake activity. However, loss of reinjectivity has been severe, with a dropoff in flow rate of about 7% per month. Periodically, the liquid is briefly pumped under pressure into the reservoir. Reinjectivity is restored for a short time, but the higher flow rate is not maintained during atmospheric reinjection. An arsenic removal system is now under construction at Otake. This chemical treatment plant will be able to remove arsenic from up to 150 Mg/h of the waste liquid. With the arsenic removed from the spent geofluid, it may be possible to reduce the amount that must be reinjected.

Since the condensate from the geothermal steam is recirculated to provide cooling water for the plant, large quantities of external water are not needed. There is a need, however, for make-up water to prevent the accumulation of acid-laden water in the cooling tower basin; 22 Mg/h (48,500 lbm/h) is drawn from the Kusu River.

There are no controls on the emissions of hydrogen sulfide; the full amount is discharged through the reciprocating vacuum pumps to the atmosphere. The specific amount is not large, however—about 542 g/MW·h, based on a concentration of 48 mg/kg of main steam, 113 Mg/h turbine steam flow, and 10 MW output.

The district water-heating program of Otake (and nearby Hatchobaru) provides for about 135 private homes and 30 hotels in the Sujiyu Hot Springs area. Waste hot water from the plant is used to heat clean river water to about 90°C (194°F) before the geofluid is reinjected. The hot water is distributed to homes and inns at no charge and arrives at the point of use at about 75°C (167/F).

6.3.6   Economic and operating data

Capital investment for the Otake plant was $3.27 million (1968 U.S. dollars) or about US $300/kW installed. At a load factor of 90%, the 1968 cost of electricity was 7.1 US mill/kW·h [Hayashida and Ezima, 1970]. Table 6.11 summarizes the operating history of the Otake plant for the first 11 years of its existence. The plant has a remarkable and consistent availability factor, averaging 95.5% from 1967 to 1977. The loss of produc-

Table 6.11—*Performance of Otake separated-steam 10-MW geothermal power plant* [a]

Year [b]	Electricity generation, MW·h		Mean load kW	Net capacity factor, percent	Availability factor, percent
	Gross	Net			
1967 [c].........	52, 238	48, 320	5963	86. 2	86. 7
1968..........	75, 824	68, 317	8632	78. 0	95. 3
1969..........	71, 311	63, 895	8141	72. 9	96. 1
1970..........	84, 885	77, 160	9690	88. 1	98. 9
1971..........	73, 538	65, 964	8395	75. 1	94. 9
1972..........	72, 128	64, 482	8211	73. 6	99. 2
1973..........	70, 318	62, 794	8027	71. 7	93. 9
1974..........	74, 758	66, 684	8534	75. 9	99. 2
1975..........	69, 216	61, 741	7901	70. 5	94. 0
1976..........	65, 761	57, 738	7486	65. 9	98. 8
1977..........	52, 403	44, 385	5982	50. 7	93. 9

[a] Source: Kyushu Electric Power Company.
[b] Year beginning April 1.
[c] Plant commissioned August 12, 1967.

tion wells is responsible for the steady decline in net capacity factor from a value of 86.2% in 1967 to 50.7% in 1977.

Maintenance is performed by the Nishi-Nippon Plant Company. Inspection must be carried out at least once every 2 years by Japanese law. Overhauls have taken place during 1968, 1969, 1971, 1973, 1975, and 1977.

The cost of the routine maintenance for Otake is 16 million yen (or $90,000). About 1 month is required to carry out routine inspection and maintenance. At Otake the strainer atop the barometric condenser must be cleaned of dirt and other debris once per month; this operation takes about 4 hours.

The production wells seem to develop a buildup of calcium carbonate, $CaCO_3$, at a point about 20 m (66 ft) below the end of the production casing in the open-hole portion of the well. The deposit thus occurs 270 m (886 ft) below the surface and most likely corresponds to the flash point of the geofluid. An attempt to clean one of the wells (O–6) by treatment with 5000 kg (11,000 lbm) of 35% HCl acid resulted in a temporary increase in total flow from 29 Mg/h (64 klbm/h) to 64 Mg/h (141 klbm/h), or a 120% improvement. The treatment caused the steam flow to double from 4 Mg/h (8.8 klbm/h) to 8 Mg/h (17.6 klbm/h). Shut-in pressure also increased by about 18%.

6.4   Onuma

The power plant at Onuma was the third completed in Japan and the second of the liquid-dominated type. The plant was completed and began operations in December 1973. Built and operated by the Mitsubishi Metal

Corporation to supply electricity to one of its metal processing mills, the plant is located in the northern part of the main island of Honshu in Akita Prefecture in Towada-Hachimantai National Park. The turbine is rated at 10 MW. A site photograph is given in figure 6.10.

6.4.1 Geology and exploration

Like its neighbor at Matsukawa, the Onuma plant is situated in a Quaternary volcanic region. The area is dotted with surface manifestations of subterranean magmatic activity such as fumaroles, active volcanoes, and hot springs. Prospecting and development began in 1965, and construction commenced in 1970.

6.4.2 Well programs

The geothermal wells are drilled from inclined central platforms so that wellhead equipment can be grouped together to minimize land usage and to avoid unnecessary replication of equipment. The wells are so inclined that the well bottoms are located 150 m (492 ft) from the drilling platform, measured horizontally at ground level [MHI, 1976c].

The plant began by using three production and two reinjection wells. Additional production wells were added in 1975 (well 3Rb) and 1977 (well 8R); during this time, two more reinjection wells also were completed. Table 6.12 lists the flow capacities of the nine wells serving the Onuma plant as of June 1977. The reinjection wells receive waste liquid at atmospheric pressure immediately after the pressure is let down in the cyclonic silencers. Depth of the five production wells averages about 1600 m (5250 ft) and for the four reinjection wells about 960 m (3150 ft). One reinjection well is slant drilled (7R) as are all production wells except one (3Ra) [Ito et al., 1977a].

6.4.3 Energy conversion system

The steam turbine is similar to Otake's except that the Onuma turbine operates at a speed of 3000 rpm and has a last-stage blade height of 500 mm (19.7 in). The inlet and exhaust conditions are essentially identical to those at Otake. The Onuma turbine is located on the ground floor of the powerhouse, with the exhaust duct directed upward and leading outdoors to the barometric condenser. Placing the turbine at ground level and the auxiliaries beneath the floor made it possible to keep the roof elevation of the powerhouse very low, a desirable feature because the station is situated in a scenic national park.

The generator is rated at 12,500 kVA, 6.6 kV, 0.8 power factor and is air cooled. The step-up transformer is rated at 12,000 kVA and feeds power to the transmission line at 66 kV. The plant schematic flow diagram is quite similar to that for the Otake plant (see figure 6.8) ; the condenser,

FIGURE 6.10—Onuma geothermal power plant. Centralized wellhead equipment is in right foreground; powerhouse, barometric condenser, and cooling tower are at left [MHI, 1976c].

Table 6.12—*Flow capacities of wells serving Onuma plant
(as of June 1977)*[a]

Well number	Flow rate, m³/h
Production wells:	
3Ra	97. 8
3Rb	75. 0
5R	60. 9
6R	90. 0
8R	72. 0
Total	395. 7
Reinjection wells:	
1R	65. 0
2R	63. 7
7R	126. 2
7T	140. 4
Total	395. 3

[a] Source: Ito et al., 1977b.

gas extractor, and cooling tower are also similar. Table 6.4 lists technical particulars for the major components of the energy conversion system.

6.4.4 Operating experience and co-utilization

The performance of the Onuma plant is summarized in table 6.13. From 1973 through 1977, Onuma operated for 36,073 out of a possible total of 38,424 for an availability factor of 93.9 percent. However, unlike Otake, the Onuma plant began at about one-half capacity and has steadily increased its output, achieving a mean output of 7.7 MW in 1977. Reinjection has proved trouble-free even though the fluid is returned to the reservoir at atmospheric pressure. The plant has been overhauled once each year from 1974 through 1977.

Table 6.13—*Performance of Onuma separated-steam 10-MW geothermal power plant*

Year	Operating hours	Availability factor, percent	Mean output, kW	Capacity factor, percent
1973 [a]	3, 044	90. 6	5, 055	0. 458
1974	7, 873	89. 9	6, 183	0. 556
1975	8, 389	95. 8	7, 248	0. 692
1976	8, 378	95. 4	6, 946	0. 664
1977	8, 389	95. 8	7, 688	0. 737

[a] Plant commissioned November 12. 1973.

At the end of 1978 the output at Onuma was 8,200 kW (with a maximum capacity of 8,600 kW). The steam cost was about 2.5 yen/kW · h, the cost of generation was about 6.5 yen/kW·h, and the electric power charge by the Tohoku Electric Power Company was 8 yen/kW · h (for midnight power demand) and 10 yen/kW·h (for normal power demand). Power at one time was sent to a copper refinery owned by Mitsubishi Metals, but that plant closed, and power is now sold to the Tohoku Electric Power Company.

Several nearby inns and hotels at Tsunodate city receive about 15 Mg/h (33 klbm/h) of water heated to 60°–70°C (140°–158°F) by the waste water from the Onuma plant. About 266 Mg/h (586 klbm/h) of geothermal liquid is circulated through heat exchangers before being reinjected [Iga and Baba, 1974].

6.5   Onikobe

The 25-MW power plant at Onikobe was the fourth to be built in Japan. In March 1975 the plant was turned over to the Electric Power Development Company, Ltd., by Kawasaki Heavy Industries, Ltd., which constructed all major plant elements except for wellhead equipment, steam transmission pipelines, and the substation. The plant is located in Kurikoma National Park in Miyagi Prefecture on the northern part of the island of Honshu, about 120 km (75 mi) south of the Matsukawa-Takinoue region. A site photograph is given in figure 6.11.

6.5.1   Geology and exploration

The Onikobe reservoir may be likened to that of Wairakei, New Zealand. Faults in a Miocene and granite basement allow high-temperature fluids to reach from depth to the overlying Pliocene beds of porous rocks, which form the hydrothermal convective reservoir containing ground water.

The area is an elliptical basin 10 km × 8 km (6 mi × 5 mi) across the axes. A major part of the subsided area was filled by a lake from the Pliocene to the Pleistocene epoch. Exposed pre-Neogene granite and a thick series of Miocene rocks make up most of the subsided area. The rocks are mostly volcanic and partly sedimentary. Total thickness of the formation is about 300 m (984 ft) and consists of a lacustrine formation in loose deposits of conglomerate, andesitic tuff-breccia and tuff, pumiceous tuff of dacitic nature, and mudstone from the base upward [Saito, 1961].

Although the region is a Quaternary volcanic zone, there are no active volcanoes; the usual surface manifestations (hot springs, fumaroles, solfataras, geysers) are scattered throughout the area. Early exploration included a topographic survey and geologic mapping in 1955 and the drilling of three test bores to depths of 255, 250, and 192 m (836, 820, and 630 ft) [H. Nakamura, 1959].

FIGURE 6.11—Onikobe geothermal power plant [Kawasaki, 1975].

The general environment of the site is quite hostile to power station operation. Altitude is 530 m (1738 ft) above sea level, winter snowfall exceeds 2 m (6.6 ft), ambient temperature ranges from $-15°$ to $+30°C$ (5° to 86°F), the natural concentration of hydrogen sulfide in the air is 0.2 ppm, and the neighboring brooks are extremely acidic with pH of 2–3. Thus the effects of corrosion and erosion place severe design constraints on the Onikobe power plant [Kawasaki, 1977].

6.5.2 Wells and productivity

The geothermal reservoir consists of two producing zones, a shallow zone at about 300 m (980 ft) and a deep zone at about 1000 m (3280 ft). The upper reservoir is vapor dominated and yields a mixture at the wellhead with a dryness fraction of about 50%, i.e., equal parts of liquid and vapor by weight. The deep reservoir is liquid dominated but contains highly acidic water that, so far, has defied utilization because of its corrosive nature. Unlike most geothermal hydrothermal reservoirs in Japan, which are characterized by sodium-chloride water, the Onikobe fluid in the deep reservoir contains a large percentage of hydrochloric acid [Ozawa, 1973].

Thus the plant is supplied with steam obtained from relatively shallow wells drilled into the upper hydrothermal reservoir. Originally, nine production wells were used; three more have since been completed. Characteristics of the geothermal fluid are given in table 6.14. Hydrogen sulfide concentration is the highest found in any geothermal power plant now in operation anywhere in the world.

Table 6.14—*Characteristics of geothermal steam and condensate at Onikobe* [a]

Percentage of noncondensable gases (by volume of steam).............................	0 5%
Composition of noncondensable gases (volume percentage):	
Carbon dioxide, CO_2.........................	58. 4%
Hydrogen sulfide, H_2S....................	36. 0%
Others (hydrogen, methane, ammonia)..........	5. 6%
Composition of condensate (ppm):	
Chloride, Cl................................	7590
Calcium, Ca................................	930
Amorphous silicon dioxide, $SiO_2 \cdot H_2O$.....	550
Magnesium, Mg.........................	340
Hydrogen sulfide, H_2S.... .	107
Sulfate, SO_4...................	24
pH of condensate....................	3. 0

[a] Source: Kawasaki. 1977.

Owing to the vapor-dominated nature of the reservoir, the geofluid arrives at the surface essentially as a saturated vapor with a small amount of liquid entrained. Approximately 400 times more steam is generated than liquid, by volume. Separators are used at the wellheads and at the powerhouse to remove the liquid droplets. Steam is collected at the station header, where its pressure is controlled by an electrohydraulic regulating valve. Two main steam lines conduct the working fluid to the turbine.

Production from the wells has been only half what was expected. Furthermore, actual well characteristics of pressure versus flow rate differ from the design values. Attempts to correct for these deficiencies are described in the next section.

6.5.3 Energy conversion system

A flow diagram of the plant is given in figure 6.12; a listing of technical particulars for the major mechanical elements can be found in table 6.4. Kawasaki Heavy Industries, Ltd., built the steam turbine, a 5-stage, impulse-reaction, single-cylinder, single-flow machine rated at 25 MW at 3000 rpm. The generator was furnished by Fuji Electric Company, Ltd., and is rated at 28,000 kVA, 11 kV, power factor of 0.9 (lagging), and is a horizontally-mounted, revolving-field, self-ventilated, enclosed-duct circulating type with air cooler [Kawasaki, 1975].

The very high levels of hydrogen sulfide in the working fluid (36% of the total noncondensable gases is H_2S) required extensive corrosion fatigue tests on the materials for the turbine rotor and blades in addition to the usual static corrosion tests. Fatigue strength decreases markedly in the presence of a corrosive atmosphere. Thus, absolutely no resonance could be allowed in the turbine, and all blades for each of the five stages were designed to avoid resonance with any excitation force. Further, the static steam-bending stress was set at half the value allowed in conventional steam turbines. Stellite plates were silver soldered to the leading edge of the first-stage blades to prevent erosion by dust particles entrained in the steam. Titanium, with its outstanding anticorrosion characteristics, was used in the rotor shaft glands because of their exposure to geothermal steam and air leakage.

The turbine was forced to shut down in May 1975, only 2 months after it began operation. A new well had been connected to the plant, and an excessive amount of scale was carried over and deposited on the nozzles and blades of the first stage. Excessive vibrational stress resulted when the leading edge of the blades came in contact with the deposited scale, breaking one blade and cracking 11 others. Following an overhaul, no repetition of this problem has occurred, and no scale buildup has been observed.

The Onikobe plant uses fairly conventional equipment to handle the problem of heat rejection—a barometric condenser, a gas extraction apparatus that consists of one set of two steam jet ejectors (in parallel) working to maintain vacuum in the barometric condenser, a motor-driven vacuum

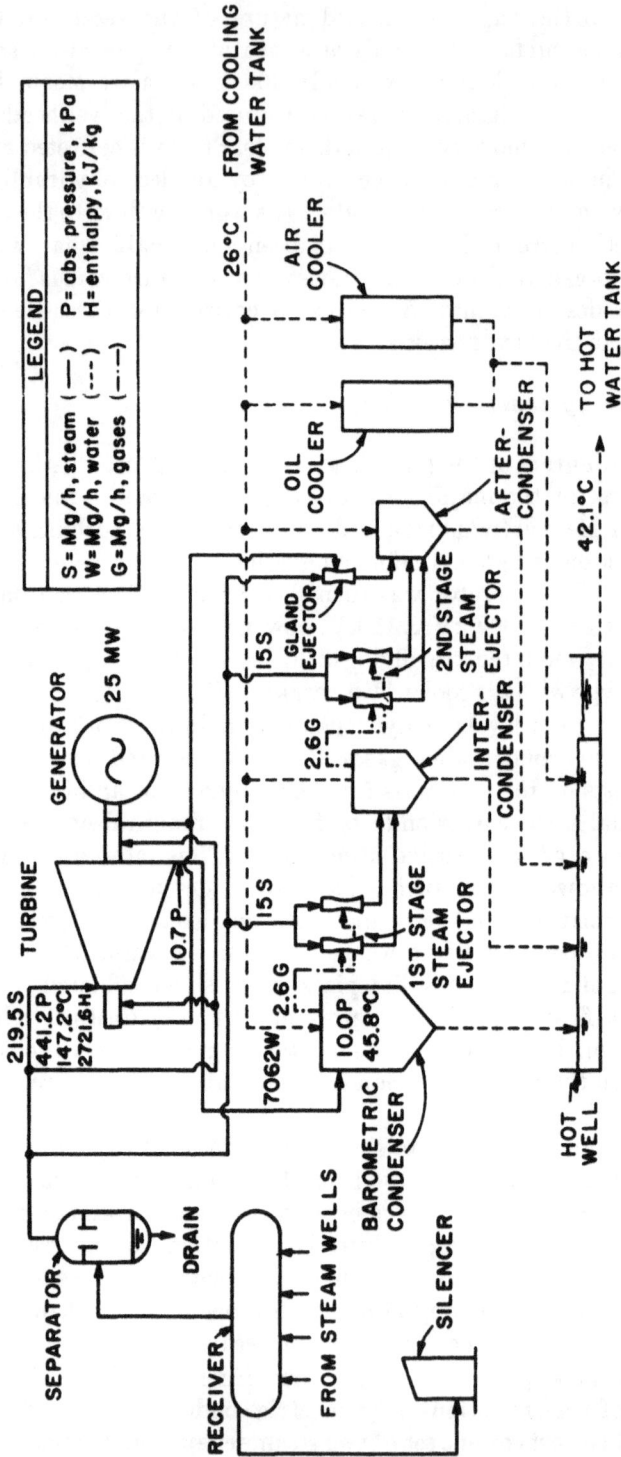

FIGURE 6.12—Schematic and heat balance diagram for Onikobe power plant [Kawasaki, 1977].

LEGEND

S = Mg/h, steam (——)   P = abs. pressure, kPa
W = Mg/h, water (---)   H = enthalpy, kJ/kg
G = Mg/h, gases (—·—)

pump that removes noncondensable gases from the intercondenser, and a counterflow, mechanically-induced-draft water cooling tower. The plant was originally designed and operated with two sets of steam jet ejectors, but the second set was replaced by the vacuum pump in October 1975.

The inner surface of the condenser is coated with a neoprene lining that is more resistant to corrosion and wears better than epoxy paint. The May 1975 overhaul revealed that part of the lining had separated from the condenser wall near the bottom. This separation was attributed to the direct impingement of hot water and a lack of adhesion. Some minor improvements solved this problem.

Some of the spray nozzles of the cooling tower had been attacked by corrosion and erosion and had fallen off. Repairs and improvements such as fixing the nozzles to the header solved this problem. In October 1975 the sleeve on the circulating water pump was replaced; the new one is fabricated from 304 stainless steel whereas the original (which had heavy pitting) was made of 403 stainless steel. A summary of the fouling experience of the plant is given in table 6.15.

Table 6.15—*Fouling experience at Onikobe* [a]

Problem	Component
Scale accumulation......	Nozzles and blades of first and second stages; turbine gland; nozzle of gland ejector; main-steam strainer.
Corrosion...............	Gland packing ring; cooling pipe of heat exchanger; pressure gauge; spray nozzle of cooling tower.
Erosion........	Valve stem used in condensate line; circulating water pump sleeve.
Separation............ .	Neoprene lining at condenser bottom; epoxy coating of exhaust pipe.

[a] Source: Kawasaki, 1977.

6.5.4 Environmental effects

All waste liquid (drains from separators, turbines, and other pieces of equipment plus overflow from the cooling water basin) is collected in a drain pond and reinjected into the formation. This amounts to a relatively small quantity because of the high dryness fraction of the geofluid at the wellhead. Hydrogen sulfide is released to the atmosphere without controls. The concentration of H_2S is 0.40 ppm in the turbine room, 0.050 ppm in the control room, and 0.20 ppm outside the plant. Since the main steam contains about 3400 mg/kg, the specific emission level is approximately 31,400 g/MW · h, an extremely high value.

Because Onikobe is located within the Kurikoma National Park, special consideration was given to design features that help the plant blend into its scenic surroundings, including colors, low height of buildings and components, and restricted sound levels.

6.5.5 Operating experience

The Onikobe plant maintained an average availability of 93.7% for the first 17 months of its operation. During this period the plant was shut down twice for improvements, twice for inspection according to the regulations of the Ministry of International Trade and Industry, and twice for periodic inspection.

Plant output started at about 9 MW and rose to about 13 MW after 17 months. Production from the steam wells has been disappointing, and as a result the plant is operating at about half its design capacity, albeit very reliably. In October 1975 two improvements were carried out to boost the output. First, the second set of steam jet ejectors was replaced by a motor-driven vacuum pump. Second, the nozzle of the first stage of the turbine was redesigned to correspond more closely with actual steam conditions. These changes increased plant output about 10%. However, steam production remains at about 120 Mg/h (0.264×10^6 lbm/h), far below the design value of 219.5 Mg/h (0.484×10^6 lbm/h).

6.6 Hatchobaru

The geothermal power plant at Hatchobaru is the most advanced steam plant in Japan, incorporating the following novel features: (1) two-phase transmission of geothermal fluid; (2) so-called double-flash operation (one separation followed by a flash process); (3) low-level jet condenser; and (4) so-called ERR system gas extractor (steam ejector followed by two radial blowers). The fifth commercial geothermal station in Japan, it came on-line in June 1977 with a power rating of 50 MW. The Kyushu Electric Power Company, Inc. owns the plant.

6.6.1 Geology and exploration

Hatchobaru lies about 1.7 km (1 mi) south of Otake, about 40 km (25 mi) north-northeast of the active volcano Mount Aso in Oita Prefecture on the island of Kyushu. Figure 6.13 is a map of the thermal region showing the locations of the two geothermal sites and the town of Sujiyu, a small hot-springs resort.

The geology of the Hatchubaru region is similar in general to that of Otake, but different in certain details. In both cases the source of the thermal anomaly is the same magma chamber. Both areas are underlain by two hydrothermal reservoirs. However, whereas at Otake the upper zone of the tuff breccia is being tapped for the power plant, at Hatchobaru the

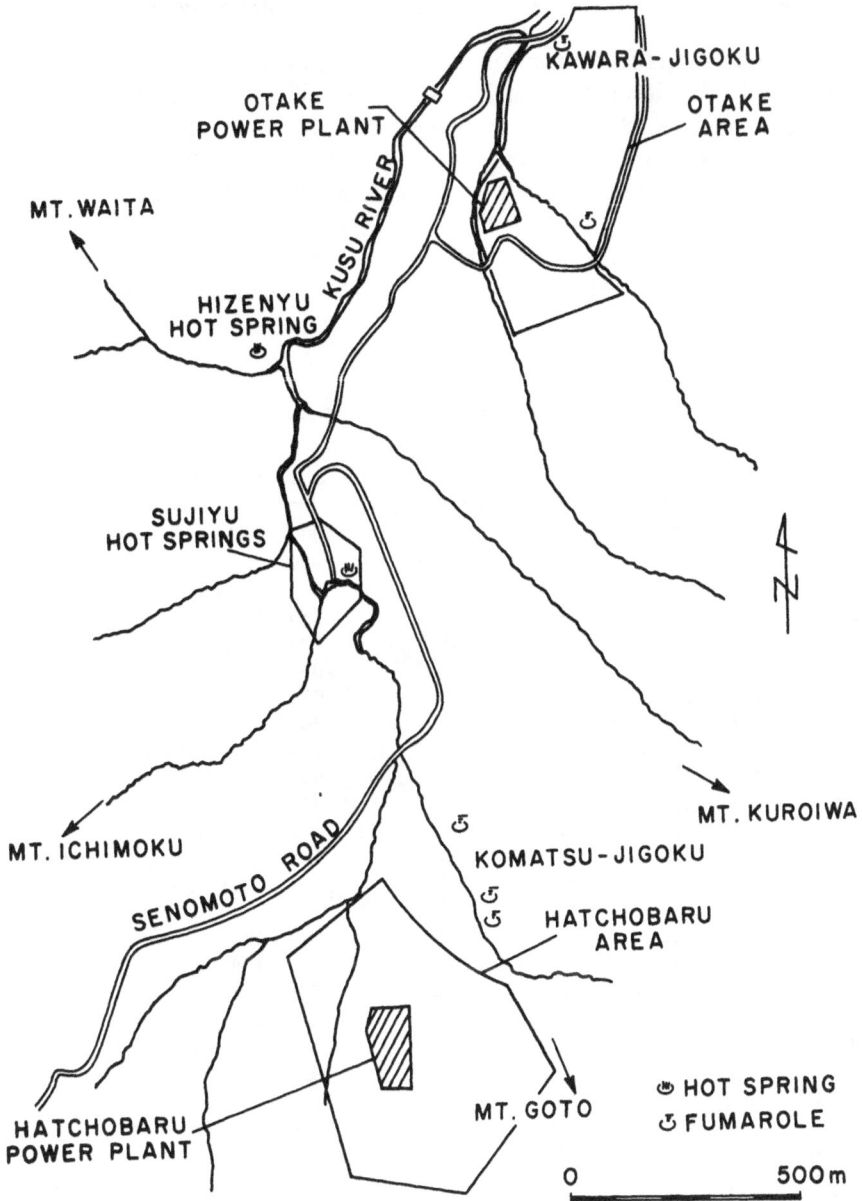

FIGURE 6.13—Location of Otake and Hatchobaru geothermal areas. Hot-springs resort town of Sujiyu lies midway between the two sites.

corresponding upper zone, uplifted relative to the Otake zone, is cased off, and the deep zone consisting of so-called Kusu and Usa groups, is being exploited. Exploration of Hatchobaru was done simultaneously with Otake (see section 6.3.1).

6.6.2 Well programs and two-phase fluid transmission system

An intensive drilling program was underway during 1978 to bring the station up to its designed capacity of 50 MW. This program includes the drilling of both production and reinjection wells. Table 6.16 gives a summary of the wells drilled at Hatchobaru as of the end of 1978. At that time five successful production wells had been drilled, four of which were delivering steam to the plant (H-4, H-6, H-7, and H-10), and one was being tested. Three additional wells were being drilled. Ten wells will be required to permit the plant to generate 50 MW. Drilling success rate for production wells was 45% through 1978.

Table 6.16—*Drilling experience at Hatchobaru*

Well No.	Drilling period	Size, mm	Depth, m	Comments
HT-1...	Aug. 1964–Mar. 1965	76. 2	1000	Failure
HT-2...	Oct. 1965–Nov. 1966	76. 2	900	Successful; stopped prod. Jan. 1967
H-1....	Jul. 1968–Feb. 1969	193. 7	785	Failure
H-2....	Feb. 1969–Jul. 1969	193. 7	739	Failure [a]
H-3....	Mar. 1970–Jun. 1970	142. 9	1089	Failure
H-4....	Jun. 1970–Oct. 1970	193. 7	1084	Successful; still in use
H-5....	Jan. 1971–Dec. 1971	193. 7	1600	Failure
H-6....	Jan. 1971–Aug. 1971	193. 7	1238	Successful; still in use
H-7....	Apr. 1976–Nov. 1976	193. 7	921	Successful; still in use
H-8....	Apr. 1976–Dec. 1976	193. 7	1450	Failure [b]
H-9....	Apr. 1977–Oct. 1977	193. 7	550	Failure [c]
H-10...	Mar. 1977–Sep. 1977	193. 7	759	Successful; still in use
H-11...	Jan. 1978–Dec. 1978	193. 7	—	Testing
H-12...	Aug. 1978–	193. 7	—	Drilling
H-13...	Aug. 1978–	193. 7	—	Drilling
H-14...	Sep. 1978–	193. 7	—	Drilling
HR-1...	Mar. 1973–Jul. 1973	142. 9	1500	Failure
HR-2...	Aug. 1973–Oct. 1973	219. 1	600	Failure
HR-3......	....	....		Interrupted
HR-4. .	Jun. 1974–Sep. 1974	193. 7	950	Failure
HR-6...	Jun. 1975–Feb. 1976	193. 7	988	Successful; still in use
HR-7...	Mar. 1976–Oct. 1976	193. 7	1018	Successful; still in use
HR-8...	Aug. 1975–Apr. 1976	193. 7	1130	Successful; still in use
HR-9...	Aug. 1975–Apr. 1976	193. 7	1130	Successful; still in use
HR-10..	May 1977–Aug. 1977	269. 9	702	Successful; still in use
HR-11..	Apr. 1977–Nov. 1977	269. 9	850	Successful; still in use
HR-12..	Dec. 1977–Jul. 1978	269. 9	1027	Successful; preparing to use
HR-13..	Mar. 1978–Jul. 1978	269. 9	986	Successful; still in use
HR-14..	Oct. 1978–	269. 9	—	Drilling
HR-15..	Aug. 1978–	269. 9	—	Drilling

[a] Converted to a reinjection well, H-2R; currently in use.
[b] Planning to be converted to a reinjection well.
[c] Converted to a reinjection well, H-9R; currently in use.

Nine wells were being used as reinjection wells; seven of these had been drilled purposely as reinjectors and two had been converted from unsuccessful production wells. One well was being prepared for use, another was in the process of being converted, and two were being drilled. The success rate for reinjection wells (drilled for that purpose) was 67%.

Figure 6.14 is a plan view of the power station and bore field. As can be seen, many of the wells are slant drilled for economic and topographic reasons. Production well H-7, for example, which has a total drilled length of 821 m (3022 ft), actually has a well bottom depth of 897 m (2943 ft) below the surface and offset from the wellhead by 167 m (548 ft). Average depth of the successful producing wells is 1001 m (3283 ft); for the reinjection wells in use, it is 899 m (2950 ft).

One of the most interesting features of the Hatchobaru plant is the method of transporting the geothermal fluid to the powerhouse. For a double-flash plant, there are essentially three ways to carry out this opera-

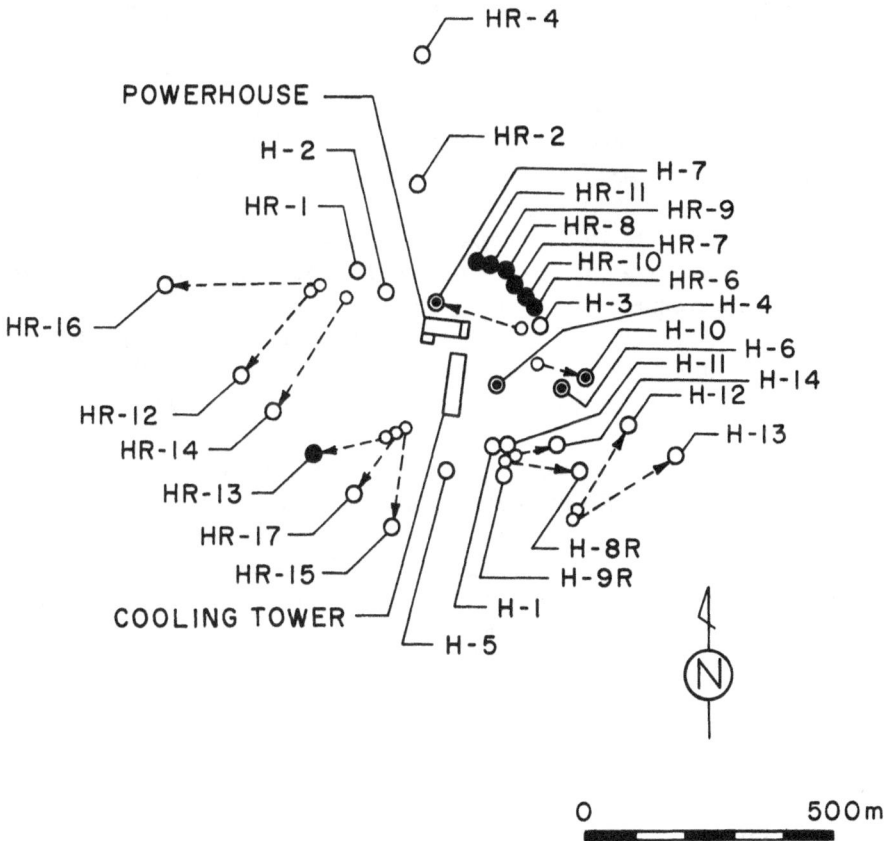

FIGURE 6.14—Location of wells serving the Hatchobaru 50-MW geothermal power plant.

tion: (1) steam-only transmission; (2) separate steam and hot water transmission; (3) two-phase mixture transmission.

In the first case, parallel main and secondary steam pipelines must run from each well to the powerhouse. Although this concept is simple, it has some serious disadvantages. Secondary steam pipelines are very large, expensive, and must extend long distances. Waste hot water must be piped from each well to the disposal site and such pipes are subject to scaling. Separators and flash tanks must be located at each well, and maintenance of scattered wellhead equipment is a problem.

In the second case the main steam is separated at the wellhead, and both steam and hot liquid are piped separately to the powerhouse, where the liquid is flashed to produce the secondary steam. While the number of flash tanks is reduced and the size of the liquid pipeline is smaller than a secondary steam line, there are still disadvantages to this method. Water-hammer occurs in the liquid pipeline. Static pressure will fall in the liquid, leading to flashing and scale deposition. Hot water pipes cannot be joined on the way to the powerhouse. Furthermore, pumping of the liquid is required unless the wellheads are located at an elevation higher than the powerhouse.

In the last case the entire steam liquid mixture is transmitted to the station in a single pipeline. This technique does not automatically eliminate the possibility of waterhammer or scale deposition, and there is a danger of large pressure losses. However, this method does offer a number of distinct advantages over the previous two methods. Only one pipeline is needed between each well and the powerhouse, and the pipelines from several wells may be joined on the way to the powerhouse. Various piping arrangements are possible, including upwardly inclined pipes. The separator and flasher are centrally located at the station for ease of maintenance, and a small number of separators and flashers serve many wells. Finally, depending on the method of disposal, only one waste hot water pipeline may be needed.

Extensive field tests were conducted beginning in 1965, including tests at Otake (well No. 5, 1966) and Hatchobaru (well No. 1, 1970). These tests established the superiority of the two-phase transmission system for the Hatchobaru plant [Takahasi et al., 1970]. Each wellhead is equipped with a simple U-bend separator and silencers to allow easy and stable operation during startup of the plant. Table 6.17 gives some of the characteristics of the geofluid from the four producing wells at Hatchobaru.

6.6.3 Energy conversion system

The turbine is of the mixed-pressure type. Main steam is admitted to the first two stages; second steam enters between the second and third stages, mixing with the main steam, and the full flow continues through the last three stages. The machine is of the single-cylinder, double-flow, impulse-reaction type.

Table 6.17—*Characteristics of the geofluid from producing wells at Hatchobaru*

	Well H-4	Well H-6	Well H-7	Well H-10
Shut-in pressure, kPa..	4207	2472	—	1560
Separator pressure, kPa..	481	481	481	481
Primary steam flow, Mg/h.. .	68. 1	11. 9	54. 0	38. 2
Primary hot water flow, Mg/h....	245. 0	71. 0	166. 4	71. 8
Total flow rate, Mg/h . .	313. 1	82. 9	220. 4	110. 0
Flasher pressure, kPa.. . . .	113. 7	113. 7	113. 7	113. 7
Secondary steam flow, Mg/h......	21. 9	6. 3	14. 9	6. 4
Secondary hot water flow, Mg/h..	223. 1	64. 7	151. 6	65. 4
Wellhead quality, %..	21. 8	14. 4	24. 5	34. 7
Noncondensable gas content [a].	0. 05	0. 08	0. 08	0. 59
Composition of NC gases· [b]				
Carbon dioxide..	75. 5	39. 6	77. 4	84. 1
Hydrogen sulfide..	14. 8	3. 2	6. 7	4. 8
Others; O_2, N_2, etc..	9. 7	57. 2	15. 9	11. 1
Analysis of hot water:				
pH..	6. 5	4. 7	6. 1	6. 4
Sodium, ppm..	2200	2860	2390	2590
Potassium, ppm.	243	300	340	383
Calcium, ppm..	34	108	68	80
Chloride, ppm..	3850	5150	4125	4475
Silica, ppm..	922	620	834	737
Arsenic, ppm..	3. 2	3. 0	3. 1	2. 9

[a] As a percentage by weight of primary steam.
[b] As a percentage by volume of total noncondensables.

The main steam is produced from two separators located just outside the turbine room. They are of the vertical cyclone variety with bottom steam outlets and yield steam with a quality in excess of 99.9%. They operate at a pressure of 481.4 kPa (89.9 lbf/in²). The secondary steam is flashed from the residual hot water from the separators. The flash tank is a horizontal drum operating at a pressure of 113.7 kPa (16.5 lbf/in²) and producing steam with a quality greater than 99.9%. A simplified plant schematic is shown in figure 6.15. Table 6.4 lists the specifications of the major mechanical components of the station. Values shown in the table are the design values when the plant is at a full output of 50 MW.

Both the primary and secondary steam inlets are fitted with two main stop valves and two control valves. All stop valves are of the swing-check type; the control valves are of the butterfly type. Turbine speed and flash tank pressure are both controlled by the butterfly valves on the secondary steam line. The generator is of the horizontal cylindrical rotating-field type; it is rated at 55,000 kVA and is equipped with hydrogen cooling. A view of the power plant as seen from well H-6 is shown in figure 6.16; the main steam line into the twin cyclone separators is shown in figure 6.17.

This is the first Japanese geothermal power plant to use a low-level jet condenser. Set below the turbine-generator within the concrete foundation,

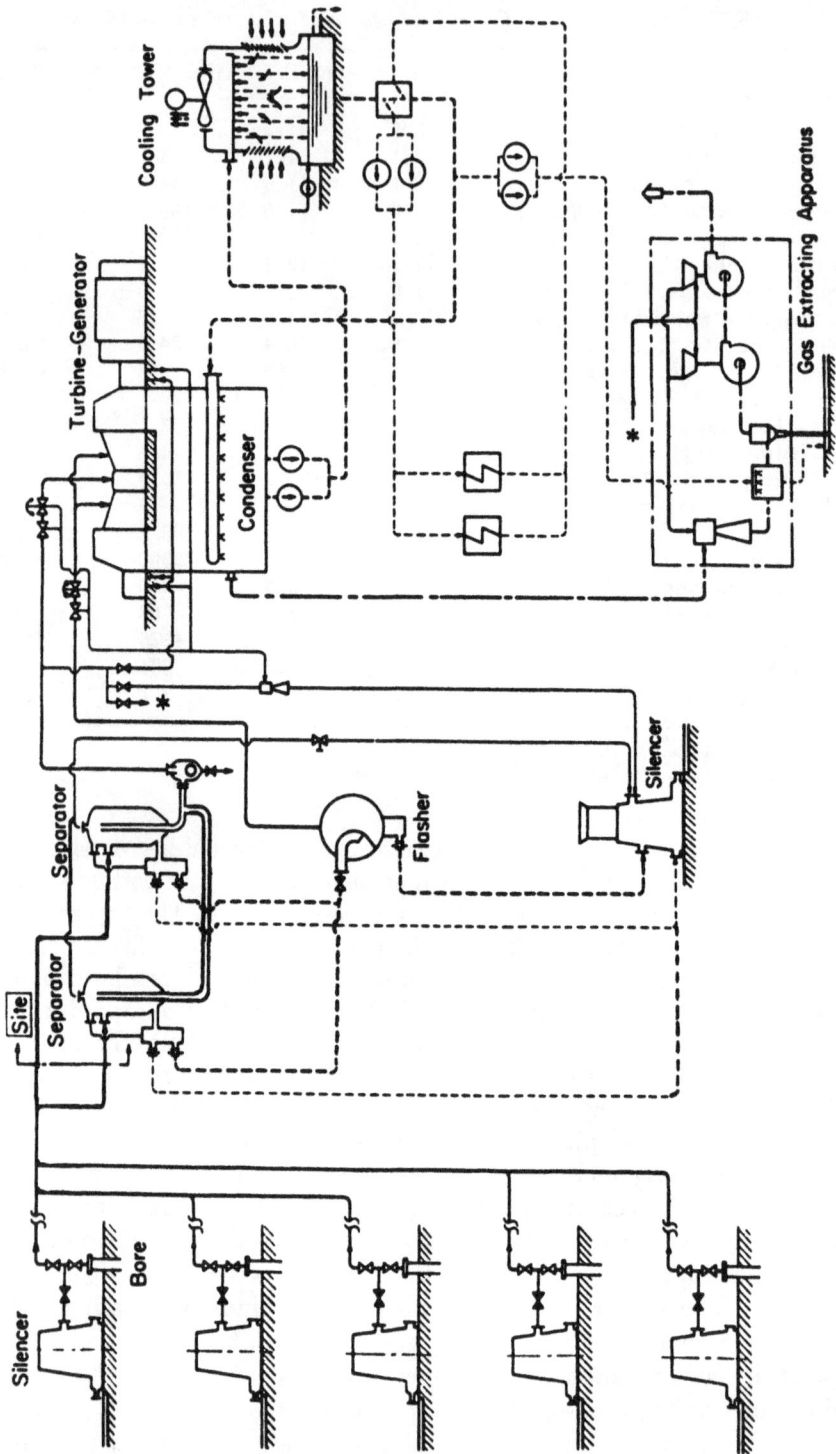

FIGURE 6.15—Schematic diagram for Hatchobaru power plant [Aikawa and Soda, 1975; MHI, 1977b].

FIGURE 6.16—Hatchobaru geothermal power plant. Powerhouse and separator/flasher equipment are at right; cooling tower and well No. 4 are at left [MHI, 1977b].

FIGURE 6.17—Main two-phase flow line and vertical cyclone separators at Hatchobaru [photo by R. DiPippo].

the compact unit allows the profile of the powerhouse to be kept low. The condenser arrangement is shown in figure 6.18. The right side of the unit is the gas cooler section leading to the gas extraction apparatus. The walls of the concrete condenser are coated with polyester resin to protect the surfaces from the corrosive action of the condensate. This new, compact design is expected to remove one of the impediments to larger-sized geothermal power units [Aikawa and Soda, 1975].

The gas extractor is another innovation of the Hatchobaru station. The system combines steam jet ejectors, which are very efficient for high-vacuum use, with radial blowers, which are better suited for low-vacuum applications. A schematic of the system is shown in figure 6.19. Three steam ejectors in parallel handle the high-vacuum noncondensable gases, while two radial blowers in series handle the low-vacuum load. The blowers are driven by small steam turbines; turbine exhaust is used to supply the steam jet ejectors. Steam consumption is about half that required for the usual two-stage ejector system.

The cooling tower is of the counterflow, mechanically-induced-draft type constructed of reinforced concrete with four cells. There is no need

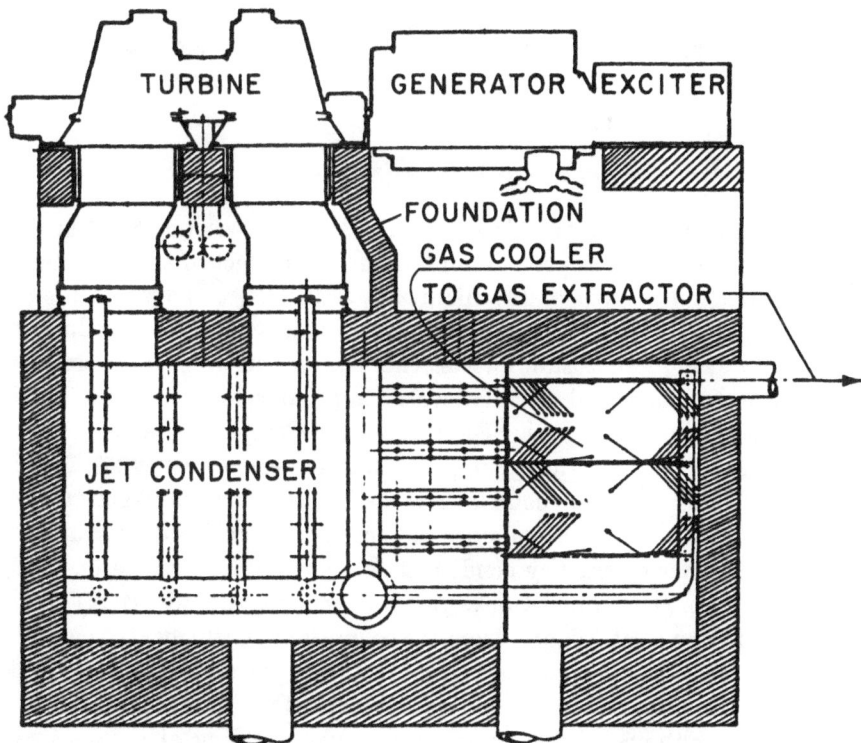

FIGURE 6.18—Low-level concrete jet condenser at Hatchobaru [Aikawa and Soda, 1975].

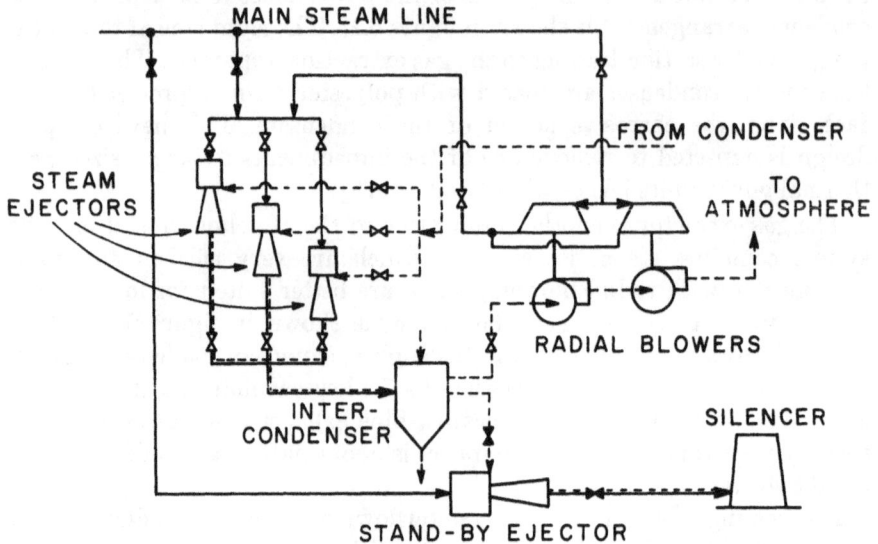

FIGURE 6.19—Schematic diagram of ERR gas extraction system at Hatchobaru
[MHI, 1977b].

for cooling water pumps because of the pressure differential between the tower basin at atmospheric pressure, and the condenser at 9.8 kPa (2.9 in Hg).

6.6.4 Environmental effects and co-utilization

Like the Otake plant, the Hatchobaru station is located in Aso National Park, and attention was paid to minimizing the impact of the plant on its surroundings. The area of the plant site is 36 ha (89 acres). All waste hot liquid is reinjected into the reservoir under gravity feed to reduce the possibility of induced seismic effects. There are no controls on the emission of hydrogen sulfide, and specific levels comparable to that of the Otake plant are likely. The Kyushu Electric Power Company, Inc. is planting trees on the site to help the plant conform to the scenic beauty of the area.

Restrictions placed on noise generation at Hatchobaru are extremely tight: 50 dB(A) under normal operating conditions at the plant boundary, 70 dB(A) under emergency conditions when the safety valves open. By comparison, in the United States the requirements are 65 dB(A) at the plant boundary or at 0.5 mile from the sound source, whichever is greater. Stringent noise restrictions have led to careful placement of potentially noisy pieces of equipment. Every attempt was made to take advantage of the natural contours of the terrain to provide shielding of noise sources.

The district water-heating system of Otake and Hatchobaru was discussed in section 6.3.5. Pure hot water is generated in a heat exchanger of the direct-contact type at Hatchobaru, with the hot geothermal waste liquid being flashed at the heat exchanger and the clean steam brought into contact with a falling stream of fresh water from the Kusu River, heating it to about 90°C (194°F). The separated geothermal waste liquid (now containing a higher concentration of dissolved solids) is disposed of by reinjection. Some trouble has been encountered with clogging of the pipelines carrying these liquids due to silicia deposition.

6.6.5   Economic data

Capital cost for the Hatchobaru plant is 12.96 billion yen (about US $73 million or US $1460/kW installed, 1978). This includes all well costs and field development. Roughly half this sum is attributed to power plant equipment. When full power is established (i.e., 50 MW expected in early 1980), electricity cost should be about 10 yen/kW·h ($\sim$ 56 US mill/kW·h). The percentage breakdown of the capital costs is as follows: equipment, 46.3%; wells and cooling tower, 25.5%; interest, 6.9%; land and buildings, 6.2%; miscellaneous, 15.1% [K. Tanaka, personal communication].

The total savings of the two-phase geofluid transmission system, compared to a single-phase system with individual wellhead separators, is estimated at about 5% of total construction cost [K. Aikawa, personal communication]. The cost of maintaining the Hatchobaru plant is 40.4 million yen ($\sim$ US $225,000). Drilling costs run over 100 million yen ($\sim$ US $500,000) per well.

6.6.6   Operating experience

The Hatchobaru plant was commissioned on June 24, 1977, with a rated output of 23 MW. In March 1978, the plant was upgraded to 27 MW. As of August 31, 1978, the plant had logged a total of 10,327 hours of operation for an availability of 99.2%. Reinjection wells HR-6 through HR-11 (see figure 6.15) were found to be interfering with production well H-3, necessitating the drilling of six additional reinjection wells HR–12 through HR-17. The first two of these have been successfully drilled, the second two are being drilled, and the last two are being planned.

The most serious problem associated with the plant is lack of sufficient steam stemming from the low success rate in drilling production wells. This difficulty will eventually be overcome as the reservoir model becomes more accurate and techniques for the selection of well sites improve. A related problem has been the lack of adequate reinjection wells. In spite of these difficulties the plant has shown an outstanding record of reliability, including a capacity factor of 90.5% based on its rated capacity for the first 13 months of operation. As of the summer of 1979 the plant was generating about 40 MW.

6.7   Kakkonda (Takinoue)

The newest Japanese geothermal power plant is a 50-MW, separated-steam (or single-flash) plant at Kakkonda in the Takinoue geothermal region. It is located about 7.5 km (4.7 mi) southwest of the Matsukawa dry-steam field and 50 km (31 mi) west of Morioka, a city with a population of roughly 200,000, in Iwate Prefecture, northern Honshu.

6.7.1   Geology and exploration

Studies by the Geological Survey of Japan (GSJ) in the 1950's established the fact that a valuable geothermal resource existed at Takinoue. From 1961 to 1964 the GSJ carried out detailed geological surveys, investigated the geochemistry of hot springs, took temperature gradients at depths of 30 m (98 ft) at 50 points, and measured electrical resistivities. The Japan Metals and Chemicals Company, Ltd. (JMC) participated by drilling several shallow wells of 70-mm (2¾-in) diameter to depths of 110 m (360 ft), cored all the way.

Beginning in 1967 deeper exploratory holes were sunk and temperatures of 200°C (392°F) were encountered at 400 m (1312 ft). Temperature profiles for well No. 203 at Takinoue [GERD, 1975] showed steep gradients—60°C/100 m (32°F/100 ft)—to depths of about 100 m (328 ft). This is about 18 times the normal gradient of 33°C/km (1.8°F/100 ft). Average gradient over 400 m (1312 ft) was about 30°C/100 m (16°F/100 ft); there was little gradient between 400 m and 800 m (1312 ft and 2625 ft), which is typical of a hydrothermal convection system.

From 1969 to 1973 a variety of geothermal exploration techniques were brought to bear on the site. Six deep exploratory wells 700–1000 m (2300–3280 ft) were drilled by JMC beginning in 1973. Geological maps show that the region is comprised of various layers of dacite, silt, sandstone, andesite, shale, and tuff. Numerous faults cut through the region, many of them nearly vertical. The production zone lies 900–1600 m (2950–5250 ft) below the surface.

6.7.2   Well programs and productivity

From August 1974 to December 1977, JMC drilled 11 production wells with diameters of 219 mm (8⅝ in) and an average depth of 1183 m (3881 ft). During the same period 15 reinjection wells were also completed; these have a diameter of 269.9 mm (10⅝ in) and an average depth of 754 m (2474 ft). Total drilled depth of the 26 wells is 24.3 km (15.1 mi). Air-drilling methods were used for all wells, and all were slant drilled from several central platforms. Table 6.18 summarizes information on all wells serving the Kakkonda plant.

Table 6.18—*Characteristics of wells at Kakkonda*

Well area	Production Wells				Reinjection Wells		
	Well No.	Depth m	Flow rate [a]		Well No.	Depth m	Flow rate [b] liquid Mg/h
			Vapor Mg/h	Liquid Mg/h			
A	A-1	1600	20	130	AR-1	570	162
	A-2	1100	125	725	AR-2	700	312
					AR-3	700	289
					AR-4	700	287
Totals			145	855			1050
B	B-1	1160	35	230	BR-1	690	243
	B-2	897	65	340	BR-2	615	212
	B-3	1265	70	313	BR-3	700	122
					BR-4	700	118
Totals			170	883			695
C	C-1	1091	38	171	CR-1	724	546
	C-2	887	80	354	CR-2	700	308
	C-3	1298	85	546	CR-3	700	100
					CR-4	954	130
Totals			203	1071			1084
D	D-1	954	27	77	DR-1	600	50
	D-2	1565	38	136	[c] DR-2	1600	85
Totals			65	212			135
E	E-1	1200	53	196	ER-1	650	196
Averages		1183				754	
Totals			636	3217			3160

[a] At a pressure of 686 kPa (99.5 lbf/in²).
[b] At a pressure of 539 kPa (78.2 lbf/in²).
[c] Drilled as a production well; used for reinjection.

JMC began building the pipeline system in April 1976 and completed the task at the end of 1977. A total of 2900 m (9515 ft) of pipeline was constructed from the wells to the powerhouse, including 750 m (2460 ft) of hot water pipes. Three types of piping are used: (1) two-phase geofluid pipes, (2) geothermal steam pipes, and (3) geothermal hot water pipes. For the first type there are two sets, a high-temperature set designed for 250°C (482°F) and ranging in diameter from 318.5 to 355.6 mm (12.5 to 14 in), and a medium-temperature set designed for 191°C (376°F) with diameters ranging from 165.2 to 863.6 mm (6.5 to 34 in). With regard to the second type—the geothermal steam pipes—there are also two sets, one designed for 191°C (376°F) with diameters of 216.3 to 762 mm (8.5 to 30 in), and one designed for 179°C (354°F) with diameters of 216.3 to 914.4 mm (8.5 to 36 in). The hot water pipes can handle 191°C (376°F) water and have diameters of 165.2 to 711.2 mm (6.5 to 28 in).

The general layout of the power station and the bore field is given in figure 6.20. A schematic of the steam gathering system is shown in figure 6.21. The wells are located in five groups, designated A, B, C, D, and E.

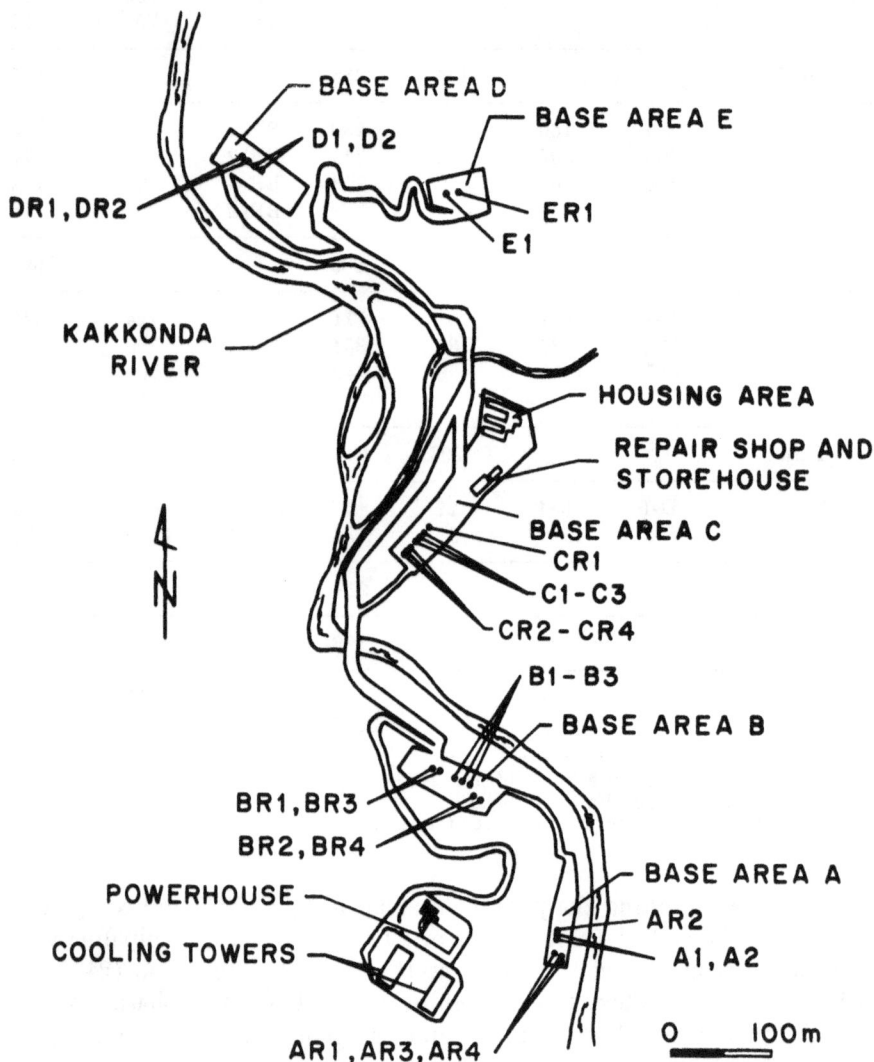

FIGURE 6.20—Arrangement of wells and powerhouse at Kakkonda (Takinoue). Production and reinjection wells are clustered in five areas. Two-phase fluid transmission pipelines connect wells to area B where main separators are located [JMC, 1978].

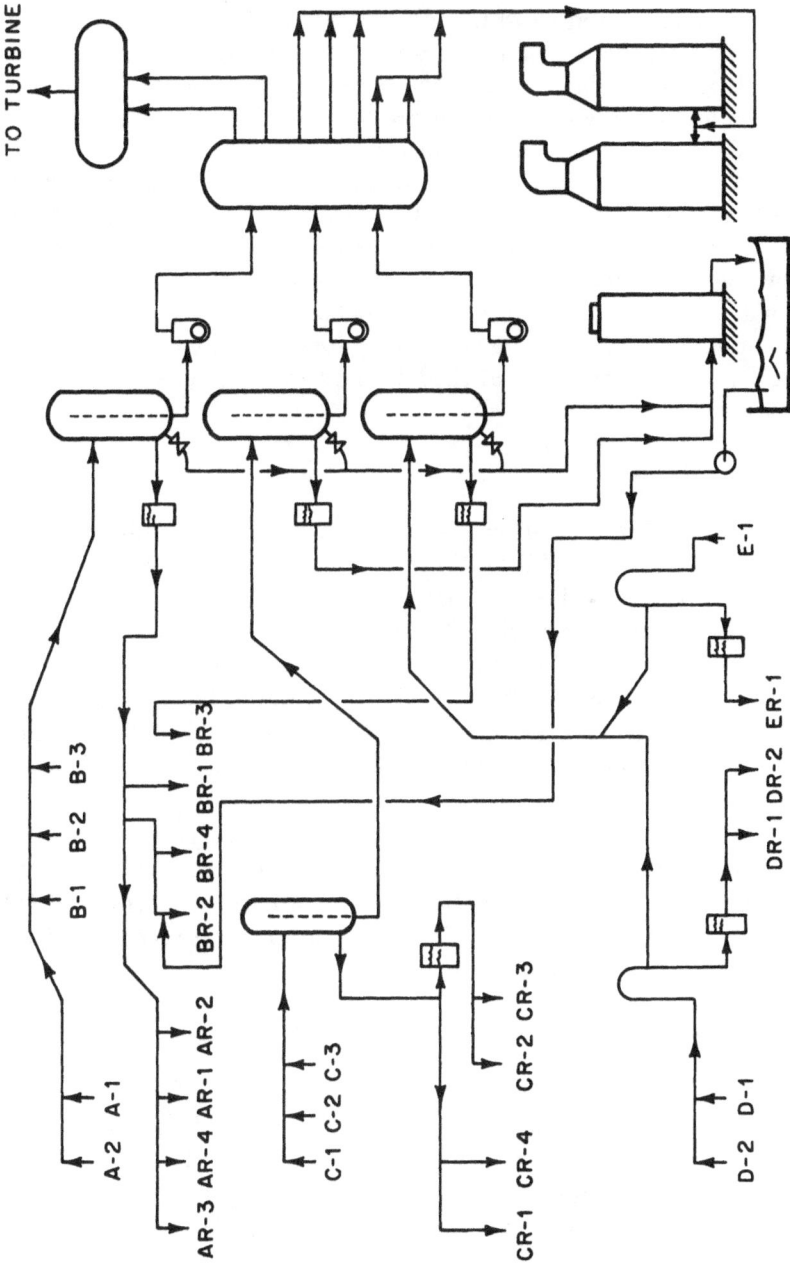

FIGURE 6.21—Steam gathering system for Kakkonda geothermal power plant. The three main separators, silencers, and steam receiver are located at area B (see figure 6.20) [H. Nakamura, personal communication].

Two-phase geofluid from the two wells at A and the three wells at B is piped to a separator located at the B area. Separated liquid is either sent to a silencer or returned either to area A or B for reinjection. Three wells at area C produces a mixture of liquid and vapor that is separated initially in a cycline separator in area C, the waste liquid being reinjected in that area, and the steam being piped to another separator in area B. The two-phase geofluid from the most distant areas, D and E, is first separated by means of U-bend devices, one at each area, the liquid being reinjected at each site. The steam is then passed through a final cyclone separator (at area B) to remove any liquid that may have condensed along the pipeline. All the steam is then sent from area B up to a steam receiver at the power-house, which is situated on the side of a hill about 60 m (200 ft) above the Kakkonda River. The photograph in figure 6.22 shows base area B in the foreground and the powerhouse and cooling towers on the hill in the back-ground. The well silencers are at the left and three cyclone separators are at the right.

6.7.3 Energy conversion system

The turbine and generator were supplied by the Tokyo Shibaura Elec-tric Company, Ltd. (Toshiba). The turbine is of the single-cylinder, double-flow type, rated at 50 MW and turning at 3000 rpm. The machine has four impulse stages in each flow. Each half of the turbine is essentially a duplicate of the single-flow machine used at Matsukawa. The generator is air cooled, rated at 55,600 kVA, 11 kV with 0.9 power factor. The turbine steam inlet conditions are 441 kPa (64 lbf/in^2), 147°C (297°F); exhaust is 13.5 kPa (4 in Hg). Complete technical specifications for the plant are given in table 6.4.

The power plant converts about 33% of the thermodynamic available work in the geofluid at the wellhead into useful electric work at the busbar. This relatively low utilization factor is a result of wastage of the liquid fraction rather than its utilization by means of a secondary flash process to generate additional steam.

6.7.4 Economic data

The capital cost for the Kakkonda single-flash 50-MW plant was 18 billion yen (or US$100 million or US$2000/kW installed). Net cost of electricity is given as 7.39 yen/kW·h, for 80% capacity factor (~42 US mill/kW·h, 1978). These figures include the cost of the wells in the capital cost of the plant [H. Nakamura, personal communication].

6.7.5 Operating experience

Unlike all other geothermal power plants in Japan where reinjection is done at atmospheric pressure, the waste liquid at Kakkonda is returned to

FIGURE 6.22—Kakkonda 50-MW geothermal power plant. Well area B is in foreground; powerhouse and cooling towers are on the hill in background [photo courtesy of H. Nakamura].

the formation at the separator pressure, about 550 kPa (80 lbf/in²). Although extensive monitoring is being carried out to check for any signs of induced seismicity, no adverse effects have yet been observed. The reinjection temperature is about 160°C; reinjection under pressure seems to promote reinjectivity. Maintaining the fluid temperature above 150°C also aids in preventing silica deposition in the reinjection wells. If this procedure proves successful, other plants in Japan may be converted, where feasible, to a similar mode of reinjection.

The performance of the plant, from the day it was commissioned (May 26, 1978) to January 20, 1979, is summarized in table 6.19. In the early stages of operation the separators were discovered to be somewhat under-sized, leading to excess carryover of mist to the turbine with a design flow rate of 400 Mg/h (882 klbm/h). To avoid the deposit of silica scale on the turbine blades, it was necessary to restrict the flow below the design value. From November 11 to December 5, 1978, another separator was constructed which allowed the plant to reach its full capacity. Otherwise there have been no problems with the operation of the plant [H. Nakamura, personal communication].

Table 6.19—*Performance of Kakkonda single-flash 50-MW geothermal power plant May 26, 1978 to January 20, 1979* [a]

Time Period	Output, MW	No. days	Generation, MW·h	Capacity factor, percent	Availability, percent
5–26–78 to 6– 7–78	37	13⎤			
6– 8–78 to 6–14–78	0	[b] 7⎬	63, 000	84. 0	92. 0
6–15–78 to 8–17–78	37	64⎦			
8–18–78 to 11–10–78	40	85	79, 800	99. 0	100. 0
11–11–78 to 12– 5–78	0	[c] 25			
12– 6–78 to 1–20–79	50	71	53, 900	90. 0	100. 0
Total		265	196, 700	[d] 74. 0	88. 0

[a] Source: H. Nakamura, personal communication.
[b] Inspection period.
[c] Plant down for construction of additional separator.
[d] Based on an effective capacity of 41.7 MW for the period 5/26/78–1/20/79.

Finally, a plan is in the works to make use of about 2000 Mg/h of hot water from the plant for a variety of heating purposes in the city of Morioka and the town of Shizukuishi. These uses will include district heating, snow melting on roads, agriculture, tourism, and other industrial applications [Mori, 1975].

6.8   Pilot binary plant at Otake

Development of binary geothermal power cycles began in Japan in 1975 in cooperation with the Sunshine Project, promoted by the Agency of

Industrial Science and Technology (AIST) of the Ministry of International Trade and Industry (MITI). Two 1-MW experimental pilot plants have been built and were tested: one of these was at the Otake site and the other at Mori on the northern island of Hokkaido. The latter plant will be described in section 6.9.[a]

6.8.1 Nature of the geothermal fluid

The pilot plant uses a portion of the geothermal steam from the steam receiver of the adjacent Otake power plant as well as the geofluid from the production well O-10 at the Otake field. The quality of the fluid at the separator is about 9%, at a temperature of 130°C (266°F). After being used as the heating medium for the secondary working fluid, about 78% of the produced fluid is reinjected into the reservoir through Otake reinjection well OR-2. The geothermal brine (liquid portion of geofluid) has a pH of 8 and contains a total of 4030 ppm of dissolved solids, mainly chloride (Cl) 1720 ppm, sodium (Na) 1060 ppm, and silicon dioxide (SiO_2) 612 ppm.

6.8.2 Energy conversion system

Otake's pilot binary plant is an ambitious design employing isobutane as the working fluid. An overview of the test site is shown in figure 6.23 and a highly simplified flow diagram is given in figure 6.24. Isobutane vapor is generated in a multistage flash heater from geothermal liquid from the separator of the Otake well O-10. A portion of the steam from the receiver at the nearby Otake power plant is used to provide the final heating needed to vaporize the isobutane. The turbine is a radial-inflow machine fitted with an extraction point to allow for feed-heating of the isobutane liquid as it returns from the air-cooled condenser before entering the multistage heater. Note that the regenerator is not shown in figure 6.24, and that the multiflash heater actually consists of 18 sections. A supplementary water spray is used in the air-cooled condenser during warm weather [MHI, 1977a].

The multistage flash heater has two principal advantages. First, it produces pure, distilled water which is then used in the dry/wet isobutane condenser during the hot season. Second, the dissolved solids in the geofluid are contained in the liquid or precipitate to the bottom of the heating vessel, leaving the heat exchanger tubes free of scale. Noncondensable gases are extracted from the dead end of the heater and pumped out to the atmosphere. The heat exchanger is of the shell-and-tube type, receiving geothermal steam and hot liquid in a ratio of roughly 1 to 10. The geothermal steam makes two passes through the heat exchanger while the liquid makes four passes. The isobutane leaves the vaporizer in a dry,

---

[a] The test programs are now completed and both plants have been dismantled. (*Note added in proof.*)

FIGURE 6.23—Pilot binary power plant at Otake. Powerhouse is at center; isobutane boiler is to right rear of the power-house; air-cooled condenser is to left of the powerhouse; at left-center background is an arsenic removal system for the adjacent Otake 10-MW flash-steam plant [photo by R. DiPippo].

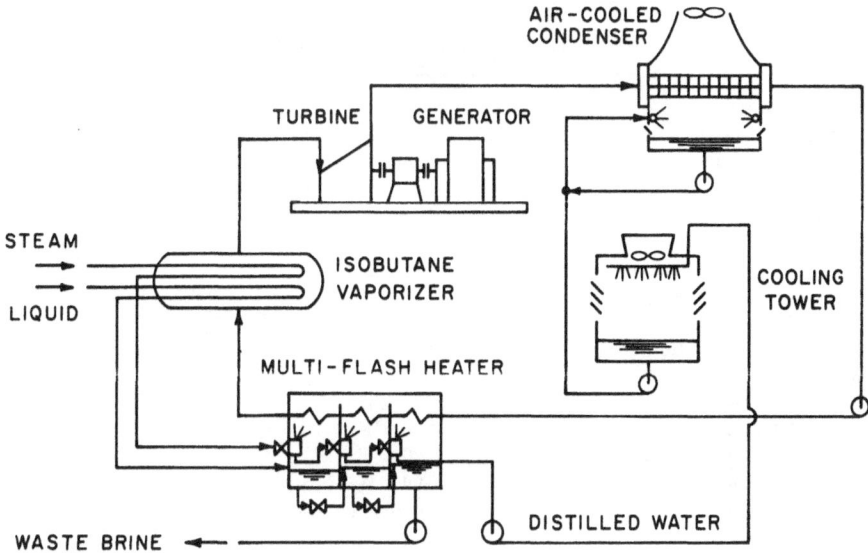

FIGURE 6.24—Schematic diagram for pilot binary plant at Otake (regenerator not shown) [MHI, 1977a].

saturated state through a cyclone separator mounted on top of the evaporator.

The turbine, generator, reduction gear, and all associated auxiliary equipment for governing, sealing, and lubrication are mounted on a common bed for ease of transportation and installation. The turbine is a radial expander with a spiral inlet, axial exhaust, and a bleed connection for regenerative heating (not currently in use). The turbine is fitted with variable inlet nozzles that serve as a control mechanism. Turbine speed is 17,000 rpm; generator speed is 1800 rpm. A planetary gearbox serves as a speed reducer. Turbine shaft sealing is achieved by means of a double-flow, balance-type oil film seal for normal running, a contact seal for shutdown, and a buffered gas seal for abnormal operation [MHI, 1977a].

The plant uses a forced-draft, air-cooled surface condenser augmented by a water spray. The water is obtained from the flash distillation process just described and from a single-cell, counterflow, mechanically-induced-draft cooling tower. The air-cooled condenser uses six bundles of circular finned tube elements with two top-mounted, induced-draft fans with variable pitch blades.

Table 6.20 lists technical particulars for the plant, which are subject to change owing to its experimental nature. As of October 1978 three test runs had been conducted and another was scheduled for the winter of 1978/79, during which the regenerator was to be tested for the first time.

Table 6.20—*Specifications for pilot binary plant at Otake* [a]

Period of operation: 1977–1979

Turbine data:

Type...........	Single-cylinder, radial-inflow, geared
Rated capacity......	1,000 kW
Speed, turbine/generator.... .	17,000/1800 rev/min
Secondary working fluid.	Isobutane, i-C_4H_{10}
i-C_4H_{10} inlet pressure... . ..	412.4 lbf/in^2
i-C_4H_{10} inlet temperature. . ..	248.0°F
i-C_4H_{10} exhaust pressure.... ..	71.1 lbf/in^2
i-C_4H_{10} exhaust temperature... .	125.6°F
i-C_4H_{10} mass flow rate.. . .	154×10^3 lbm/h

Geothermal fluid data.

Inlet pressure.	39.1 lbf/in^2
Inlet temperature...... .	266.0°F
Outlet temperature..	122.0°F
Mass flow rate, liquid	106×10^3 lbm/h
Mass flow rate, vapor	10.36×10^3 lbm/h

Condenser data:

Type [b]..	Forced-draft, air-cooled with water spray, surface type
Cooling water inlet temperature. .	78.8°F
Cooling water flow rate..... ...	25.4×10^3 lbm/h

Gas extractor data:

Type....	Water-sealed rotary pump
Number....	1
Suction pressure.. . .	3.62 in Hg
Capacity....	353.1 ft^3/min
Power consumption....	37 kW

Heat rejection system data:

Type......	Counterflow, mechanically-induced-draft cooling tower
Number of cells..	1
Water inlet temperature..	122°F
Water outlet temperature..... . .	78.8°F
Design wet-bulb temperature.. ...	59.0°F
Water flow rate...	121×10^3 lbm/h
Draft fan type...	Vertical, axial type
Air flow rate	28.2×10^3 ft^3/min
Fan motor power consumption. .	30 kW

[a] Sources: MHI, 1977a; MHI, 1978.
[b] Water spray is used only during the hot season.

6.9 Pilot binary plant at Mori (Nigorikawa)

The other binary test facility is located at the Mori geothermal field in the Nigorikawa district of Hokkaido and is being developed under the auspices of the Geothermal Energy Research and Development Company, Ltd. (GERD), in conjunction with the Tokyo Shibaura Electric Company, Ltd. (Toshiba).

The plant uses Refrigerant-114 ($CClF_2$ $CClF_2$) as the working fluid in the power cycle. Shell-and-tube heat exchangers are used for the vapor generators and condensers. The turbine is an axial flow machine rated at 1,000 kW. A summary of the technical specifications for the plant is given in Table 6.21. In the case of the preheaters for the R-114, of which there are five, and the evaporator, the geothermal fluid is passed through the tubes of the heat exchanger while the R-114 flows through the shell side. This presumably allows for easy cleaning of the tubes in case of fouling. On the other hand, the main condensers operate with the R-114 inside the tubes with the cooling water on the shell side.

Table 6.21—*Specifications for pilot binary plant at Mori* [a]

Period of operation: 1977–1979

Turbine data:

Type..........	Axial-flow, impulse, single-flow, geared
Rated capacity.	1,000 kW
Speed, turbine/generator..... .. .	3521.7/1500
Secondary working fluid..	Refrigerant-114, $(CClF_2)_2$
$(CClF_2)_2$ inlet pressure..	194.8 lbf/in²
$(CClF_2)_2$ inlet temperature.......	209.3°F
$(CClF_2)_2$ exhaust pressure	34.1 lbf/in²
$(CClF_2)_2$ exhaust temperature....	(b)
$(CClF_2)_2$ mass flow rate	471.8×10^3 lbm/h

Geothermal fluid data:

Inlet pressure..........	(b)
Inlet temperature.........	284°F
Outlet temperature.............	197.6°F
Mass flow rate, total	396.8×10^3 lbm/h

Main condenser data:

Type.......................	Liquid film flowing type
Cooling water inlet temperature...	63.9°F
Cooling water outlet temperature..	71.6°F
Cooling water flow rate.....	2.43×10^6 lbm/h
$(CClF_2)_2$ flow rate..............	471.8×10^3 lbm/h

Auxiliary condenser data:

Type................	Vertical, shell-and-tube, surface type
Cooling water inlet temperature...	59°F
Cooling water outlet temperature..	63.9°F
Cooling water flow rate..........	2.43×10^6 lbm/h
$(CClF_2)_2$ flow rate...	183×10^3 lbm/h

[a] Source: Sunshine Project, 1977.
[b] Not available.

The experimental station is equipped with a 1260 kVA generator rated at 3.3 kV at 50 Hz with a power factor of 0.9. The generator is of the 3-phase synchronous type, drop-proof, air cooled, horizontal, and connected to the turbine through a gearbox with a 2.348 reduction ratio. No information has been reported on the performance of this pilot plant.

6.10   Other promising areas in Japan

The ultimate geothermal electric generating capacity in Japan may be as high as 100,000 MW [MITI, 1976]. A great many areas are excellent sites for geothermal development, including:

Akinomiya: a liquid-dominated reservoir at which a few wells have been drilled to 800–900 km (2625–2950 ft).

Fushime: the highest salinity (35,000 ppm) liquid-dominated reservoir in Japan, located in southernmost Kyushu.

Kirishima: deep exploratory drilling is underway.

Kumamoto: a 55-MW double-flash plant is in the planning stages.

Mori: a 55-MW single-flash plant is under construction on this liquid-dominated reservoir in southwestern Hokkaido; wells range in depth from 1000 to 1200 m (3280 to 3940 ft); geofluid has high gas content.

Oyasu: possible dry steam reservoir; located about 30 km (19 mi) north of Onikobe; wells range from 800 to 1200 m (2625 to 3940 ft) in depth; bottom-hole temperatures of 300°C (572°F), highest in Japan, have been observed.

Sumikawa: located about 4 km (2.5 mi) from Onuma but outside the national parks; 30-MW plant is being contemplated; wells are highly promising with one 50.8-mm-diameter (2-in) well producing 16 Mg/h (35,000 lbm/h) of steam and 32 Mg/h (70,000 lbm/h) of liquid under wide-open wellhead conditions.

Yakedake: liquid-dominated reservoir with bottom-hole temperatures of about 180°C (356°F), but low permeability.

Information on these sites has been obtained from a variety of sources [Mori, 1975; GERD, 1975; Iga and Baba, 1974; MITI, 1976; K. Aikawa, personal communication; K. Baba, personal communication].

REFERENCES [a]

Aikawa, K. and Soda, M., 1975. "Advanced Design in Hatchobaru Geothermal Power Station," *San Francisco, 1975*, Vol. 3, pp. 1881–1888.

Akiba, M., 1970 "Mechanical Features of a Geothermal Plant," *Pisa, 1970*, Vol. 2, pp. 1521–1529.

GERD, 1975. *Harnessing the Earth's Thermal Energy*, Geothermal Energy Research and Development Company, Ltd., Tokyo.

Hayashida, T. and Ezima, Y., 1970. "Development of Otake Geothermal Field," *Pisa, 1970*, Vol. 2, pp. 208–220.

Iga, H. and Baba, K., 1974. "Annual Information on Development and Utilization for Geothermal Energy in Japan, 1973," *Geothermics*, Vol. 3, No. 3, pp. 122–124.

Ito, J., Kubota, Y., and Kurosawa, M., 1977a. "On the Geothermal Water Flow of the Onuma Geothermal Reservoir," *Geothermal Energy*, Vol. 14, No. 3, pp. 139–151. (In Japanese.)

[a] See note on p. 24.

Ito, J., Kubota, Y., and Kurosawa, M., 1977b. "On the Silica Scale of the Onuma Geothermal Power Plant," *Geothermal Energy*, Vol. 14, No. 4, pp. 173–179. (In Japanese.)

JGEDG, 1978. *Present Status of Geothermal Development and Utilization in Japan*, Japan Geothermal Energy Development Center, Tokyo.

JMC, 1978. *Kakkonda Geothermal Power Plant*, Japan Metals and Chemicals Co., Ltd., Tokyo. (In Japanese.)

Kawasaki, 1975. *Geothermal Power Plant*, Kawasaki Heavy Industries, Ltd., Tokyo, Cat. No. GE 3847 EC.

Kawasaki, 1977. *Outline of Onikobe Geothermal Power Plant*, Kawasaki Heavy Industries, Ltd., Kobe, Dwg. No. 83GZ1–GT99–02.

Kubota, K. and Aosaki, K., 1975. "Reinjection of Geothermal Hot Water at the Otake Geothermal Field," *San Francisco, 1975*, Vol. 2, pp. 1379–1383.

MHI, 1976a. *Geothermal Power Development by Mitsubishi*, Mitsubishi Heavy Industries, Ltd., Tokyo, Brochure JA–237, (1.0) 76–11, M.

MHI, 1976b. *Geothermal Power Generation by Mitsubishi*, Mitsubishi Heavy Industries, Ltd., Tokyo, Brochure JA–243, (0.5) 76–11, H.

MHI, 1976c. *Mitsubishi Geothermal Power Plant*, Mitsubishi Heavy Industries, Ltd. Tokyo, Brochure JA–212, (2.0) 76–4.

MHI, 1977a. *Binary Cycle Geothermal Power Generation Systems with Multistage Flash Heating*, Mitsubishi Heavy Industries, Ltd., Tokyo.

MHI, 1977b. *Mitsubishi 50,000 kW Hatchobaru Geothermal Power Plant*, Mitsubishi Heavy Industries, Ltd., Tokyo, Brochure JA–327, (1.0) 78–6 M.

MHI, 1978. *Otake Binary Cycle Geothermal Generation System*, Mitsubishi Heavy Industries, Ltd., Tokyo.

MITI, 1976. *Present Status and Future Prospect of Developing Geothermal Resources in Japan*, Ministry of International Trade and Industry, Agency of Natural Resources and Energy, Tokyo.

Mori, Y., 1970. "Exploitation of the Matsukawa Geothermal Area," *Pisa, 1970*, Vol. 2, pp. 1150–1156.

Mori, Y., 1975. "Geothermal Resources Development Program in Northeastern Japan and Hokkaido," *San Francisco, 1975*, Vol. 1, pp. 183–187.

Nakamura, H., 1959. "Geothermal Conditions in the Onikobe Basin, Miyagi Prefecturer, Japan," *J. Japan Assoc. Mineral. Petrol. Econ. Geol.*, Vol. 43, No. 3.

Nakamura, H., Sumi, K., Katagiri, K., and Iwata, T., 1970. "The Geological Environment of Matsukawa Geothermal Area, Japan," *Pisa, 1970*, Vol. 2, pp. 221–231.

Nakamura, S., 1970. "Economics of Geothermal Electric Power Generation at Matsukawa," *Pisa, 1970*, Vol. 2, pp. 1715–1716.

Ozawa, T., Kamada, M., Yoshida, M., and Senemasa, I., 1973. "Genesis of Hot Spring. Part I: Genesis of Acid Hot Spring," *Geothermal Energy*, Vol. 10, No. 2, pp. 31–40. (In Japanese.)

Saito, M., 1961. "Known Geothermal Fields in Japan," *Rome, 1961*, Vol. 2, pp. 367–372.

Sakakura, S., 1975. "Sunshine Project of the Japanese Government," *Proc. 2nd Energy Tech. Conf.*, Government Institute, Washington, pp. 105–121.

Sato, H., 1970. "On Matsukawa Geothermal Power Plant," *Pisa, 1970*, Vol. 2, pp. 1546–1551.

Sunshine Project, 1977. *Japan's Sunshine Project: Summary of Geothermal Energy R&D*, Ministry of International Trade and Industry, Industrial Science and Technology, Tokyo.

Takahashi, Y., Hayashida, T., Soezima, S., Aramaki, S., and Soda, M., 1970. "An Experiment on Pipeline Transportation of Steam-Water Mixtures at Otake Geothermal Field," *Pisa, 1970*, Vol. 2, pp. 882–891.

Usui, T. and Aikawa, K., 1970. "Engineering and Design Features of the Otake Geothermal Power Plant," *Pisa, 1970*, Vol. 2, pp. 1533–1545.

CHAPTER 7

MEXICO

7.1 General remarks

The first exploration for sources of geothermal energy in Mexico took place in 1955 west of the city of Pachuca at Pathé. This geothermal field is situated along the Neovolcanic axis trending east-west across the country. The region consists of upper Tertiary and Quaternary basaltic, andesitic, rhyolitic, and pyroclastic rocks [Alonso, 1975].

As of 1975 the total installed electric capacity of Mexico was 7500 MW, with 48% supplied by hydroelectric plants, 51% from oil- and gas-fired thermal power plants, and the remaining 1% by coal and geothermal. It is unlikely that expansion in hydroelectric capacity will amount to more than about 12,000 MW. The discovery of extensive petroleum reserves in Mexico has allowed that country to become an exporter of crude oil and refined petroleum products. The proven reserves of crude oil, natural gas, and condensate as of 1977 were 16 billion barrels, with potential reserves of 120 billion barrels [Diaz Serrano, 1978]. Thus Mexico's energy situation is extremely favorable for the next several decades, well into the twenty-first century.

Geothermal energy will, nevertheless, play an important role in meeting the growing demand for electricity in Mexico. More than 130 geothermal regions dot the country, in 24 of the 32 states. The largest concentration of geothermal sites lies in the states of Michoacán (22), Jalisco (16), Baja California (15), and Guanajuato (9). Owing to their wide geographic distribution and their potential as an inexpensive source of local power, these geothermal regions will be taken into account in national plans to meet the expected future demand for electricity in Mexico.

7.2 Pathé

Mexico's first geothermal power plant was installed at Pathé, a geothermal field located in the municipality of Tecozaulta in the state of Hidalgo, about 80 km (50 mi) north-northeast of Mexico City. The plant began operations in 1959 but is now shut down [G. Cuéllar, personal communication]. The Pathé unit had a capacity of 3.5 MW and employed a noncondensing turbine supplied with steam separated, most likely, from one well.

Very little information exists on this plant in the literature. Figure 7.1 shows a view of the powerhouse at Pathé.

7.3 Cerro Prieto

The major geothermal development in Mexico is taking place at Cerro Prieto in the state of Baja California, roughly 35 km (22 mi) south of the city of Mexicali and the international boundary between the United States and Mexico. The geothermal field is located between 114°50' and 115°48' west longitude and between 31°55' and 32°44' north latitude. The general location of the area is shown in figure 7.2. Since 1973 Cerro Prieto has been generating 75 MW of power on a highly reliable and economic basis, recently achieving the highest capacity factor of any power plant in Mexico. This experience has been so successful that work has recently been completed on an extension of the plant, duplicating the first two power units and bringing the installed capacity to 150 MW. The full potential of the field is estimated to be at least 400 MW.

7.3.1 Geology

The Cerro Prieto geothermal field, located on a plain in the Mexicali-Imperial rift valley, covers an area of about 3000 ha (7400 acres). The most prominent feature of the flat desert area is the volcano of Cerro Prieto ("Black Hill"), which has an elevation of 260 m (853 ft), covers an area of 400 ha (988 acres), and lies about 6 km (4 mi) west-northwest of the main geothermal steam production area [Reed, 1975]. The field is bounded on the west by the Sierra de los Cucapahs and contains numerous faults of the San Jacinto fault zone. These faults strike northwest and have resulted in considerable downthrows on the northeast side. This can be seen in figure 7.3, for example, where the vertical separation between the plutonic rock at the crest of the Sierra de los Cucapahs and the same type of rock beneath the geothermal field is about 5.5 km (18,000 ft).

Exploration studies, undertaken in the 1950's and motivated by the large number of surface manifestations of thermal activity, have revealed a good deal about the nature of the geothermal field. Figure 7.3 is a highly simplified geologic cross-section of it. The reservoir is capped by a layer of plastic, impermeable clays with a thickness of 600–700 m (1970–2300 ft) over the main portion of the field. These are sedimentary rocks and deltaic deposits of the quaternary period deposited by the Colorado River [Paredes, 1975].

Underlying the cap clays is the main reservoir, which was formed in the Tertiary period. Shales and sandstones of considerable porosity and permeability were deposited in a large graben as a result of intensive erosion of igneous granitic and metamorphic rocks. Basement faulting contributed to the permeability of this formation. The pores of these rocks became filled with water during the late Tertiary from the Colorado

FIGURE 7.1.—Powerhouse and substation for the Pathé geothermal unit (now abandoned) [CFE, 1971].

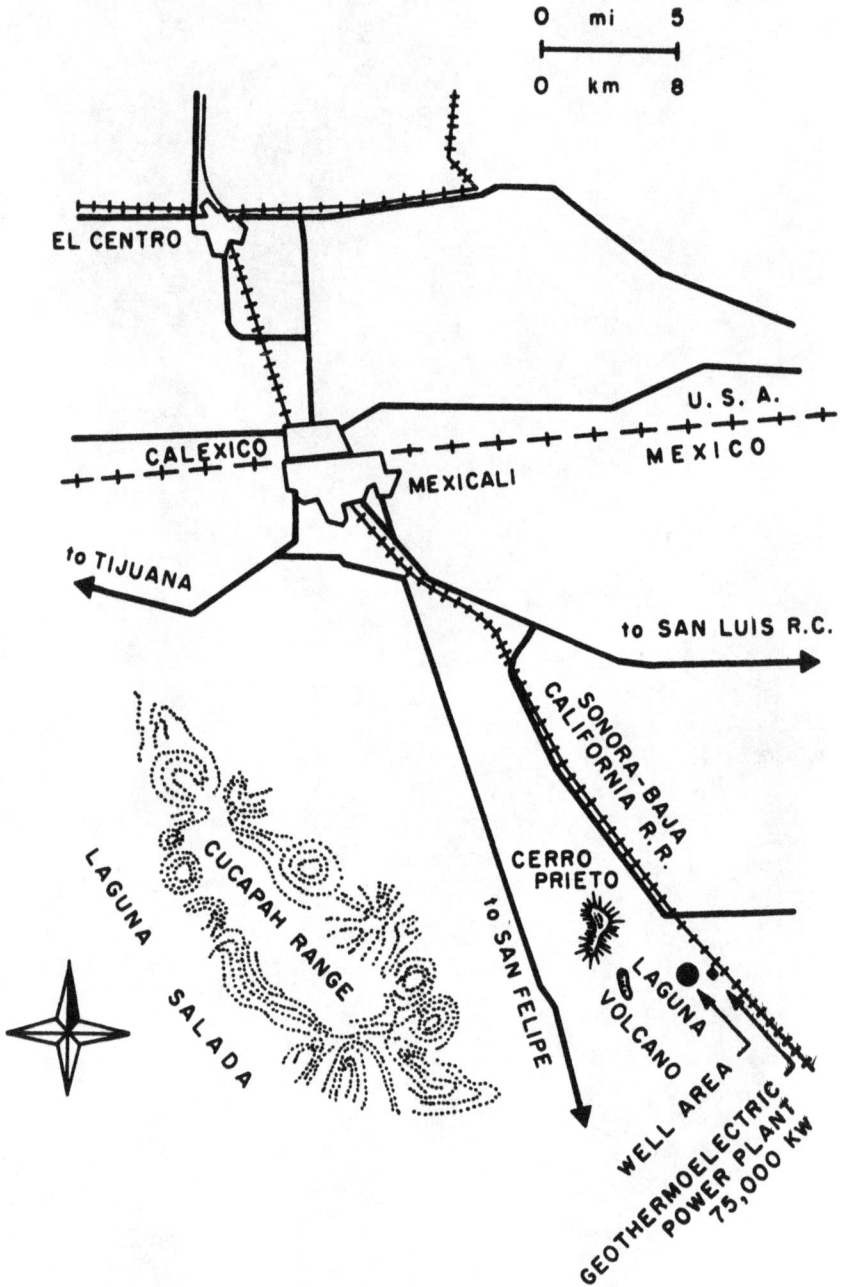

FIGURE 7.2—Geographical location of Cerro Prieto geothermal field [after CFE, 1971].

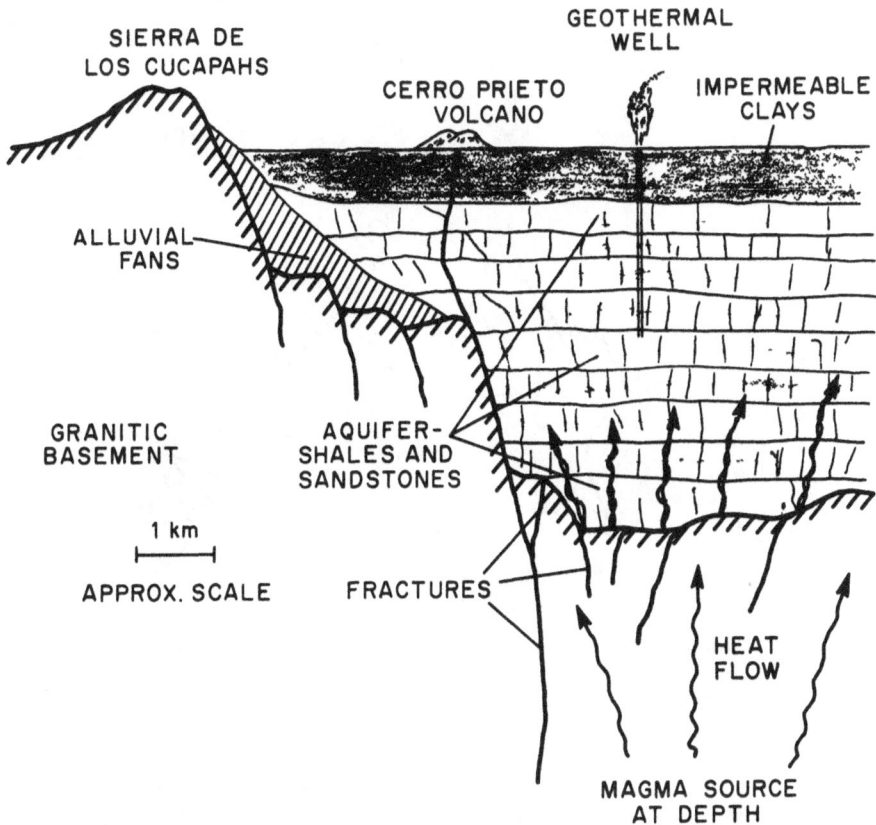

FIGURE 7.3—Schematic cross-section of Cerro Prieto geothermal reservoir [after CFE, 1971].

River. The aquifer is believed to consist of alternating layers of nondeltaic lutites and sandstones, with the sandstones saturated with connate water at a pressure in excess of the hydrostatic saturation pressure. Total thickness of the nondeltaic sediments is about 2 km (6500 ft). The basement rock is granitic and may be seen in large outcroppings in the Sierra de los Cucapahs. These are of the Mesozoic period, probably of the Cretaceous age.

7.3.2 Well programs and gathering system

Information on the drilling program and reservoir development at Cerro Prieto has been reported at various stages during the project [Dominguez and Vital, 175; Guiza, 1975; Mercado, 1974, 1975a, 1976]. A total of 41 wells had been drilld through 1974 including exploration wells, step-out wells, and deep production wells. Locations of these wells are shown in figure 7.4, which also illustrates the power plant and evapo-

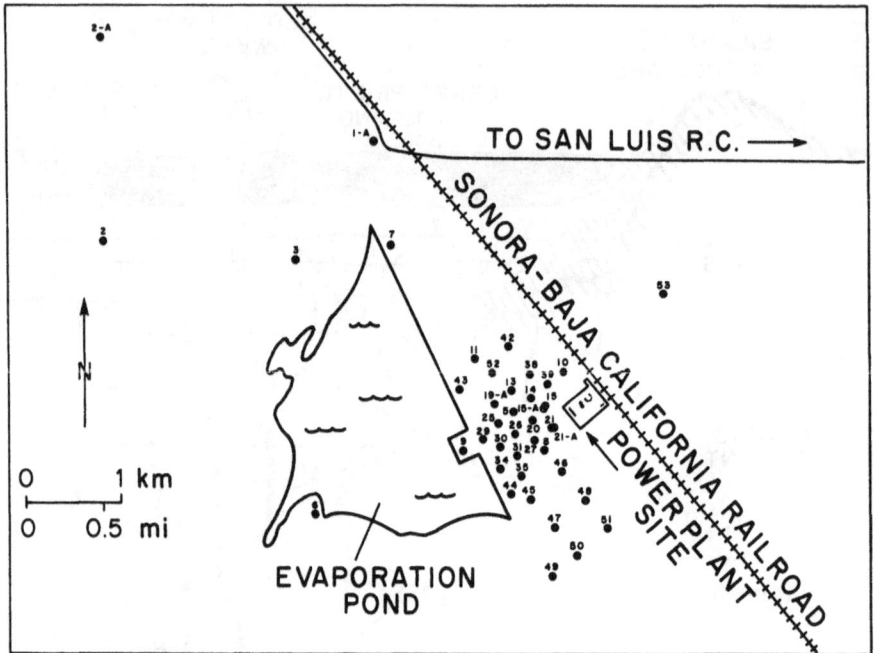

FIGURE 7.4—Location of wells at Cerro Prieto [after Dominguez and Vital, 1975].

ration pond sites. Eighteen wells are connected to the first two units of the power plant, nine for each unit. Of these, 15–16 are needed to generate 75 MW; the others are held on reserve. Figure 7.5 shows the piping layout for the steam-gathering system. Four main steam-gathering lines run from the wells to the steam receivers at the powerhouse. There are more than 6 km (20,000 ft) of steam pipelines with diameters greater than 406 mm (16 in).

The wells are drilled with rigs rated for 2200 m (7200 ft) using a drawworks powered by two 250-hp engines. The drill string consists of FH 114-mm (4.5-in) drill pipe and 165-mm (6.5-in) drill collars. Drilling mud is of the bentonitic type, emulsified with 4%–9% diesel oil and having a density between 1.09 and 1.24 Mg/m^3 (68.05 and 77.41 lbm/ft^3). The mud has a solids content of 8–15% and a pH of 7.5–10.5 [Guiza, 1975].

The wells were completed during three periods as follows:

    1964–1967: M-5, 8, 9, 10, 15
    1967–1968: M-11, 13, 15A, 19A, 20, 21, 21A, 25, 26, 30, 35
    1972–1974: M-29, 31, 34, 38, 39.

From the first group, M-10 is not used, and M-15 needed extensive repairs and was replaced with a new well, M-15A. From the second group,

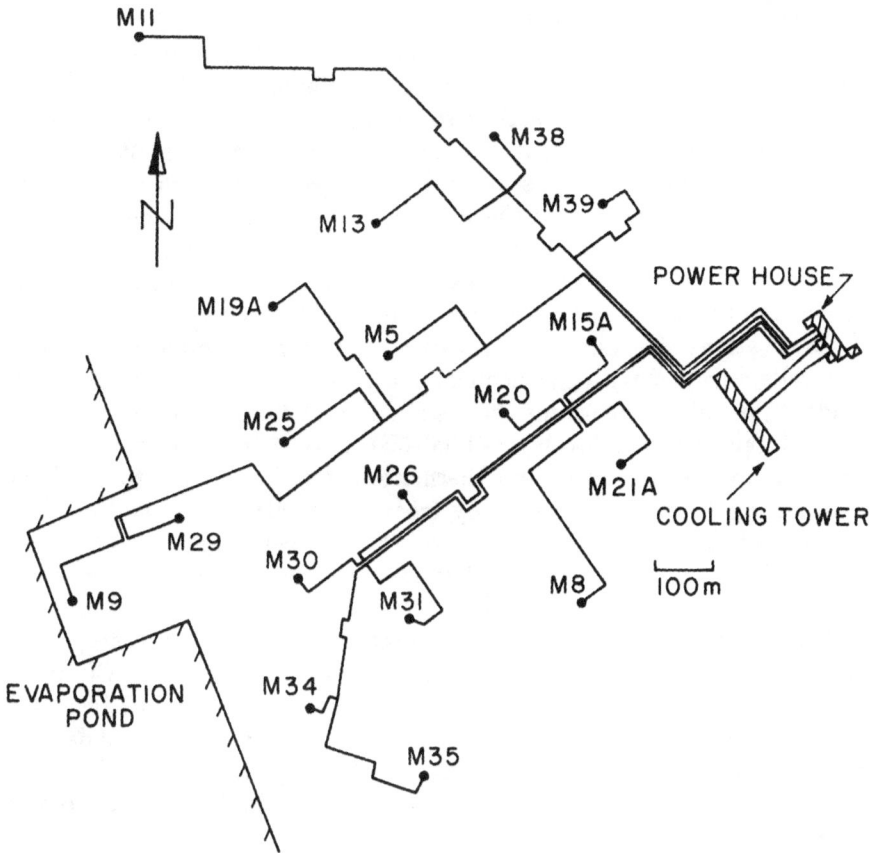

FIGURE 7.5—Steam pipeline gathering system for Units 1 and 2 at Cerro Prieto [after CFE, 1971; Mercado, 1975].

five wells suffered casing failures or other problems and needed to be repaired or replaced: M-13, 19A, 20, 21, 26. Wells M-13, 20, and 26 were successfully repaired, M–19A was plugged at the bottom to seal off a zone of cooler water, and M-21 was permanently sealed and replaced with M-21A. From the last group, only M-29 did not need repairs or reworking. The causes of the problems include mechanical accidents during drilling, inadequate materials or materials incapable of withstanding the high temperatures and stresses encountered, and improper well development and stimulation techniques that subjected the casings to abrupt changes in conditions [Dominguez and Vital, 1975].

A schedule of gradual heating is now followed to avoid casing failure. The water in the well is allowed to reach its maximum temperature in not less than 15 days, during which period gases and steam produced are vented to the atmosphere. In this way casing and cement expand simultaneously, without excessive temperature differentials between the casing and the formation [Guiza, 1975]. The mean lifetime of a well at Cerro Prieto is considered to be 15 years; some wells that are 14 years old are still in good condition and produce steam [Mercado, 1976]. Casing profiles for the 18 wells that serve the two units are shown in figures 7.6 and 7.7.

The two-phase geofluid is processed conventionally in Webre-type centrifugal separators; the separated steam passes through a ball-check valve before entering one of the four main-steam transmission lines (see figure 7.8). The main steam is collected outside the powerhouse in a set of receivers (see figure 7.9) and passes through a final stage of moisture separation (see figure 7.10) before entering the turbines.

Observations of pressure, temperature, and flow measurements taken in the wells, together with geochemical analyses of the fluid produced, have made it possible to determine the distribution and movement of hot water through relatively shallow layers (100–500 m, 330–1640 ft) from the southeast toward the northwest. Furthermore, hot fluid appears to be rising and flowing horizontally, generally from the eastern and central portions of the field toward the western portion. The temperature gradient is most gradual in the eastern part of the field (in the vicinity of well M-53), where the magma intrusion is believed to lie at depth.

The reservoir is probably being recharged from two sources: (1) the highly pervious and saturated alluvial fans of the Sierra de los Cucapahs (see figure 7.3), and (2) meteoric water from the Colorado River. The first of these feeds the reservoir from the west, whereas the second delivers water from the east. The second source is relatively unimportant, however, since the field has only moderate permeability in the eastern region [Mercado, 1975a].

On the basis of these studies, it has been determined that future wells should be drilled to the following approximate depths, depending on location in the field: 800–1400 m (2625–4590 ft) for wells west of the powerhouse, and 1400–2600 m (4590–8530 ft) for those east of the powerhouse. The depth required to reach the aquifer increases from west to east across the field [Mercado, 1975a].

7.3.3 Geofluid characteristics

The liquid-dominated reservoir produces a mixture of liquid and vapor at the wellhead. The fluid in the reservoir is a compressed liquid that partially flashes to vapor during its ascent through the wells. Under high flow rates there is annular flow in the well bore, i.e., liquid on the walls and vapor in the core [Reed, 1975]. Table 7.1 lists average pressure, temperature, mass flow rate, and dryness fraction for wells supplying the power plant during January and February 1974. As pointed out in note b of the table, an average wellhead pressure of 729 kPa (106 lbf/in²) is more representative than the average value of 1119 kPa (162 lbf/in²) because of the unusually high pressures in wells M-11 and 31. Tempera-

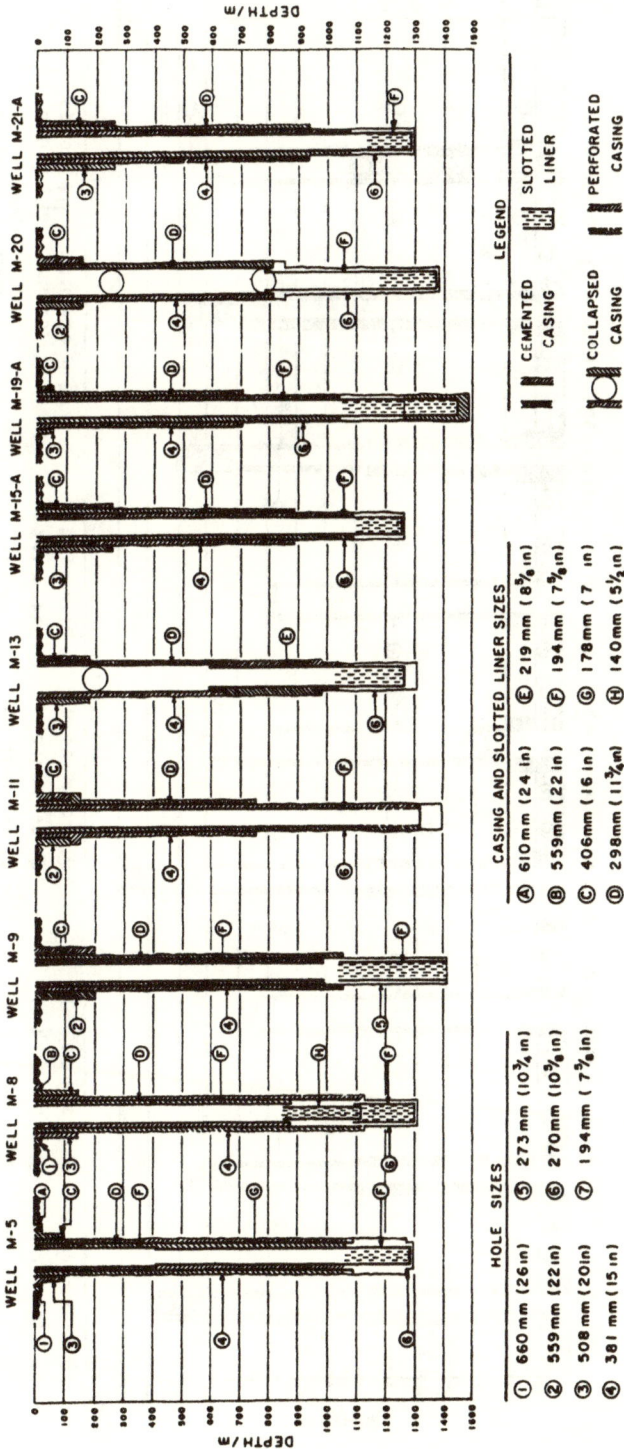

FIGURE 7.6—Well profiles for wells M-5, 8, 9, 11, 13, 15A, 19A, 20, and 21A at Cerro Prieto [after Dominguez and Vital, 1975].

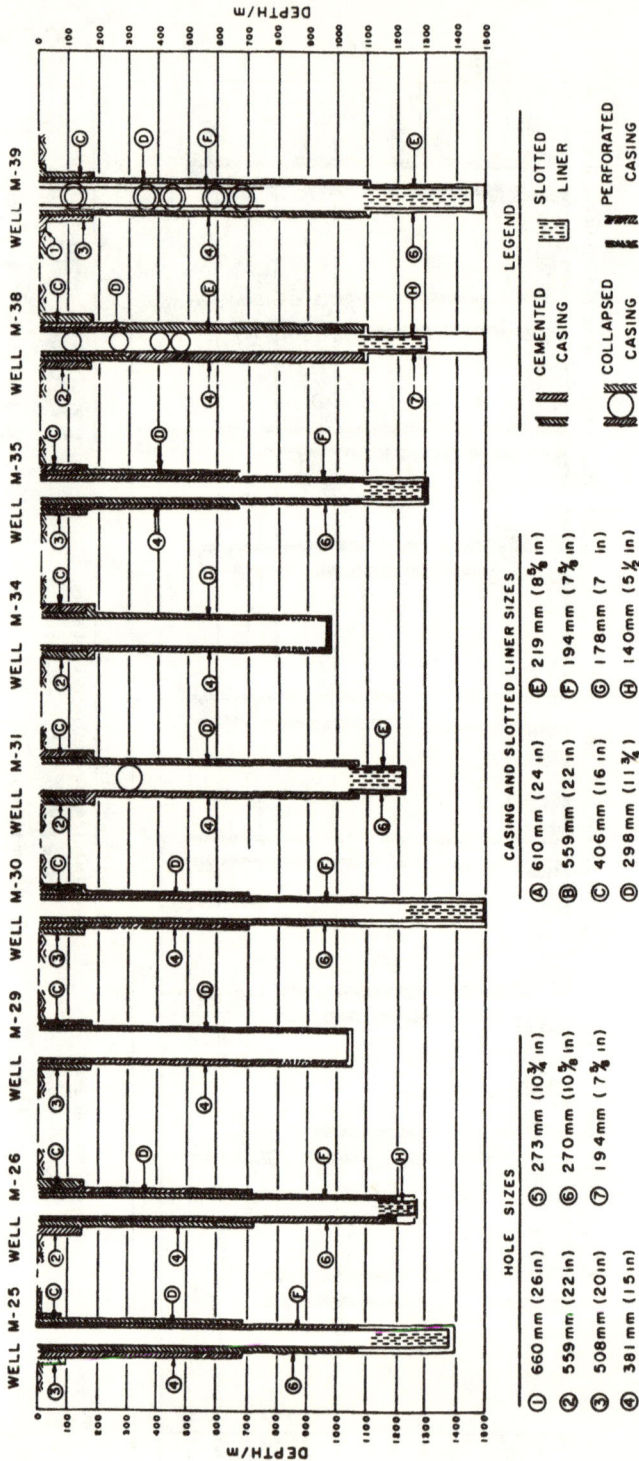

FIGURE 7.7.—Well profiles for wells M-25, 26, 29, 30, 31, 34, 35, 38, and 39 at Cerro Prieto [after Dominguez and Vital, 1975].

FIGURE 7.8—Well M-8 at Cerro Prieto showing wellhead, separator, ball-check valve, and twin silencer [photo courtesy of D. J. Ryley].

FIGURE 7.9—Steam receivers at Cerro Prieto. Steam lines from the field enter from left (four large-diameter pipes); receivers are at right; smaller pipe at top carries noncondensable gases from powerhouse back to the field for discharge at evaporation pond [photo by R. DiPippo].

FIGURE 7.10—Final moisture separator at Cerro Prieto [photo by R. DiPippo].

Table 7.1—*Wellhead and separator conditions averaged over 12 wells at Cerro Prieto* [a]

	Pressure kPa	Tempera- ture °C	Mass Flow Rate Mg/h	Dryness Fraction percent
Wellhead..........	[b] 119±982	—	—	—
Separator..........	710±72	165±4	[c] 47. 5±22. 7	24±6
Silencer...........	atmos.	—	[d] 130. 0±42. 8	—
Silencer...........	atmos.	—	[e] 16. 0±9. 2	—

[a] Source: Reed, 1975; samples taken from wells M-5, 8, 9, 11, 20, 25, 26, 29, 30, 31, 34 and 39 during January and February 1974.
[b] Wells M-11 and M 31 showed wellhead pressures of 4170 and 1960 kPa, respectively; average wellhead pressure for the other 10 wells was 792 kPa with a standard deviation of 66 kPa.
[c] Separated steam.
[d] Liquid.
[e] Steam.

ture of the geofluid at the separator was 165°C (329°F), and quality averaged about 24%.

The geothermal steam contains about 1% by weight of noncondensable gases, mainly carbon dioxide and hydrogen sulfide. Average concentrations of these two constituents for eight wells (M-5, 8, 9, 11, 20, 29, 31, 34) were reported to be 7320 and 1564 ppm, respectively [Reed, 1975]. The samples were taken for an average steam flow rate of 54.5 Mg/h (120.2 klbm/h) and at an average separator pressure of 790 kPa (114.6 lbf/in²). The impurities in the steam as it enters the turbine were quoted by Mercado as the following: CO_2, 14,100 ppm; H_2S, 1500 ppm; NH_4, 110 ppm; Cl, 0.8 ppm; Na, 0.4 ppm; and SiO_2, 0.2 ppm [Mercado, 1976].

Table 7 2—*Chemical composition of separated water samples from 12 wells at Cerro Prieto* [a]

Constituent	Concentration mg/l	Standard Deviation mg/l
Chloride, Cl..	14, 370	2, 220
Sodium, Na..	7, 760	1, 060
Potassium, K..	1, 660	430
Silica, SiO_2...	850	185
Calcium, Ca...	545	110
Bicarbonate, HCO_3....................................	50	10
Lithium, Li..	18	3
Boron, B..	18	4
Sulfate, SO_4..	17	14
Magnesium, Mg...	1. 4	1. 0
pH @ 25°C..	8. 2±0. 1	

[a] Source: Reed 1975; samples taken from wells M-5, 8, 9, 11, 20, 25, 26, 29, 30, 31, 34, and 39 during January and February 1974.

The composition of the separated liquid is given in table 7.2. Total dissolved solids amount to roughly 25,200 ppm or 2.5%. The main impurities are chloride, sodium, and potassium, which together constitute 94% of the total. Since all this liquid is discharged to the environment, either to the atmosphere through the silencers after flashing or to the ground and atmosphere by means of the settling/evaporation pond, attention must be given to the effects of the dissolved solids on the environment. The concentrations of the principal chemical components in the pond (on July 9, 1974) were reported as follows: Cl, 47,462 ppm; Na, 20,412 ppm; K, 4950; Ca, 2169; and Li, 68 ppm [Mercado, 1975b].

7.3.4 Energy conversion system

The power plant is of the separated-steam (or "single-flash") type with two units, each of 37.5 MW rated capacity. It is owned and operated by the Comisión Federal de Electricidad (CFE). The units were supplied by the Tokyo Shibaura Electric Company Ltd., (Toshiba), and are of the single-cylinder, double-flow variety with six stages of impulse-reaction blades in each flow. Unit 2 and a spare rotor are shown in figures 7.11 and 7.12, respectively. A simplified flow diagram/heat balance schematic is shown in figure 7.13; technical specifications are given in Table 7.3. Units 3 and 4 are essentially identical to the first two units.

The condenser is of the barometric, direct-contact type and is located next to the powerhouse. It is 25.35 m (83.2 ft) high, with a shell diameter of 6.7 m (22 ft), shell height of 9.6 m (31.5 ft), and tail pipe 2 m (6.6 ft) in diameter and 12 m (39.4 ft) in length. The exhaust from the turbine is conveyed to the condenser by means of a duct that is 3.6 m (11.8 ft) in diameter and about 40 m (131 ft) in length, including three right-angle bends. Noncondensable gases are removed from the top of the condenser shell through a 0.7-m-diameter (2.3 ft) pipe. Figure 7.14 shows the condensers for Units 1 and 2.

The gas extraction system consists of a two-stage steam ejector with an inter- and aftercondenser. There are three first-stage steam ejector nozzles operating in parallel, presumably for redundancy. Each is connected to a separate intercondenser; three second-stage steam ejectors are also arranged in parallel, as are three aftercondensers. The gas extraction system requires 24.2 Mg/h (53.4 klbm/h) of motive steam from the main-steam line and 373 Mg/h (822 klbm/h) of cooling water.

The generator is rated at 44,200 kVA at 13.8 kV and 60 Hz with a power factor of 0.85. Speed of rotation is 3600 rev/min, and both the rotor and the stator are conventionally hydrogen cooled. The geothermal resource utilization efficiency η_u of the plant is 40%, based on wellhead flow conditions assuming 24% quality and a sink temperature of 26.7°C (80°F). This value is typical of geothermal plants of this type.

FIGURE 7.11—Unit 2, 37.5 MW, in turbine hall at Cerro Prieto [photo by R. DiPippo].

FIGURE 7.12—Spare 6×2 impulse-reaction turbine spool at Cerro Prieto [photo by R. DiPippo].

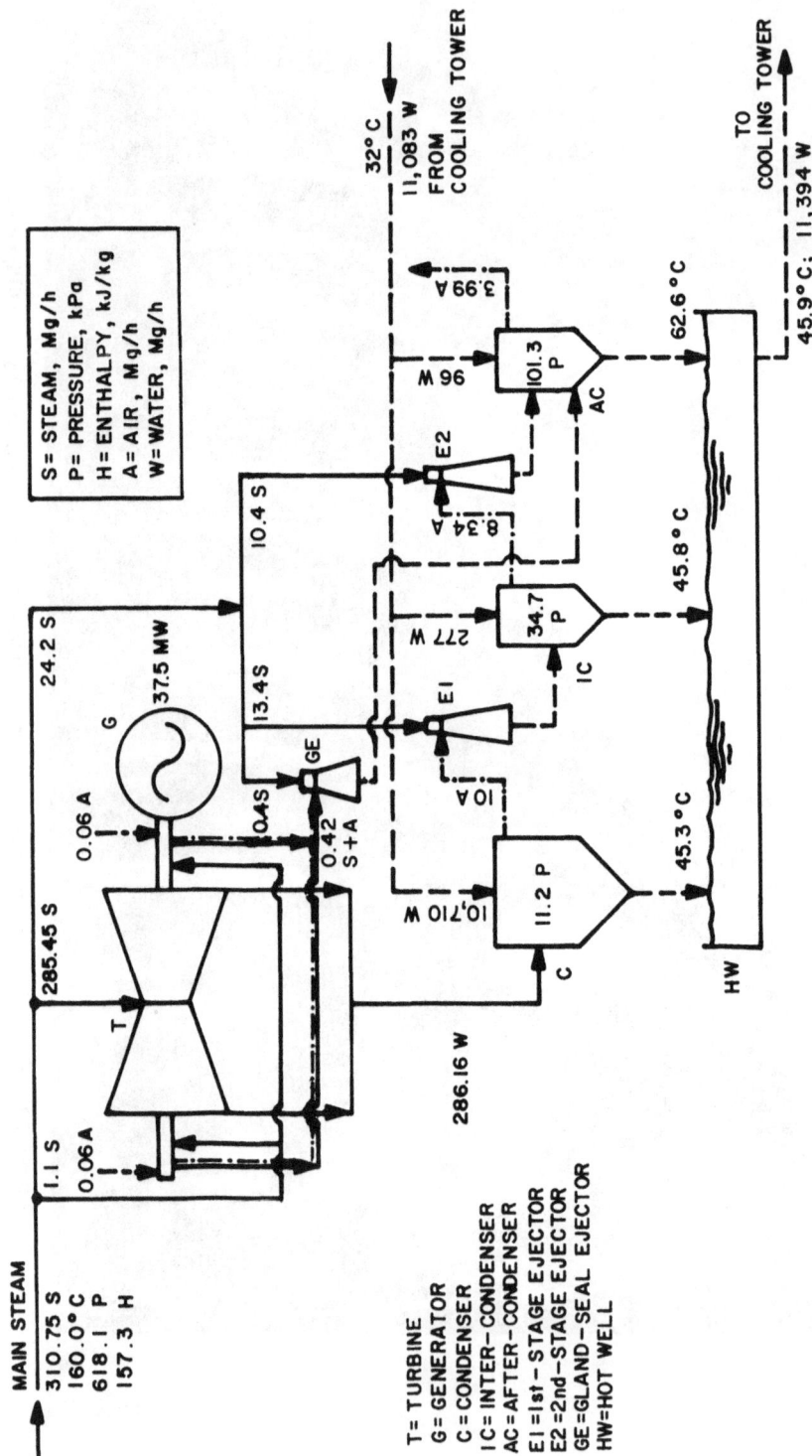

FIGURE 7.13—Simplified flow diagram for energy conversion system for each unit at Cerro Prieto [after Akiba, 1970; Mercado, 1976].

FIGURE 7.14—Barometric condensers for Units 1 and 2 at Cerro Prieto [photo by R. DiPippo].

Table 7.3—*Technical specifications for Cerro Prieto geothermoelectric power plant, Units 1 and 2* [a]

Year of startup: [b] 1973	

Turbine data:

Type	Single-cylinder, double-flow, 6×2, impulse-reaction
Installed capacity	37.5 MW, each
Speed	3600 rev/min
Steam inlet pressure	89.6 lbf/in²
% (wt.) noncondensable gases	~1.0% of steam
Exhaust pressure	3.5 in Hg
Steam flow rate	629×10³ lbm/h, each

Condenser data:

Type	Elevated, direct-contact, barometric condenser
Pressure	3.3 in Hg
Cooling water temperature	89.6°F
Outlet water temperature	113.5°F
Cooling water flow rate	23.6×10⁶ lbm/h, each

Gas extractor data:

Type	Steam jet ejectors, 2-stage
Suction pressure	3.1 in Hg
Steam consumption	53×10³ lbm/h, each

Heat rejection system data:

Type	Crossflow, mechanically-induced-draft cooling tower
Number of cells	6, each
Water inlet temperature	114.6°F
Water flow rate	25.5 x10⁶ lbm/h, each
Water pump power	835 kW

[a] Source: Mercado, 1976; Toshiba, 1977.
[b] Units 3 and 4 began operating on April 24, 1979 and are essentially identical to Units 1 and 2.

7.3.5 Construction materials

Prior to selection of materials for the various critical parts of the plant, 1000 tests were conducted on a number of candidate materials [Mercado, 1975b]. Samples were exposed to geothermal steam, both aerated and non-aerated, and to condensate for a period of 150 days. The steam was obtained from well M-8 and the condensate from steam from wells M-3 and M-5 [Tolivia et al., 1975]. In this section we shall describe the materials used in the construction of the wells, silencers, piping, turbines, condensers, cooling towers, and electrical equipment.

Well casings are fabricated from J–55 API standard-weight pipe with buttress couplings. Extra-heavy wall thickness may be required in future wells. The cement consists of API type G with silica flour, perlite, and retarders as additives [Guiza, 1975]. Silencers are made of concrete with wooden stacks in a twin-silo design [Mercado, 1975b].

Carbon steel is used for pipes carrying nonaerated steam. The corrosion rate follows a parabolic law and is about 0.04 mm/yr (0.0016 in/yr). When the steam is in contact with air; the corrosion rate is 0.11 mm/yr (0.0043 in/yr); aerated-steam pipelines are, therefore, allowed extra wall thickness. The worst case occurs in condensate lines where the corrosion rate for carbon steel is 0.66 mm/yr (0.026 in/yr). Condensate lines are provided with a corrosion allowance and also coated with epoxy resin [Tolivia et al., 1975].

The turbine rotor is fabricated from a 1 Cr/1 Mo/¼ V alloy steel forging machined to form a solid unit composed of shafts, wheels, bearing journals, and coupling flanges [Akiba, 1970]. Alloy steels containing nickel are not use because of their poor corrosion resistance. The turbine blades are machined from 12 Cr alloy steel bar stock and are enclosed with a shroud that is hand riveted in place with tenons on the outer edges of the blades. The blades of the last (sixth) row are fitted with stellite erosion shields and fastened together with lashing wire to minimize vibrations [Akiba, 1970]. Nozzle partitions are of 12 Cr/0.2 Al alloy steel, and labyrinth strips are of 15 Cr/1.7 Mo alloy steel. The turbine outer and inner casing are made of carbon steel according to ASTM specification A285 [Mercado, 1976].

The shell of the condenser is carbon steel with a coating of epoxy resin; the barometric pipe is made of naval brass [Mercado, 1976]. The structural members of the cooling tower are constructed from AISI 4140 steel with redwood and fiberglass packing [Mercado, 1976].

Since electrical equipment is susceptible to corrosive attack by hydrogen sulfide, special precautions are taken for the protection of this equipment. Most switchboards, including the main one, are installed in rooms that have air-conditioning systems fitted with activated carbon filters. These filters are filled with activated alumina beads impregnated with potassium permanganate [Mercado, 1974]. The electrical contacts on the high-voltage side at the substation are gold plated, although platinum may have been more appropriate.

7.3.6 Effluent and emissions handling systems

Liquid separated from the geofluid at the wellhead separators is either sent to silencers located at each wellhead site or piped directly to the evaporation and settling pond (see section 7.3.3). Eventually all waste liquid (liquid from the separators or the silencers and excess steam condensate) is discharged into the pond (see figure 7.15). The present pond has an estimated capacity sufficient to support 180 MW of generation, i.e., through Unit 5. Beyond that, another means must be found for disposal of the waste liquid. Among the options are (1) reinjection, (2) construction of a channel to the Laguna Salada, a dried-up lake, and (3) construction of a channel to the Sea of Cortez [Guiza, 1975; Mercado, 1975b].

FIGURE 7.15—Discharge of waste liquid into evaporation pond. Wellhead M-9 is in background and volcano Cerro Prieto is on horizon [photo by R. DiPippo].

The noncondensable gases removed from the condenser by means of the steam ejector system are discharged to the atmosphere through 475-mm-diameter (18-in) fiberglass pipes (one for each unit) that extend to a height of 40 m (131 ft) above the ground. Since the prevailing winds blow from either the northwest or the southeast, these gases should be swept away from the plant. The composition of the stack gases sampled on July 9, 1973, when Unit 2 was generating 32 MW, are given in table 7.4. The specific discharge of H_2S is more than 11,000 g/MW·h; for comparison, note that the U.S. Environmental Protection Agency (EPA) is suggesting a maximum specific H_2S emission level of 200 g/MW·h for geothermal plants in the United States [Hartley, 1978].

Table 7.4—*Composition of stack gases, sampled July 9, 1973* [a]

Gas	Percent by wt	Discharge kg/h
Carbon dioxide, CO_2............ ...	62. 5	3440
Hydrogen sulfide, H_2S............	6. 5	355
Air and other gases..............	6. 7	365
Water (condensate).............	24. 3	1340

[a] Source: Mercado, 1975b.

On windless days the concentration of H_2S may reach dangerous levels in certain areas. An alarm system is connected to a series of H_2S detectors to protect personnel in and around the powerhouse. An additional vent line was constructed from the powerhouse at the base of the gas extraction stacks to the evaporation pond. The resin-lined steel vent duct is 584 mm (23 in) in diameter and 1250 m (4100 ft) in length. Since H_2S is also emitted from the cooling tower stacks, it is impossible to vent all the H_2S away from the powerhouse, but the prevailing winds tend to carry the cooling tower plumes away from the powerhouse. The cooling tower is 100 m (328 ft) from the plant and is aligned with the direction of the prevailing winds (see figure 7.5).

The effects of other forms of pollution from the plant (for example, thermal and acoustic) are felt to be insignificant owing to the nature of the area and its sparse population. If it were deemed necessary, steps could be taken to curtail or further control plant emissions; however, it is believed that the best course for now is to provide special materials and protective devices and procedures where needed [Mercado, 1975].

7.3.7 Economic factors

The only economic data available were reported by CFE in 1971 on the anticipated costs of Units 1 and 2, then under construction. Capital investment was $14.4 million (US 1971), of which $7.2 million was for plant equipment, $2.08 million was for wells, and the remaining $5.12 million was for labor and overhead [CFE, 1971]. This report contained an estimate of $1.44 million for the annual operating expenses of the plant. The approximate annual savings in fuel (petroleum) costs were estimated at $170,000 million, a figure calculated before the price of oil underwent its dramatic increase in 1973 and thus vastly underestimating the savings in the present market.

7.3.8   Operating experience

The Cerro Prieto geothermoelectric powerplant has operated very reliably since it was brought on-line in April 1973. At that time Unit 2 began generating electricity; Unit 1 was started in September of the same year. The figures in table 7.5 give the annual electricity generation and capacity factors for the first 4 years of operation. Highest monthly capacity factor was 94% in September 1975. The capacity factor of 87% in 1976 is the highest of any Mexican powerplant.

Table 7.5—*Performance of Cerro Prieto plant, 1973–1976* [a]

Year	Generation GW h	Capacity factor percent
1973.............................	205	63
1974.............................	445	68
1975.............................	518	79
1976.............................	570	87

[a] Source: Dominguez, 1977; Mercado, 1976.

Regular maintenance is performed on the units at 2-year intervals; wells are cleaned once every 1 to 4 years. No scaling has been observed in the waste water disposal pipes when the liquid is maintained at the separator pressure (when it is *not* flashed down to atmospheric pressure at the wellhead silencers) and sent directly to the evaporation pond. No significant subsidence of the geothermal field has occurred, even though no reinjection has taken place. Reinjection is expected to be adopted when the field reaches full exploitation. Approximately 8×10^6 Mg/yr (2×10^{10} lbm/yr) of geofluid must be withdrawn from the field to supply each 37.5-MW unit; about 80×10^6 Mg (180×10^9 lbm) of fluid have been produced during the operating lifetime of the plant.

7.3.9   Expansion of Cerro Prieto plant

The ultimate electric generating capacity of the Cerro Prieto field is not known with certainty; 400 MW seems a conservative estimate. Units 3 and

4, each to be rated at 37.5 MW and essentially duplicates of Units 1 and 2, have been completed and came on-line in 1979. Following these, the next unit will be a double-flash, low-pressure unit rated at 30 MW, which will use steam flashed from the waste liquid produced from the first four units. Additional units will probably be standardized 55-MW designs. Although time schedules are difficult to predict accurately, table 7.6 gives the plan for near-term development of Cerro Prieto.

Table 7.6—*Planned expansion at Cerro Prieto*

Year	Site	Unit No.	Unit Capacity	Cum. Capacity
1979.......	Cerro Prieto I.......	3...........	37.5 MW.......	112.5 MW
1979.......	Cerro Prieto I.......	4...........	37.5 MW.......	150.0 MW
1982.......	Cerro Prieto I.......	5...........	30.0 MW.......	180.0 MW
1983.......	Cerro Prieto II......	1...........	55.0 MW.......	235.0 MW
1984.......	Cerro Prieto II......	2...........	55.0 MW.......	290.0 MW

7.4   Potential geothermal sites in Mexico

A large number of hydrothermal areas in Mexico are described by Alonso [1975]. In the 130 sites discovered, only a few wells have been drilled. Using conservative estimates, the ultimate geothermal power potential of the country has been placed at 4000 MW. Only the most prominent sites are discussed below.

7.4.1   Ixtlán de los Hervores

This field is located in the state of Michoacán within the area known as the Neovolcanic axis. Numerous thermal manifestations exist in the valley, where five small- and three large-diameter wells have been drilled. A satisfactory evaluation of the site has not been accomplished owing to problems with completion of the large-diameter wells [Alonso, 1975].

7.4.2   Los Negritos

The geology of this field has been studied by a variety of methods including seismic refraction, electric resistivity, gravimetric, magnetic, and geochemical. The field is located in the municipality of Villamar in the state of Michoacán, also in the Neovolcanic axis. One well, drilled to a depth of 1000 m (3280 ft), produces a mixture of liquid and vapor on an intermittent basis. The rocks are believed to be basalts and andesites with numerous fractures along the east-west axis [Alonso, 1975].

7.4.3 Los Azufres

Another promising region in the state of Michoacán, Los Azufres has been explored by means of geophysical, geochemical, geologic, and electrical resistivity techniques. All indications are favorable for the existence of a hydrothermal reservoir, but no wells had been drilled as of 1975 [Alonso, 1975].

7.4.4 La Primavera

This volcanic caldera containing a large number of hot springs and steam vents is located west of the city of Guadalajara in the state of Jalisco. The usual exploration techniques have been applied and several small-diameter wells have been drilled [Alonso, 1975].

7.4.5 San Marcos

Steam vents and hot springs, some with flows as large as 300 kg/s (661 lbm/s), exist in this field located about 80 km (50 mi) southwest of Guadalajara in Jalisco. Besides the geologic, resistivity, and geochemical surveys carried out, six small-diameter wells have been sunk to depths of 50–300 m (169–934 ft) [Alonso, 1975].

REFERENCES [a]

Akiba, M., 1970. "Mechanical Features of a Geothermal Plant," *Pisa, 1970*, Vol. 2, pp. 1521–1529.

Alonso, H., 1975. "Geothermal Potential of Mexico," *San Francisco, 1975*, Vol. 1, pp. 21–24.

CFE, 1971. "Cerro Prieto: Underground Power," Comisión Federal de Electricidad, G. V. Caravantes, Dir. Gen., Ródano 14, México 5, D.F.

Diaz Serrano, J., 1978. "Pemex Director General Reports on Mexico's Outlook," *Ocean Industry*, Vol. 13, No. 5, pp. 42–44.

Dominguez A., B., 1977. "Cerro Prieto, Mexico," *Geothermal Resources: Survey of an Emerging Industry*, Geothermal Resources Council Spec. Short Course No. 6, Houston.

Dominguez A., B., and Vital B., F., 1975. "Repair and Control of Geothermal Wells at Cerro Prieto, Baja California, Mexico," *San Francisco, 1975*, Vol. 2, pp. 1483–1499.

Guiza L., J., 1975. "Power Generation at Cerro Prieto Geothermal Field," *San Francisco, 1975*. Vol 3, pp. 1976–1978.

Hartley, R. P., 1978. *Pollution Control Guidance for Geothermal Energy Development*, EPA Rep. No. 600/7-78-101, Ind. Env. Res. Lab., Cincinnati.

Mercado G., S., 1974. "The Geothermal Plant of Cerro Prieto, B.C., Mexico, and Problems Encountered During Its Development," *Geothermics*, Vol. 3, No. 3, pp. 125–126.

Mercado G., S., 1975a. "Movement of Geothermal Fluids and Temperature Distribution in the Cerro Prieto Geothermal Field, Baja California, Mexico," *San Francisco, 1975*, Vol. 1, pp. 487–494.

---

[a] See note on p. 24.

Mercado G., S., 1975b. "Cerro Prieto Geothermoelectric Project: Pollution and Basic Protection," *San Francisco, 1975,* Vol. 2, pp. 1385–1398.

Mercado G., S., 1976. "Cerro Prieto Geothermal Field, Mexico: Wells and Plant Operation," *Proc. Int. Cong. on Thermal Waters, Geothermal Energy and Vulcanism of the Mediterranean Area, Geothermal Energy,* Vol. 1, pp. 394–408.

Paredes A., E., 1975. "Preliminary Report on the Structural Geology of the Cerro Prieto Geothermal Field," *San Francisco, 1975,* Vol. 1, pp. 515–519.

Reed, M. J., 1975 "Geology and Hydrothermal Metamorphism in the Cerro Prieto Geothermal Field, Mexico," *San Francisco, 1975,* Vol. 1, pp. 539–547.

Tolivia M., E., Hoashi, J., and Miyazaki, M., 1975. "Corrosion of Turbine Materials in Geothermal Steam Environment in Cerro Prieto, Mexico," *San Francisco, 1975,* Vol. 3, pp. 1815–1820.

Toshiba, 1977. "Toshiba Experience List on Geothermal Power Plant," Tokyo Shibaura Electric Company, Tokyo.

CHAPTER 8

NEW ZEALAND

8.1 Overview

New Zealand pioneered in the use of liquid-dominated hydrothermal resources for the generation of electricity. The use of geothermal energy was first considered in the 1930's. Early investigations led to the conclusion that geothermal energy could be used to produce electric power. A 5-year program established in 1949 had as its objective the winning of sufficient steam to support a 20-MW plant in an area of the Waiora Valley at Wairakei. Consulting engineers Merz and McLellan of England designed a 26-MW plant in 1953, based on the proven geothermal steam supply. Construction began in 1956, the first turbine was commissioned on November 15, 1958, and Stage One of the program was completed in March 1960. This was followed by the installation of 12 additional machines, the last brought on-line on October 7, 1963. The total installed capacity of the Wairakei geothermal power station was then 192.6 MW [NZED, 1975].

Multipurpose use of geothermal energy is made at Kawerau, where process heating, clean steam generation, and electricity production take place. At Ohaki (Broadlands), the New Zealand Electricity Department (NZED) is planning to build a geothermal power plant that will employ a two-stage steam generation process, i.e., a primary separation followed by a secondary flash process. This plant will eventually have a capacity of 150 MW, but is being built in 50-MW stages.

8.2 Wairakei

8.2.1 Geology

The Wairakei geothermal field lies in an extensive thermal area on New Zealand's North Island. Wairakei is situated about 8 km (5 mi) north of the northeast corner of Lake Taupo, roughly in the middle of a thermal belt 50 km (31 mi) wide and 250 km (155 mi) long that trends northeast-southwest across North Island from a central group of volcanic mountains to the White Island volcano in the Bay of Plenty. Figure 8.1 shows the location of Wairakei (insert) and the general arrangements of the bore field and power station [Armstead, 1961].

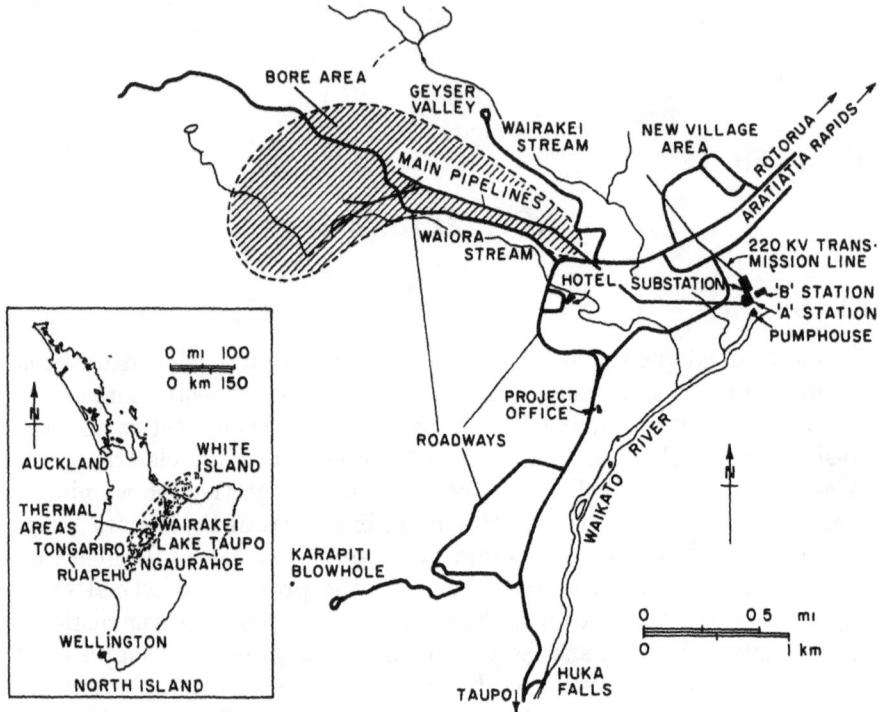

FIGURE 8.1—Location of Wairakei geothermal field, North Island, New Zealand
[Armstead, 1961].

The first bore at Wairakei delivered steam in 1951; the first electricity
was generated in 1958. This field has been exploited for a longer time than
any other liquid-dominated geothermal reservoir in the world. The reservoir has passed its peak of production. Whereas the installed capacity is
192.6 MW, the output in 1974 was only 150 MW [Bolton, 1977]. It is expected, however, that production will stabilize at between 125 and 140
MW for an indefinite period of time.

The Wairakei thermal area is located in a region of Cenozoic subsidence
and forms a part of the Taupo volcanic zone. A geologic map of the
thermal region is given in figure 8.2 [Grindley, 1961]. Large, active andesitic volcanoes are located at each end of the zone, and the wider central
portion is dominated by acid igneous activity. These include rhyolite
domes, pyroclastic pumice deposits, and ignimbrites. During the late Miocene and Quaternary periods, a huge volume of lava and pyroclastic rocks
[16,000 km³ (3840 mi³)] were erupted from the Taupo volcanic zone and
led to the formation of grabens and calderas constituting the zone of
subsidence.

A geologic cross-section taken east-west across the thermal belt at
Wairakei is shown in figure 8.3(a) [McNitt, 1965], which illustrates the

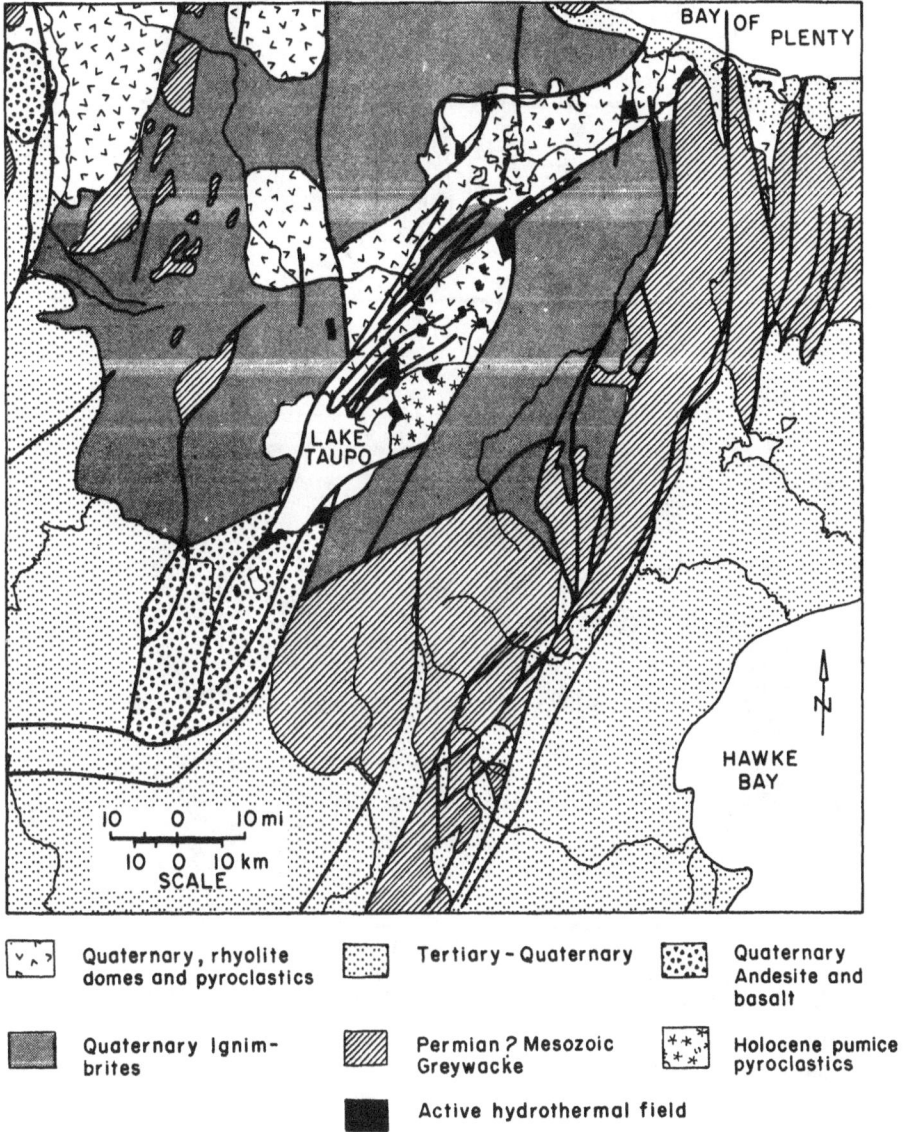

FIGURE 8.2—Generalized geologic map of Taupo volcanic zone, New Zealand
[Grindley, 1961].

magmatic intrusion at a depth of about 8 km (26,000 ft) that is the source
of the thermal anomaly. The Wairakei field is situated on a horst block
that has been elevated some 1200 m (3940 ft) relative to the surrounding
rocks. The horst is flanked by steep faults that act as feeders for the hot-
water aquifer.

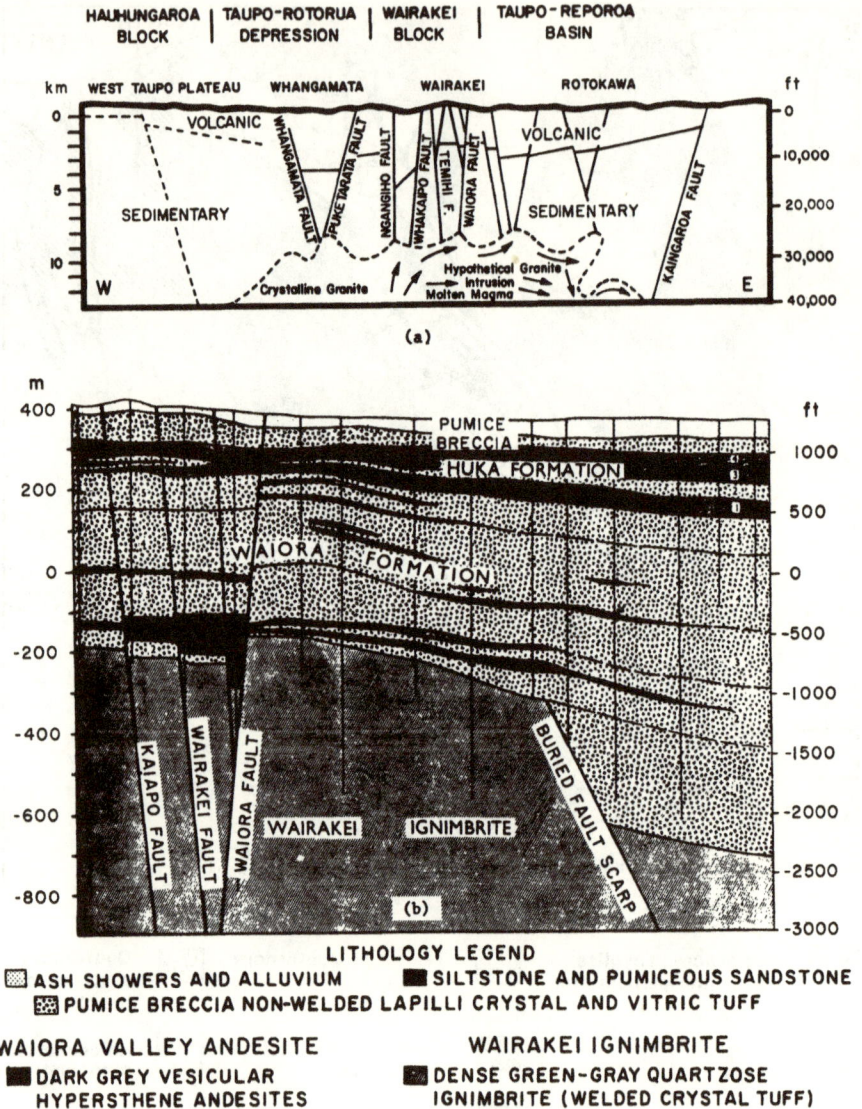

FIGURE 8.3—(a) Geologic cross-section of Taupo thermal zone at Wairakei [McNitt, 1965]; (b) Detailed cross-section of Wairakei reservoir [Grindley, 1961].

A detailed cross-section is given in Fig. 8.3(b) [Grindley, 1961], which shows a portion of the field at Wairakei. This view covers a width of about 1.9 km (6100 ft) to a depth of about 900 m (3000 ft) below sea level, or 1300 m (4300 ft) in total depth. The major faults (e.g., Waiora, Wairakei, Kaiapo) are shown at the left; the thin vertical lines represent drilled wells. The reservoir consists of a pumice breccia aquifer (Waiora formation) varying in thickness from 460 to 900 m (1500 to 3000 ft). The res-

ervoir is capped by layers of relatively impermeable lacustrine mudstones
(Huka formation) which range in thickness from 60 to 150 m (200 to 500
ft) and lie from 180 to 300 m (600 to 1000 ft) below the surface. The
surface formations comprise mainly loosely consolidated breccias (Wairakei breccia) and a top layer of recently deposited pumice cover. These
surface layers extend to a depth of about 125 m (410 ft) [Bolton, 1977].

A large proportion of the field's production comes from the permeable
Waiora aquifer. The main production zone, however, is the contact area
between the Waiora formation and the underlying Wairakei ignimbrite,
which by itself has a low production capacity. This producing interface
lies at a depth of between 570 and 680 m (1870 and 2230 ft) over the western and central portions of the cross-section, but dips sharply to about 1100
m (3600 ft) in the eastern region. The reservoir has not been drilled to its
full depth in this area [Bolton, 1977].

8.2.2 Steam wells and gathering system

The steam wells at Wairakei produce at two pressure levels: high-
pressure bores at pressures of about 1030 kPa (150 lbf/in²) and above, and
intermediate-pressure bores with pressures ranging from 620 to 860 kPa
(90 to 125 lbf/in²). A considerable reduction in field pressure has occurred
during the lifetime of the project. The high-pressure wells originally produced at pressures in excess of 1400 kPa (200 lbf/in²) [Bolton, 1970]. The
dramatic decrease in field pressure may be seen in figure 8.4, which

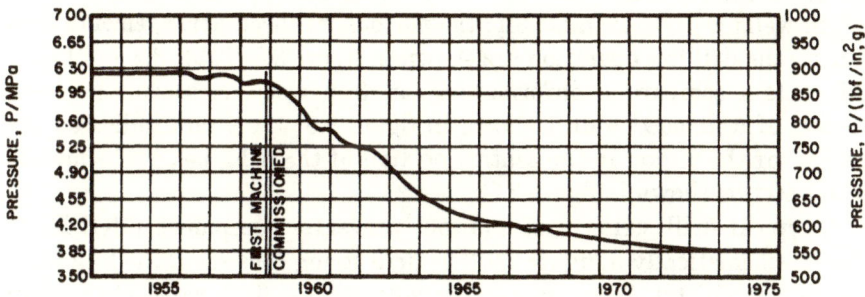

FIGURE 8.4—Reservoir pressure history at Wairakei [after Bolton, 1970, 1977; Hunt,
1977].

covers the period 1953–1975. The loss in pressure amounts to about 38%
over the 23-year span, with nearly all of it occurring since the commission-
ing of the first machine. The pressure appears to be approaching a stable
value, having lost only 6% during the last 7 years.

Note that no reinjection of the withdrawn fluid has ever taken place at
Wairakei. In addition to playing a role in the pressure loss within the
reservoir, this has led to the subsidence effects discussed later (see section

8.2.5.). Furthermore, the slight recovery in pressure that occurred during a test conducted in early 1968 was attributed to a curtailment of fluid drawoff (flow reduced to about one-third the normal value). The implication is that the reservoir is being influenced in some way by an inflow.

It has been reported recently [Hunt, 1977] that the amount of recharge of water into the geothermal field has, in fact, increased markedly over the past several years. Recharge is defined as the ratio of the mass of fluid replaced to that withdrawn over a specified time interval. On the basis of repeat gravity measurements, Hunt indicates the following values of recharge: 1958–1961, 30%; 1961–1967, 35%; 1967–1974, 90%. These values are probably accurate to about ±15%. During the last period (1967–1974), 400×10^9 kg (822×10^9 lbm) of fluid were withdrawn and were replaced by 364×10^9 kg (802×10^9 lbm). Over the full history of the field, about 695×10^9 kg (1532×10^9 lbm) have replaced 965×10^9 kg (2127×10^9 lbm) of withdrawn geofluid. If the recharge continues at 90%, it is likely that the field will not be depleted for a long time.

The practices employed for drilling, casing, operations, and maintenance of the steam wells at Wairakei have been documented extensively [Craig, 1961; Fisher, 1961; Fooks, 1961; Smith, 1961a and 1961b; Stilwell, 1970; Woods, 1961]. Figure 8.5 shows a typical drilling rig and drilling fluid circulation layout. The cellar provides a solid base on which to mount the rig and accommodates the wellhead equipment used during production. Early cellars were 3.0 m (10 ft) deep, but in 1966 they were redesigned for 2.1 m (7 ft), for medium-depth wells, i.e., wells of 900 m (3000 ft) nominal depth; cellars for deep wells, i.e., wells of 2300 m (7500 ft) nominal depth, are 3.4 m (11 ft) deep [Stilwell, 1970]. The drilling program for medium-depth wells is given in table 8.1; the various wellhead arrangements used during the drilling operation are shown in figure 8.6. Table 8.2 and figure 8.7 contain comparable information for deep wells. A schematic cross-section of a finished well is shown in figure 8.8; the strata indicated are typical of the region in the central portion of the field, i.e., in the middle of the section shown in figure 8.3b.

A typical drill string with blowout preventer equipment (BOPE) is shown schematically in figure 8.9. The drill string consists of the following elements: drill bit (usually tri-cone roller type), float valve (to control flow of drilling mud), drill collar (similar to drill pipe except somewhat larger and heavier), drill pipe, and the Kelly (square piece of pipe that allows torque to be transmitted from the rotary table through the square Kelly bushing). Three additional elements enable the well to be closed off during the drilling operation should this be necessary. A drill-through valve may be used to close the well once the drill string has been removed. Next are the Shaffer gates, two independent horizontal gates, each of which is split in half. The inner surfaces of the semicircular sections are fitted with rubber rams capable of closing around the drill pipe, drill collar, or well casing. The gates are driven by an air-motor-actuated screw with a

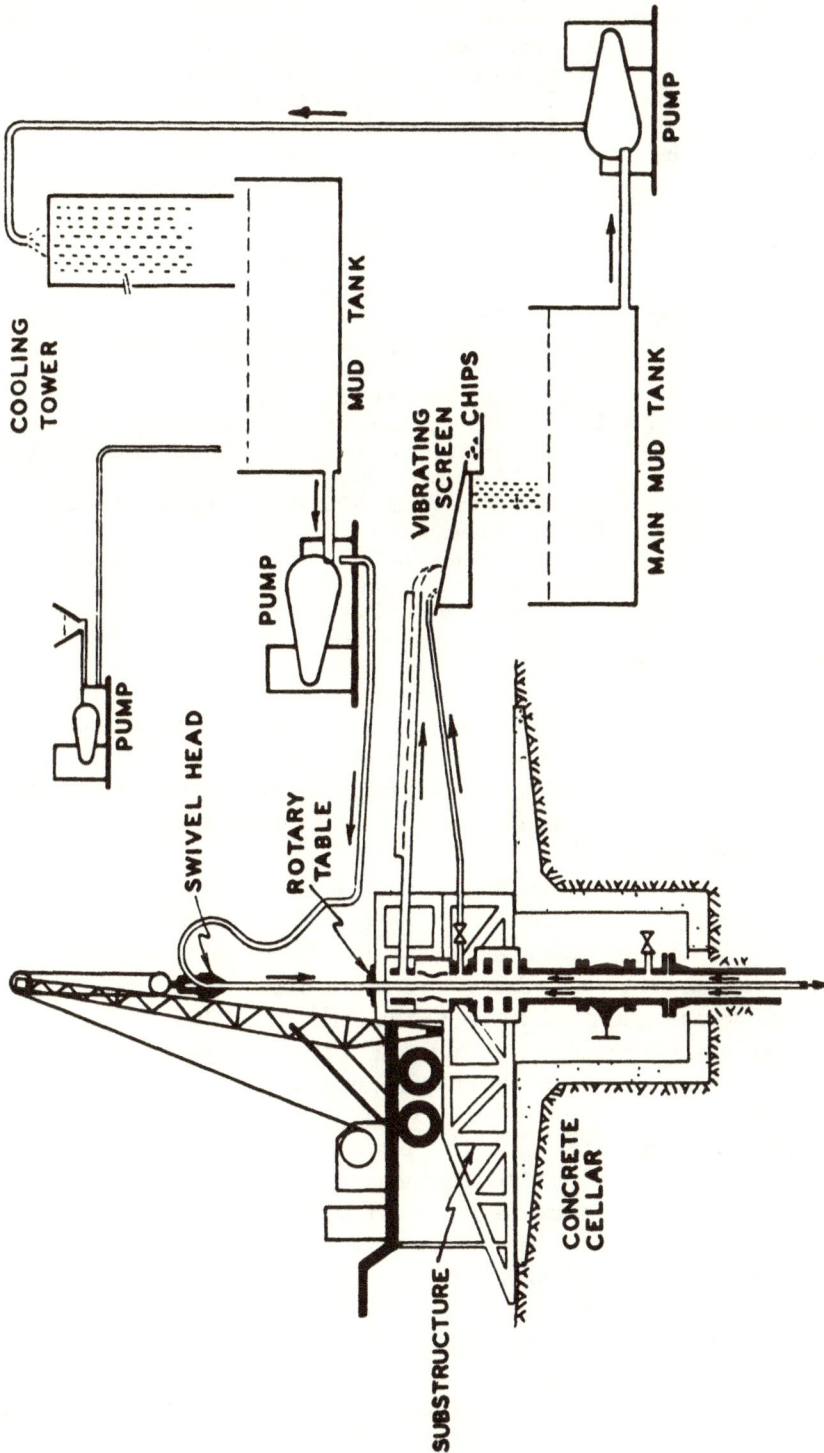

FIGURE 8.5—Typical drilling rig and drilling fluid circulation layout at Wairakei [Craig, 1961].

Table 8.1—*Drilling program for medium-depth wells at Wairakei* [a]

For 559-mm (22-in) O.D. casing.	Drill 311-mm-diameter (12.25-in) pilot hole. Open out to 660-mm-diameter (26-in). Use 184-mm (7.25-in) O.D. drill collars. Run and cement casing; set up Stage 1 wellhead.[b]
For 406-mm (16-in) O.D. casing.	Drill 311-mm-diameter (12.25-in) pilot hole. Open out to 533-mm-diameter (21-in) with expanding hole opener. Use 184-mm (7.25-in) O.D. drill collars and 168-mm (6.625-in) O.D. drill pipe. Run and cement casing; set up Stage 2 wellhead.[b]
For 298-mm (11.75-in) O.D. casing.	Drill 381-mm-diameter (15-in) hole. Use 184-mm (7.25-in) O.D. drill collars and 168-mm (6.625-in) O.D. drill pipe. Run and cement casing; set up Stage 3 wellhead.[b]
For 219-mm (8.625-in) O.D. casing.	Drill 270-mm-diameter (10.625-in) hole. Use 184-mm (7.25-in) O.D. drill collars and 89-mm (3.5-in) O.D. drill pipe. Run and cement casing; set up Stage 4 wellhead.[b]
For 168-mm (6.625-in) O.D. liner.	Drill 194-mm-diameter (7.625-in) hole. Use 146-mm (5.75-in) O.D. drill collars and 89-mm (3.5-in) O.D. drill pipe: Run liner with "J" slot.

[a] Source: Stilwell, 1970.
[b] See figure 8.6.

manual backup. Finally, the hydraulically operated blowout preventer (BOP) is able to shut off the well completely by means of a rubber packing element that can fit around any item in the well. The driller has immediate access to the control for the Shaffer gates and the BOP in case of emergency [Craig, 1961].

The arrangement of wells in the steam field is shown in figure 8.10 (see page 221) [after Haldane and Armstead, 1962; Bolton, 1977; Grindley, 1961]. Recent operations have extended the drilled area beyond the borders of this plan view. There have been a total of 102 wells drilled, of which 68 have supplied steam to the turbines [Bolton, 1977].

The gathering system is a complicated one involving three pressure levels. The complexity arose because the original plans for development of the area included a plant to produce heavy water for the United Kingdom Atomic Energy Authority. This proposal was made in 1953 and the steam pressures were selected to accommodate the requirements of the distillation plant. The proposal for the heavy-water plant was withdrawn in 1956, but only after the design of the steam system had been frozen and turbines were on order. Thus the resulting design is unnecessarily complex, and will not be repeated [Bolton, 1975]. The present gathering system is shown schematically in figure 8.11 (see page 222) [Bolton, 1977]. Two high-pressure wells are shown. The one on the left supplies fluid to a typical flash plant that produces steam at three pressure levels: high pressure (HP), intermediate pressure (IP) and intermediate-low pres-

Table 8.2—*Drilling program for deep wells at Wairakei* [a]

For 559-mm (22-in) O.D. casing.	Drill 311-mm-diameter (12.25-in) pilot hole. Open out to 660-mm-diameter (26-in). Use 229-mm (9-in) and 203-mm (8-in) O.D. drill collars. Run and cement casing; set up Stage 1 wellhead.[b]
For 457-mm (18-in) O.D. casing.	Drill 311-mm-diameter (12.25-in) pilot hole. Open out to 533-mm-diameter (21-in) with expanding hole opener. Use 229-mm (9-in) and 203-mm (8-in) O.D. drill collars and 114-mm (4.5-in) O.D. drill pipe. Run and cement casing; set up Stage 2 wellhead.[b]
For 340-mm (13.375-in) O.D. casing.	Drill 311-mm-diameter (12.25-in) pilot hole. Open out to 419-mm-diameter (16.5-in) with hole opener. Use 229-mm (9-in) and 203-mm (8-in) O.D. drill pipes and 114-mm (4.5-in) O.D. drill pipe. Run and cement casing; set up Stage 3 wellhead.[b]
For 244-mm (9.625-in) O.D. casing.	Drill 311-mm-diameter (12.25-in) hole. Use 203-mm (8-in) O.D. drill collars and 114-mm (4.5-in) O.D. drill pipe. Run and cement casing; set up Stage 4 wellhead.[b]
For 194-mm (7.625-in) slotted liner.	Drill 216-mm-diameter (8.5-in) hole. Use 159-mm (6.25-in) O.D. drill collars and 114-mm (4.5-in) O.D. drill pipe. Run liner with "J" slot.

[a] Source: Stilwell, 1970.
[b] See figure 8.7.

sure (ILP). The one at the right produces only high-pressure fluid by means of a simple cyclone separator. The figure does not show the intermediate-pressure wells that also produce intermediate-pressure (IP) steam and additional water for the flash plant.

Wellhead separators are of two types, as shown in figure 8.12(a) and 8.12(b) [Haldane and Armstead, 1962; Bolton, 1977]. Early wells were fitted with top-outlet cyclonic (TOC) separators as illustrated in (a); recent wells use bottom-outlet cyclonic (BOC) separators as shown in (b) [Hunt, 1961]. The former type incorporated a U-bend upstream of the admission point to the separator, which removed about 80%–90% of the liquid. A baffle arrangement inside the separator trapped the remaining liquid and allowed the steam to emerge with a dryness fraction of about 99%. The latter type, often called a Webre separator, is much simpler and has been shown to be capable of yielding steam with a dryness fraction in excess of 99.9% [Usui and Aikawa, 1970]. The hot water pump shown in (b) was used on the original pilot plant that pioneered the use of the liquid fraction of the geofluid; it has since been replaced by a simple water-collection tank similar to that shown in (a), but without the U-bend.

The essentially dry, saturated steam is transmitted from the separators to the main steam lines by means of branch pipelines. The main transmis-

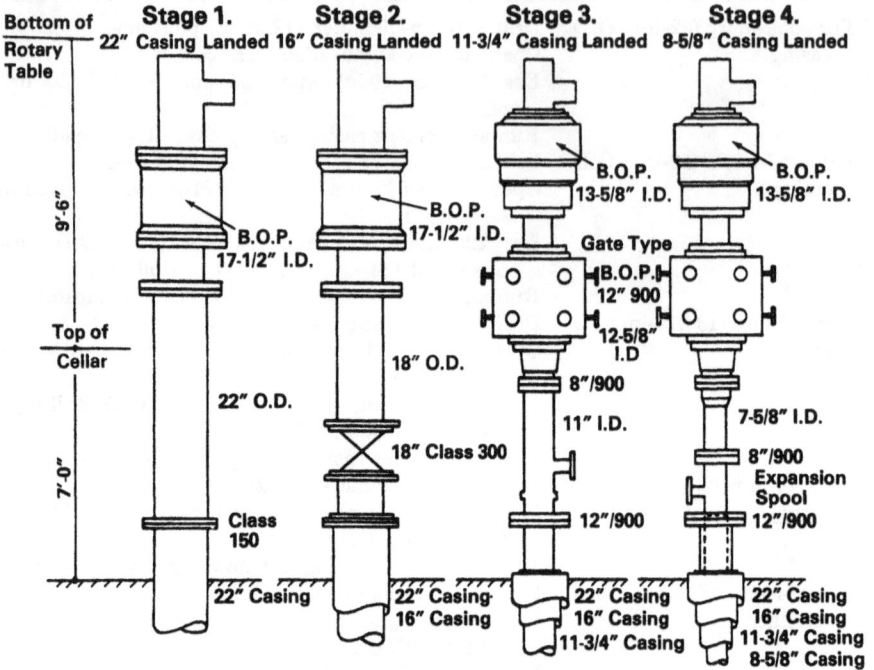

FIGURE 8.6—Wellheads for various stages of drilling program: medium-depth wells at Wairakei [Stilwell, 1970].

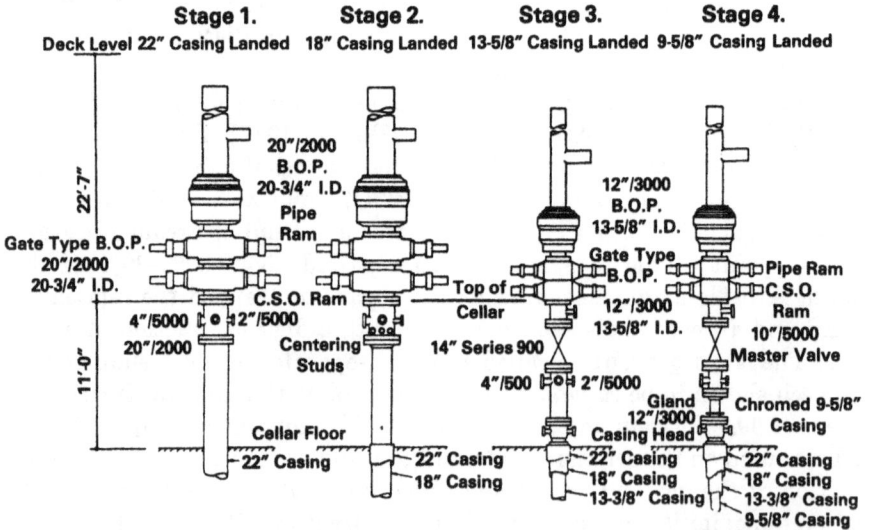

FIGURE 8.7—Wellheads for various stages of drilling program: deep wells at Wairakei [Stilwell, 1970].

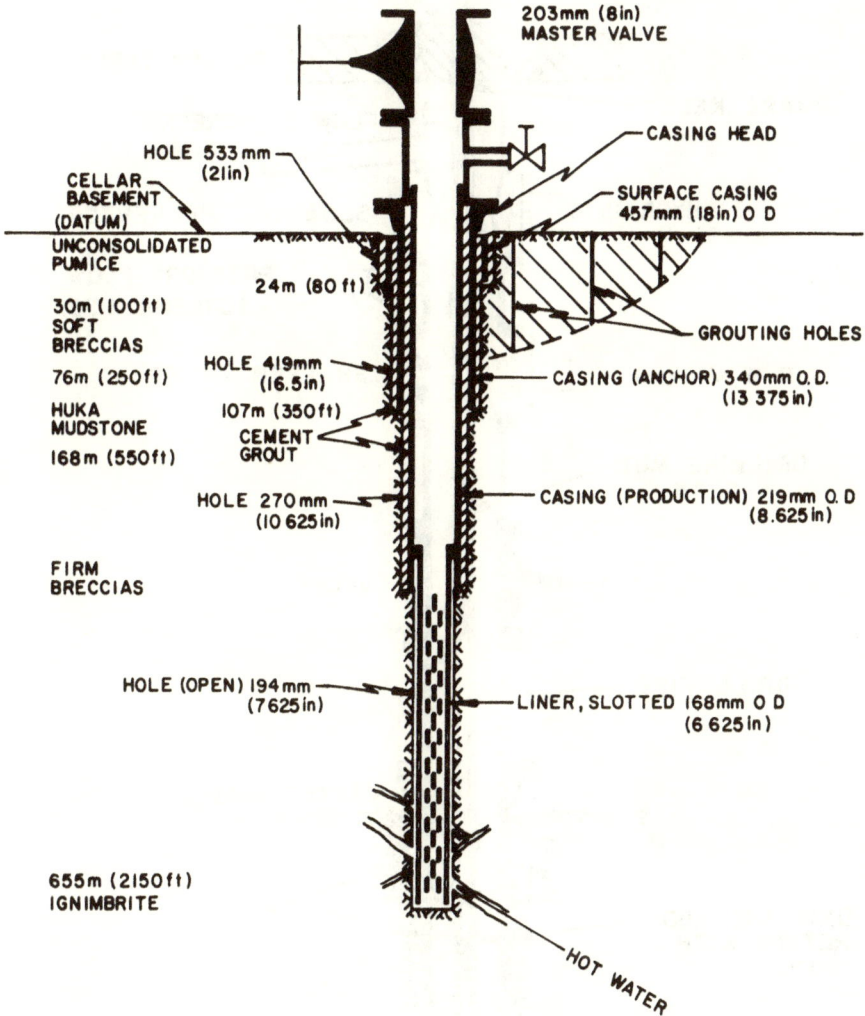

FIGURE 8.8—Typical completed well at Wairakei [after Craig, 1961; Stilwell, 1970].

sion pipelines are 508, 762, 1067, and 1219 mm (20, 30, 42, and 48 in) in diameter (nom.); the branch lines vary from 150 to 300 mm. (6 to 12 in) in diameter. The two largest steam mains are associated with the ILP steam system in which the specific volume is large. Expansion loops are located every 300 m (1000 ft) roughly; each loop can absorb about 762 mm (30 in) of pipe movement. Condensate that forms during transmission is removed by means of drain pots located about 150 m (500 ft) apart. Since the drains serve to remove impurities, the steam arrives at the powerhouse in a highly purified condition, the pipeline acting as a very efficient scrubber.

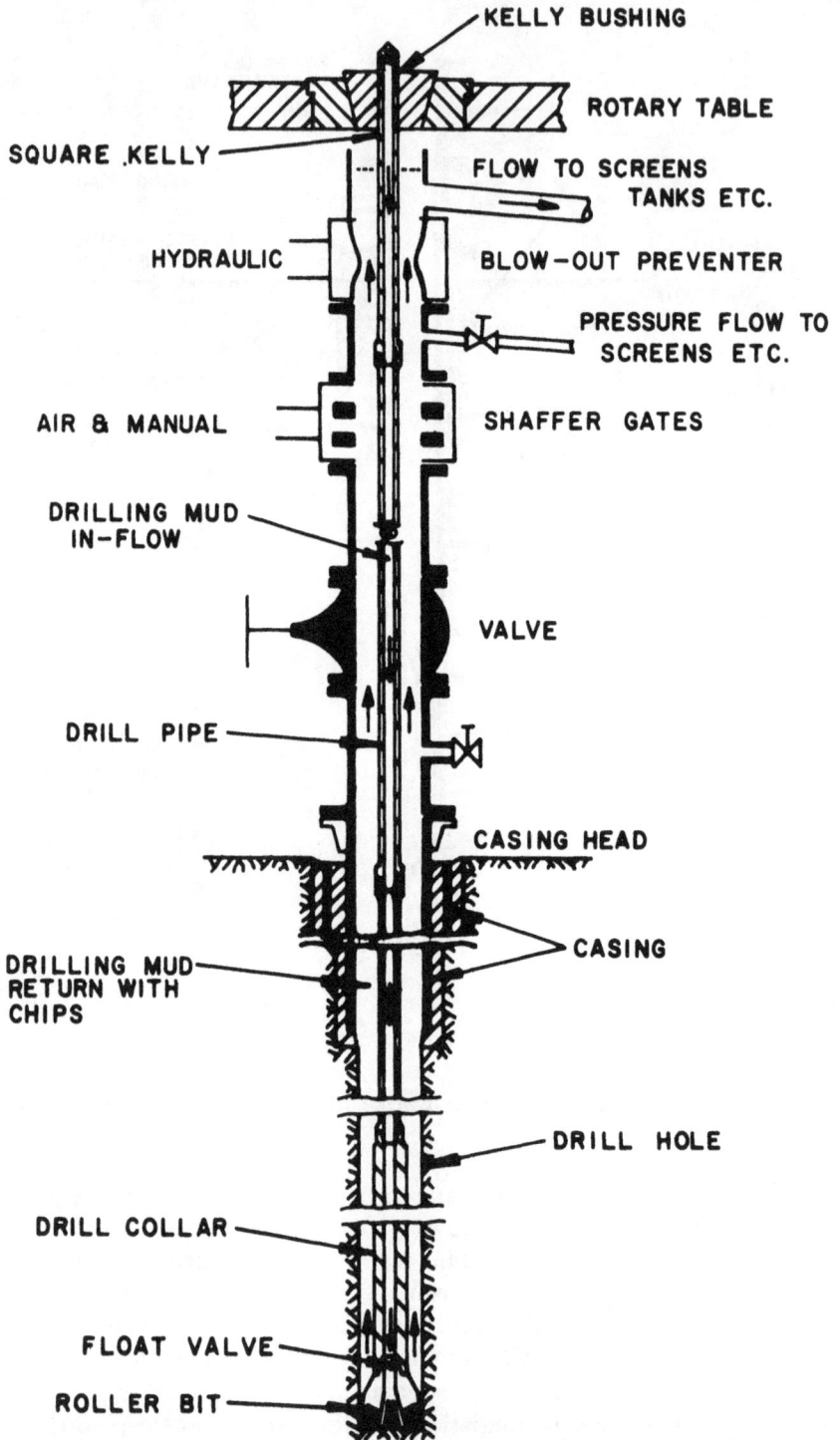

FIGURE 8.9—Typical drill string and blowout preventer equipment at Wairakei
[Craig, 1961].

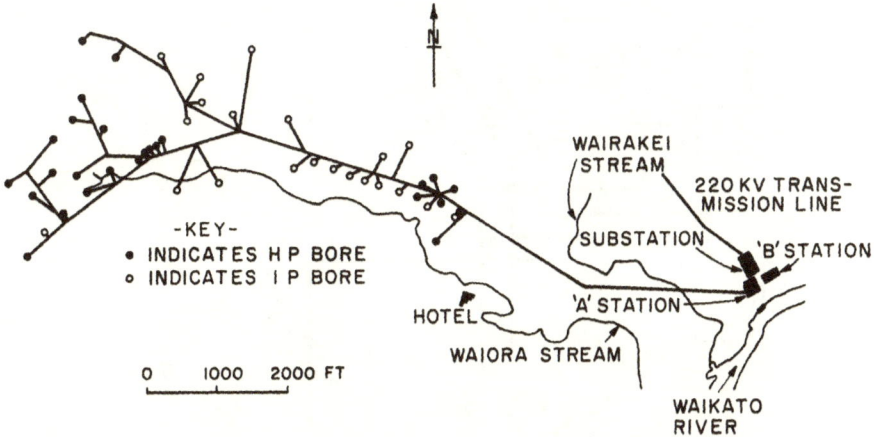

FIGURE 8.10—Arrangement of wells at Wairakei [after Haldane and Armstead, 1962; Bolton, 1977; Grindley, 1961].

8.2.3 Energy conversion system

The energy conversion system at Wairakei may be described as a multi-pressure, separated-steam, double-flash power plant. On the average, about 80% (by weight) of the high-pressure geothermal fluid at the wellhead is liquid. Steam is separated from high- and intermediate-pressure wells and sent to the powerhouses. Hot water separated from a number of HP wells is flashed to produce IP steam. An additional flash produces ILP steam, which is transmitted to the power plant and subsequently let down for use in the lowest-pressure turbines.

The arrangement of the 13 power turbines is shown in figure 8.13. Two 6.5-MW and two 11.2-MW back-pressure machines are supplied with HP bore steam; two 11.2-MW back-pressure turbines receive a mixture of IP bore steam, IP flash steam, and exhaust steam from the HP units; four 11.2-MW condensing units operate on LP steam obtained from the exhaust of the IP machines and the let-down flashed steam from the second-stage flash tanks; and three 30-MW dual-admission, condensing units are supplied with the same steam that feeds the IP turbines and receive pass-in, LP steam let down from the second-stage flash vessels.

Tables 8.3 and 8.4 list technical particulars of the several sets of power generating equipment. Since the plant has been undergoing modifications to compensate for the loss of pressure in the geothermal reservoir, some of the values shown in the tables are approximate or may not be current. They nevertheless represent the best information available at the time of writing. Figure 8.14 is a simplified plant flow diagram showing the typical wellhead separator/flasher arrangements. It is expected that eventually all high-pressure wells will be adapted to a multiflash setup; there are now

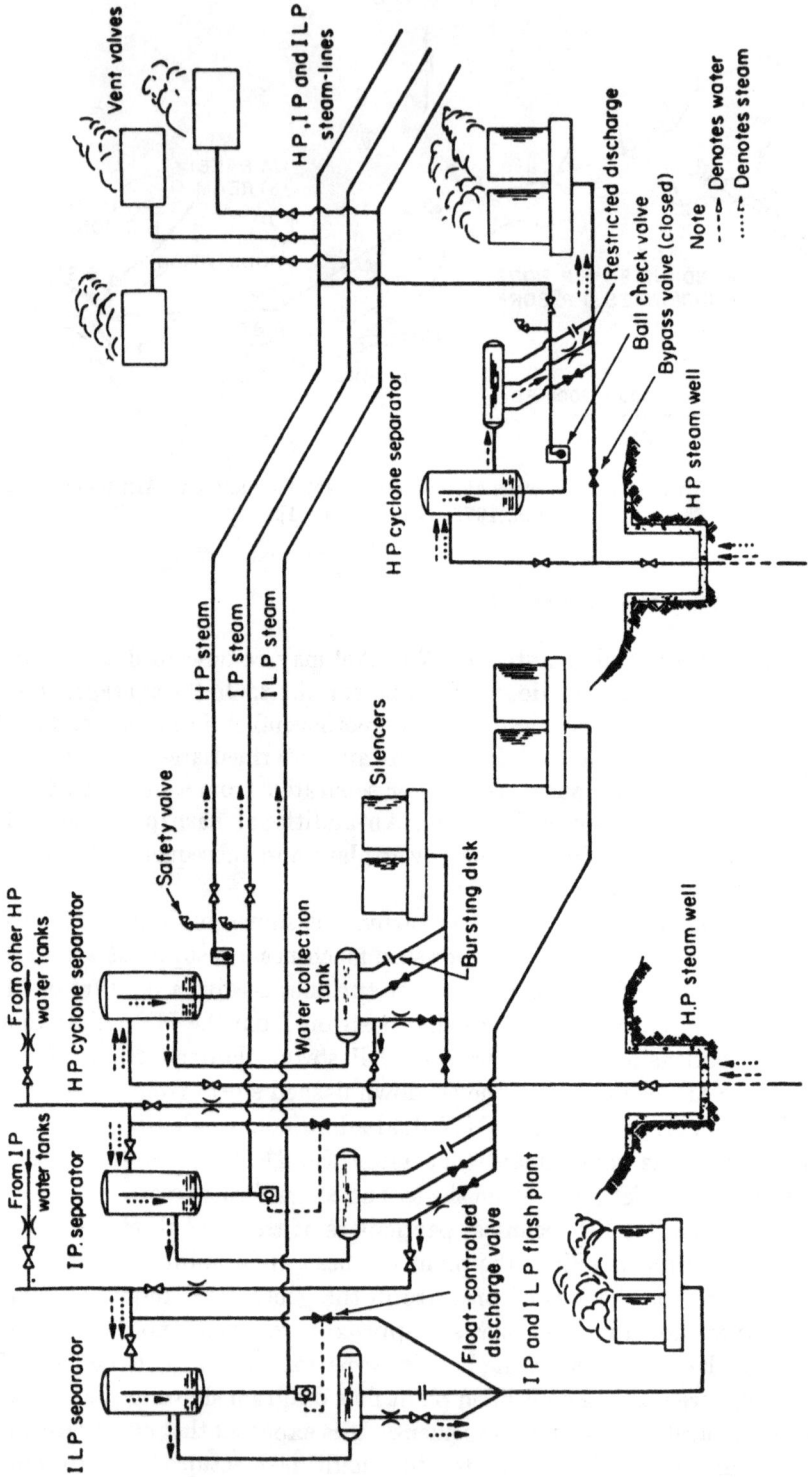

FIGURE 8.11—Steam separation equipment at Wairakei [Bolton, 1977].

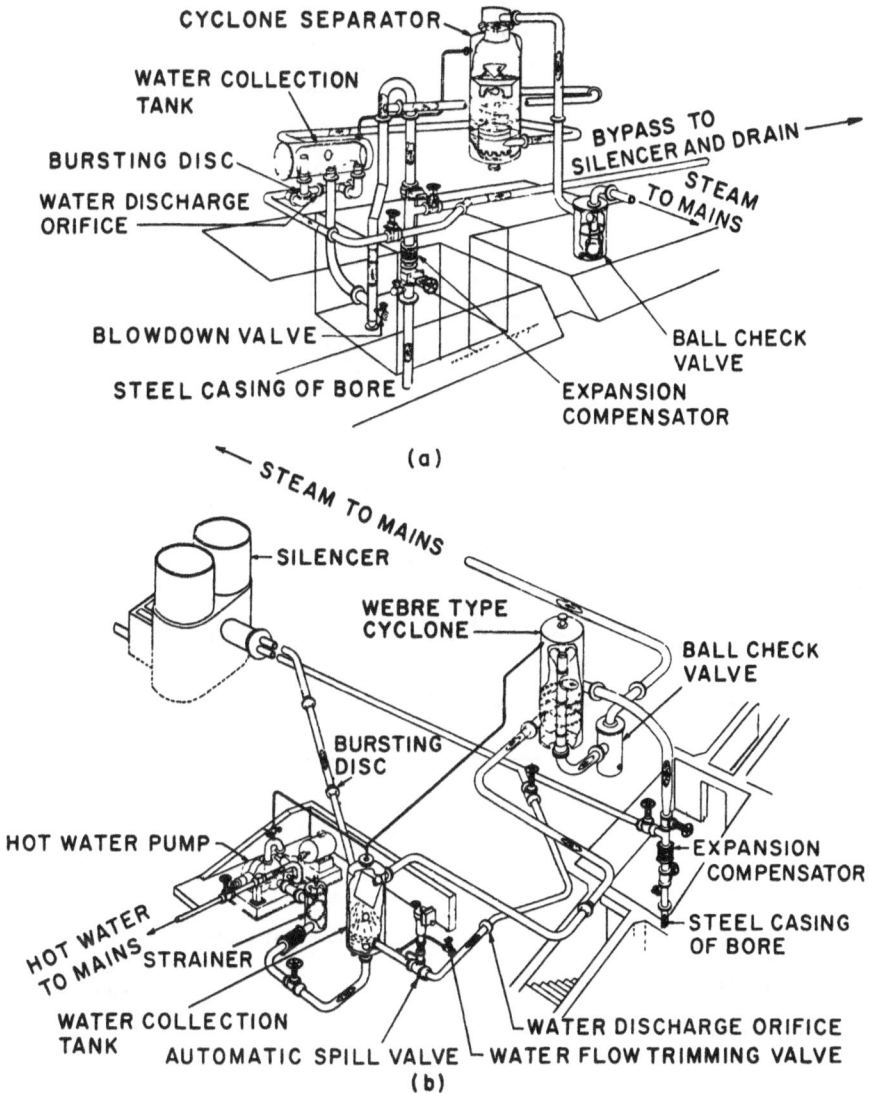

CYCLONE SEPARATOR

WATER COLLECTION
TANK

BURSTING DISC

WATER DISCHARGE
ORIFICE

BYPASS TO
SILENCER AND DRAIN

STEAM
TO MAINS

BLOWDOWN VALVE

STEEL CASING OF BORE

BALL CHECK
VALVE

EXPANSION
COMPENSATOR

(a)

STEAM TO MAINS

SILENCER

WEBRE TYPE
CYCLONE

BALL CHECK
VALVE

BURSTING
DISC

HOT WATER PUMP

EXPANSION
COMPENSATOR

HOT WATER
TO MAINS

STRAINER

STEEL CASING
OF BORE

WATER COLLECTION
TANK

AUTOMATIC SPILL VALVE

WATER DISCHARGE ORIFICE

WATER FLOW TRIMMING VALVE

(b)

FIGURE 8.12—Wellhead steam separation equipment at Wairakei; (a) top-outlet
separator with U-bend; (b) bottom-outlet separator with hot-water pump (pump
no longer used) [Haldane and Armstead, 1962; Bolton, 1977].

seven such stations. Figure 8.15 is a Mollier diagram (enthalpy-entropy
coordinates) showing the expansion portions of the cycle; it is not drawn
to scale and is intended for illustration only. The photograph in figure
8.16 shows several wellheads and associated separators, flash vessels, and
silencers.

The geothermal resource utilization efficiency, η_u, has been calculated
approximately. This factor is the ratio of the actual power delivered by

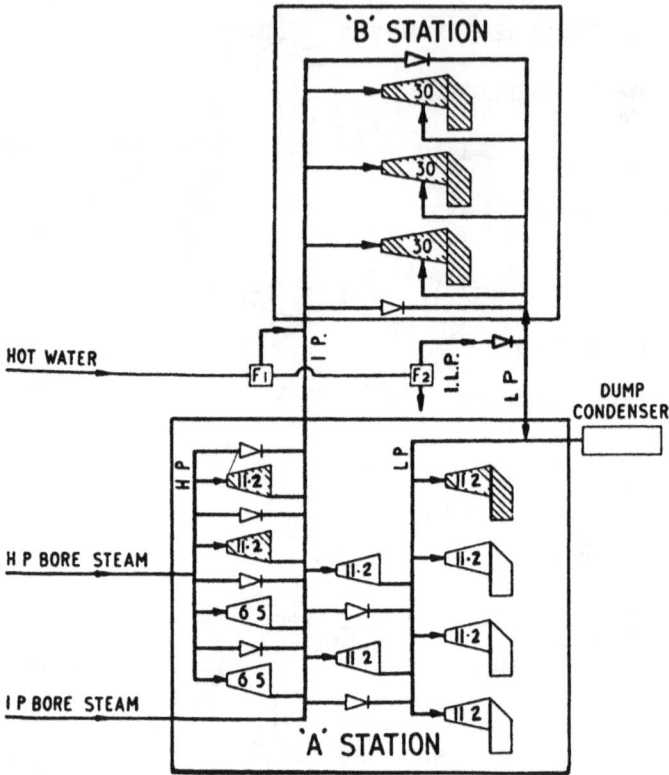

FIGURE 8.13—Turbine arrangements at Wairakei [after Armstead, 1961].

the plant to the ideal power available in the geofluid as it flows from the wellhead. For purposes of the calculation, the actual power was taken equal to the installed capacity of the plant, i.e., 192.6 MW. Actual output at the present time is considerably below this value, and ranges from 140 to 150 MW. Since flow rates will probably also be less than design values, there may be compensating effects on resource utilization efficiency.

Table 8.3—*Turbine specifications for noncondensing units at Wairakei* [a]

	Unit designation [b]		
	HP	HP	IP
Number of units.........................	2	2	2
Year of startup..	1958, 1959	1962	1959
Turbine type....	Single-cylinder, single-flow		
Installed capacity, MW [c]................	6. 5	11. 2	11. 2
Speed, rev/min.......................	3000	3000	3000
Steam inlet pressure, lbf/in².............	132	132	64
Steam inlet temperature, °F..	348. 5	348. 5	297
%(wt) noncondensable gases [d]...........	~0. 5	~0. 5	~0. 4
Exhaust pressure, lbf/in².	~64	~64	15. 5
Steam flow rate, 10³ lbm/h..............	[e] 319	[e] 569	~460

[a] Source: Bolton, 1977.
[b] HP=high pressure; IP=intermediate pressure.
[c] Per unit.
[d] As a percentage of steam.
[e] Original design flow rates when inlet pressure was 194 lbf/in².

Table 8.4—*Technical specifications for condensing power units at Wairakei*

	Unit designation [a]	
	LP	MP
Number of units................................	4	3
Year of startup................................	1959, 1960	1962, 1963
Turbine data:		
Type [b].................................	SCSF	SCSFPI
Installed capacity, MW......................	11.2, each	30, each
Speed, rev/min............................	3000	3000
Main steam pressure, lbf/in².................	15	64
Main steam temperature, °F.................	213	297
Secondary steam pressure, lbf/in².	—	15
Secondary steam temperature, °F.............	—	213
% (wt) noncondensable gases [c]................	~0.4	~0.4
Main steam flow rate, 10³ lbm/h..............	~287, each	~400, each
Secondary steam flow rate, 10³ lbm/h...........	—	~100, each
Condenser data:		
Type.....................................	Low-level direct-contact, barometric condenser	
Cooling water temperature, °F................	59	59
Outlet water temperature, °F:................	[d] 85.1	[d] 85. 1
Cooling water flow rate, 10⁶ lbm/h.............	[d]11.1, each	[d]16.4, each
Gas extraction system:		
Type.....................................	Steam jet ejectors, 3-stage	
Heat rejection system:		
Type................	Once-through, Waikato River	

[a] LP=low pressure; MP=mixed pressure .
[b] SCSF=single-cylinder, single-flow; SCSFPI=single-cylinder, single-flow, pass-in.
[c] As a percentage of steam.
[d] Average values.

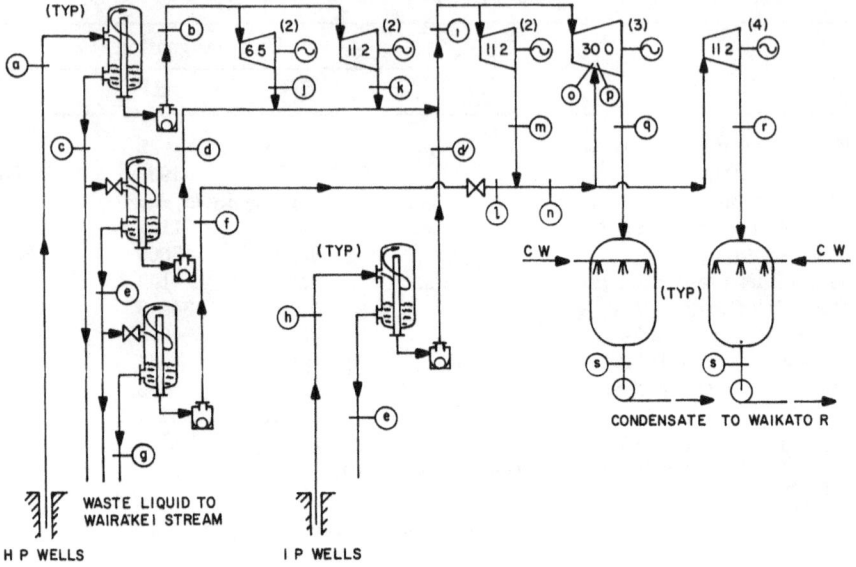

FIGURE 8.14—Simplified flow diagram for Wairakei power plant.

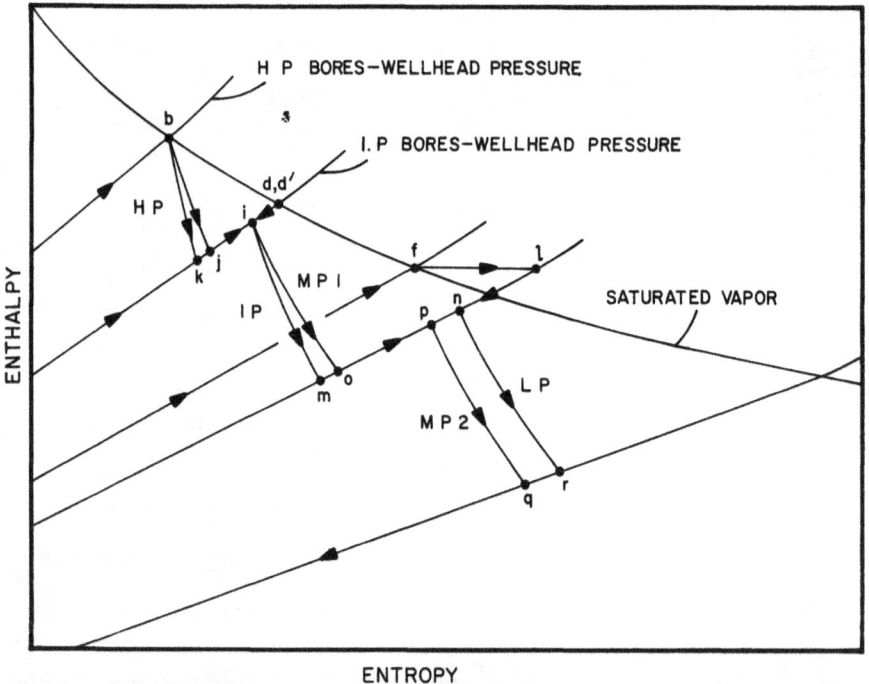

FIGURE 8.15—Mollier diagram showing expansion processes for Wairakei power cycle (not to scale).

FIGURE 8.16—Portion of the steam field at Wairakei showing wellheads, separators, flashers, and steam pipelines [photo courtesy of M. F. Conover].

With reference to figure 8.14, the ideal available power of the geofluid is composed of two contributions, one from the HP fluid at state a and one from the IP fluid at state h. Since mass flow rates were not available in the literature at states a and h, it was necessary to work backwards from the turbine main stop-valve steam flows and employ assumptions about the average dryness fraction at the HP and IP wellheads. Fair approximations for these were obtained from Hunt [1961] and Wigley [1970], making allowances for lower wellhead pressures. Table 8.5 summarizes these calculations.

Table 8.5—*Geothermal resource utilization efficiency calculation* [a]

Mass flow rates, lbm/h:	
Steam flow rate per 6.5-MW HP unit	319, 000
Steam flow rate per 11.2-MW HP unit	569, 000
Total HP steam flow rate	1, 776, 000
Total HP liquid flow rate	[b] 7, 104, 000
Steam flow rate at state d	[c] 62, 660
Steam flow rate per 11.2-MW IP unit	460, 000
Steam flow rate per 30-MW MP unit	[d] 400, 000
Total IP steam at state d	281, 340
Total IP liquid at state h	[e] 422, 010
Enthalpy, Btu/lbm:	
State a	494. 56
State h	540. 34
State o	[f] 28. 06
Entropy, Btu/lbm·°R:	
State a	0. 71704
State h	0. 80170
State o	0. 05550
Availability rate, 10^5 Btu/h:	
State a	1087. 8
State h	110. 0
Total availability entering plant, 10^6 Btu/h	1197. 8
Power output, 10^6 Btu/h	657. 2
Resource utilization efficiency, η_u	54. 9%

[a] See figure 8.14 for state designations.
[b] Dryness fraction at state a assumed to be 20%.
[c] Assumes 15% of HP geofluid is used with multiflash units.
[d] Main steam flow rate.
[e] Dryness fraction at state h assumed to be 40%.
[f] Sink condition; T_o=289 K (520 R).

The resource utilization efficiency of the Wairakei plant under conditions of maximum operation (i.e., in its original design state) is about 55%. This value is likely somewhat, but not much, lower at the present time, owing to the general deterioration of the reservoir characteristics and the corresponding mismatch between the geofluid and the energy conversion equipment. However, conversion of more and more wells to the multiflash arrangement shown in figure 8.14 will tend to maintain the level of utilization in the face of reservoir decline. Interestingly, an earlier report [Wigley, 1970] stated that the plant had an "efficiency" of about 9%, based on the "heat content" of the geofluid. Such an assessment does not take into account the thermodynamic level of the thermal energy contained in the geofield. A more meaningful measure of system performance is now being applied through the use of the so-called Second Law efficiency, i.e., one based on the thermodynamic available work (or availability, or exergy) [Milora and Tester, 1976; Wahl, 1977; Kestin, 1978].

8.2.4 Construction materials

The choice of materials for construction is largely influenced by the composition of the geothermal fluid, including impurities both in the form of dissolved solids and noncondensable gases in the steam. The fluid produced at Wairakei is reasonably "clean" in both regards, as can be seen from the data in table 8.6 [Armstead, 1961]. The total noncondensables reported at that time amounted to about 0.50% and 0.36% (by weight) of the bore steam for the HP and IP wells, respectively. Furthermore, judging from more recent investigations [Glover, 1970], it appears likely that the level

Table 8.6—*Impurities in geothermal fluid at Wairakei* [a]

Noncondensable gases in steam	Concentration, ppm (wt)	
	HP wells [b]	IP wells [c]
Carbon dioxide, CO_2	4857	3467
Hydrogen sulfide, H_2S	132	70
Nitrogen, N_2	7	17
Methane, CH_4	3	5
Hydrogen, H_2	1	1
Total	5000	3560

Constituents in hot liquid	Concentration, ppm (wt)
Chloride, Cl	2318
Silica, SiO_2	300
Metaboric acid, $H_2B_2O_4$	116
Bicarbonate, HCO_3	39
Sulphate, SO_4	34
Fluoride, F	10
pH of condensate	8.6

[a] Source: Armstead, 1961.
[b] For pressure of 1.48 MPa (214 lbf/in²).
[c] For pressure of 0.58 MPa (84 lbf/in²).

of noncondensable gases is presently lower than that quoted by Armstead. In fact, during the period from 1960 to 1969 the total noncondensable gas concentration decreased by 50%, although the hydrogen sulfide fraction remained essentially constant [Axtmann, 1975b]. The total amount of dissolved solids in the hot liquid from the bores is of the order of 3800 ppm (by weight), but increases to about 4150 ppm and 4550 ppm following the first- and second-stage flash vessels, respectively [Haldane and Armstead, 1962]. Since the presence of salinity in the fluid entering the turbine will cause stress-corrosion cracking in the blades, scrubbers are used to ensure that the total saline content of the flash steam does not exceed 10 ppm.

Mild steel is used for wellhead equipment, steam and hot water pipes, and flash vessels. There are no corrosion problems with the geothermal fluid as long as oxygen is avoided. The steam transmission lines are made of seamless steel piping, rolled and butt-welded mild steel pipes, and spiral-welded steel pipe for the largest-sized pipes. The newer, larger pipes are insulated with 38 mm (1.5 in) of fiberglass and covered with aluminum sheathing. The older, smaller pipes have 38-mm (1.5 in) slabs of 85% magnesia that are wired on, covered by bituminous roofing felt and wire netting, and painted with bituminous aluminum [NZED, 1974].

The powerhouses are steel-framed buildings finished in aluminum and glass. Owing to the relatively low load-bearing capability of the ground, it was necessary to implant a massive raft foundation 4.4 m (14.5 ft) deep, containing about 15,000 Mg (33×10^6 lbm) of concrete. Figures 8.17 and 8.18 show, respectively, an aerial view of the power station and an overall view of the plant, including a portion of the steam field.

The condensing steam turbines are particularly susceptible to erosion in the last stages owing to condensation of the steam. Since the blades are made of soft stainless steel the problem of erosion will be intensified. Brazing on erosion shields (e.g., stellite inserts) was ruled out because of the possibility of local hardening and the resulting vulnerability to stress-corrosion cracking. Instead, it was decided that blade tip speed would be kept below 274 m/s (900 ft/s), even though this limits the capacity of the individual condensing turbine units [Haldane and Armstead, 1962]. The photograph in figure 8.19 is a view of "A" station showing the turbine hall that houses Units 1–6 (foreground to background) [NZED, 1974]. The reader may refer to Haldane and Armstead [1962] for detailed plan layouts and elevation views of both powerhouses.

8.2.5 *Environmental effects*

When discussing the effects of the Wairakei geothermal power plant on the environment, it is important to keep in mind that the plant was designed and built at a time when environmental issues were regarded as far less important than they are today. The fact that an environmental impact

FIGURE 8.17—Aerial view of Wairakei power station. Steam lines enter from right, switchyard is at left center, cooling water pumphouse is located behind "A" Station on the Waikato River, with outfall seen to left of the plant [NZED, 1974].

FIGURE 8.18—Overall view of Wairakei power plant. The Waikato River is in foreground, Wairakei Village is behind the power-houses, and a portion of the steam field lies in center background [NZED, 1974].

FIGURE 8.19—Turbine hall in "A" Station at Wairakei. Six turboalternator sets are shown: Unit 1 (foreground), 11.2 MW, IP; Units 2 and 3, 6.5 MW, HP; Unit 4, 11.2 MW, IP; Units 5 and 6, 11.2 MW, HP. The LP manifold is seen at right behind the turbine gauge boards. The LP units are located along right background on the elevated deck [NZED, 1974].

report was not required for the construction of the Wairakei plant stands as evidence. Add to this the facts that Wairakei was the first liquid-dominated geothermal resource to be exploited for electrical power, and that the state of the art in geothermal technology was in its infancy at that time (1958). Although the plant has operated successfully for over 20 years with a minimum of unpleasant impact on the human population living near the plant, there are nevertheless many areas of concern relative to the general environment. A detailed report on the impact of the Wairakei plant on its environment has been published [Axtmann, 1974]; we call the reader's attention to a summary of this study that appeared in the open literature [Axtmann, 1975a]. This section relies heavily on the latter reference.

The following effects will be considered here: chemical effluents in the liquids and gases flowing from the plant, physical effects including thermal discharge and subsidence, general ecological effects, and "visual pollution" or esthetics. The waste liquid from the bore field is transported in open concrete trenches to the Wairakei Stream, which flows into the Waikato River at a point roughly 1 km (0.6 mi) upstream of the power plant. The liquid discharge contains a number of constituents, principally sodium and chloride, but also such potentially harmful substances as arsenic, mercury, hydrogen sulfide, and carbon dioxide. Table 8.7 lists the constituents of the discharged fluid and shows the increase they cause in the concentration of each element in the water of the Waikato River, assuming complete mixing without precipitation or absorption, at average flow conditions of the river, i.e., $127 \text{ m}^3/\text{s}$ $(2 \times 10^6 \text{ gal/min})$.

The arsenic content of the Waikato River has been studied [Reay, 1972], and it was concluded that about 75% of the arsenic in the river comes from the Wairakei plant. According to table 8.7, the concentration of arsenic should be about 0.04 ppm under average conditions, which is just below the allowable concentration for drinking water in the United States (0.050 ppm). However, samples taken at the inlet of the water supply for Wairakei Village have shown arsenic concentrations as high as 0.07 ppm. Furthermore, in periods of drought when the flow in the river is much less than normal, the concentration of arsenic could reach values as high as 0.25 ppm [Axtmann, 1975a].

The mercury found in the Waikato River is partly caused by the Wairakei plant, although several other geothermal areas in the vicinity are contributors. Mercury ores are known to be associated with hot springs, and several areas, including Broadlands, Waiotapu, and Orakeikorako. discharge into the Waikato. Examination of trout taken from the Waikato about 75 km (47 mi) downstream of the power plant revealed about 4.4 times the "normal" concentration of mercury, which is about 0.12 mg/kg of axial muscle tissue (wet weight basis) [Weissberg and Zobel, 1973]. These fish weigh 1.29 kg (9.4 oz) on average, and since the accepted upper limit for mercury in fish for human consumption is 0.50 mg/kg,

Table 8.7—*Effect of liquid discharge from Wairakei power plant
on chemical nature of Waikato River* [a]

Constituent	Increment in river concentration, ppm
Chloride, Cl	20. 8
Sodium, Na	12. 0
Silica, Si	6. 3
Potassium, K	1. 9
Boron, B	0. 27
Sulfate, SO$_4$	0. 24
Calcium, Ca	0. 17
Lithium, Li	0. 13
Fluoride, F	0. 077
Bromide, Br	0. 055
Arsenic, As	0. 039
Rubidium, Rb	0. 029
Cesium, Cs	0. 026
Iodide, I	0. 0047
Ammonium, NH$_4$	0. 0014
Magnesium, Mg	0. 000047
Mercury, Hg	0. 0000015

[a] Source: Axtmann, 1975a: discharge flow rate=6500 Mg/h
$(14.3 \times 10^6 \text{ lbm/h})$; river flow rate=127 m^3/s $(2 \times 10^6 \text{ gal/min})$.

trout weighing more than about 1.25 kg (9.1 oz) might be expected to be unsuitable for human consumption, since mercury concentration is known to be proportional to the weight of the fish. While reinjection of waste liquid has not been practiced at Wairakei, it is presently being considered because of its potential advantages as a disposal technique and as a means of managing the geothermal reservoir [R. S. Bolton, personal communication].

The hydrogen sulfide (H$_2$S) in the geothermal steam becomes divided between the condensate and the noncondensable gas stream, with about 80% of it going into solution. The rate of discharge in the liquid is about 54 kg/h (24 lbm/h). Assuming that no oxidation to sulfur dioxide (SO$_2$) occurs in the direct-contact condensers, the concentration of H$_2$S in the river would be 0.1 ppm (average flow) or 0.9 ppm (lowest flow). Even at average flow conditions, such a concentration exceeds by a factor of about 15 the safe limit for the eggs and fry of rainbow trout. Even though the solubility of carbon dioxide (CO$_2$) in water is quite low at the conditions prevailing at the condenser, nevertheless, about 1 Mg/h (2200 lbm/h) of CO$_2$ is discharged to the Waikato River in the condenser effluent [Axtmann, 1975a].

The condensed geothermal fluid that enters the Waikato contributes to the production of electricity from the system of hydrothermal power stations located along the river. This may be regarded as a positive or compensating environmental effect. The total rate of liquid discharged from the plant is 1.3 m³/s (21,000 gal/min); the enhanced evaporation rate (owing to the thermal effect) is about 0.3 m³/s (4800 gal/min), leaving a net increase of 1 m³/s (16,200 gal/min) in the flow of water in the Waikato. This is equivalent to about 2.4 MW of electrical power from the hydroelectric plants, or the generation of 20 GW·h of electricity per year.

The important gaseous effluents consist of the following: H_2S, CO_2, and water vapor; the first of these constitutes the most serious problem, owing to its hazardous nature. Hydrogen sulfide gas is easily detected by the human olfactory sense at extremely low concentrations (i.e., 0.020 ppm threshold). It causes eye irritation at 10 ppm, lung irritation at 20 ppm, and death in 30 minutes at 30 ppm [Miner, 1969]. Roughly 14 kg/h (6.4 lbm/h) or 93 g/MW·h (0.2 lbm/MW·h) are discharged into the atmosphere from the gas ejectors through four stacks on the roof of the plant. The tops of the stacks are 30 m (98 ft) above ground level; concentration of H_2S in the stack gas is about 5000 ppm. The odor of H_2S is not detectable in Wairakei Village, about 2 km (1.2 mi) north of the plant, although there have been some reports of the blackening of silverware and brass [Axtmann, 1975a]. The gaseous emission of CO_2 at Wairakei is less by a factor of about 60 than the emission from a conventional fossil-fuel-burning power plant per megawatt of electricity produced.

The amount of water vapor released to the atmosphere during a total plant shutdown (i.e., all wells venting to the atmosphere) would be equivalent to the amount discharged from the wet cooling towers of a 750 MW conventional power plant, i.e., 1.9 Gg/h (4.2×10^6 lbm/h). Under normal circumstances, however, the amount is 0.84 Mg/h (1.0×10^3 lbm/h), or about 0.04% of the maximum possible discharge. The presence of large amounts of water vapor can, of course, lead to the formation of fog, which is frequently seen in the area around the power plant.

The physical effects of the Wairakei plant on the environment are related to the drawoff of vast quantities of hot subsurface fluid and to the discharge of these fluids at the surface (i.e., without reinjection into the reservoir). As a result, an area of about 6500 ha (16,000 acres) shows the effects of subsidence and horizontal land movement. The maximum total drop in elevation is in excess of 4.5 m (14.8 ft) over the 10-year period from 1964 to 1974 at a spot removed from the bore field but within about 500 m (1640 ft) of the steam pipelines [Stilwell et al., 1975]. Subsidence appears to be progressing at the rate of 400 mm/y (16 in/y); subsidence volume is likely related to the volume of fluid withdraw from the field, but a precise correlation is not available. A study of subsidence from 1967 to 1971 by Glover showed that V (subsidence)/V (drawoff) =0.0076, on average [Axtmann, 1975a]. Figure 8.20 shows the area of ground subsidence relative to a bench mark, TH7, in the powerhouse (a), and the subsidence that has occurred along the main steam pipelines (b) for the period 1964–1974 [Stilwell et al., 1975]. Also shown in figure 8.20 (b) is the depth

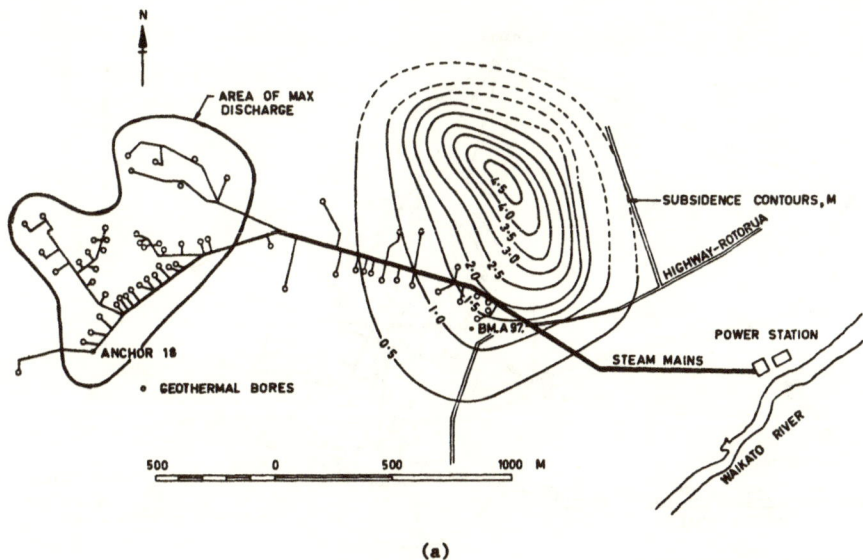

(a)

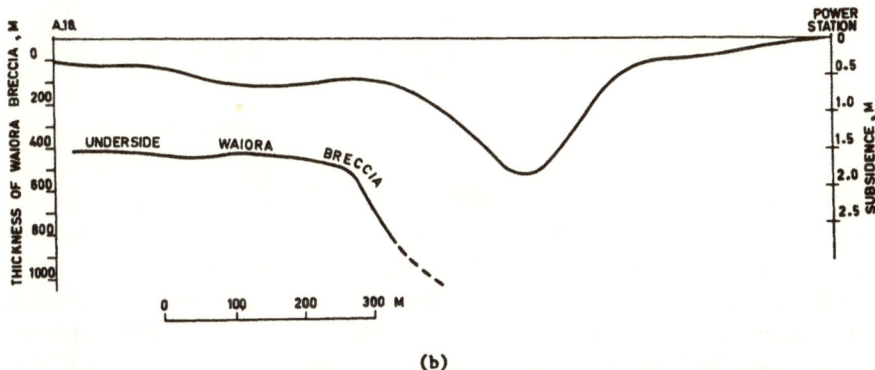

(b)

FIGURE 8.20—(a) Total subsidence at Wairakei (relative to BM TH7 in the powerhouse) for the period 1964–1974; (b) Total subsidence and Waiora breccia thickness
measured along the main steam pipelines from 1964–1974 [Stilwell et al., 1975].

of the bottom of the underlying Waiora breccia illustrated earlier in figure
8.3(b). The most drastic subsidence corresponds to the region in which the
Waiora breccia falls off along a buried fault scarp. From about 1957 to
1965 there appeared to be a correlation between pressure loss in the reservoir and field subsidence. Since that time, however, pressure has tended
to level off (see figure 8.4) while subsidence has tended to increase.

Horizontal ground movement has been traced since 1966; figure 8.21
shows a vector diagram of the movement of a number of ground stations

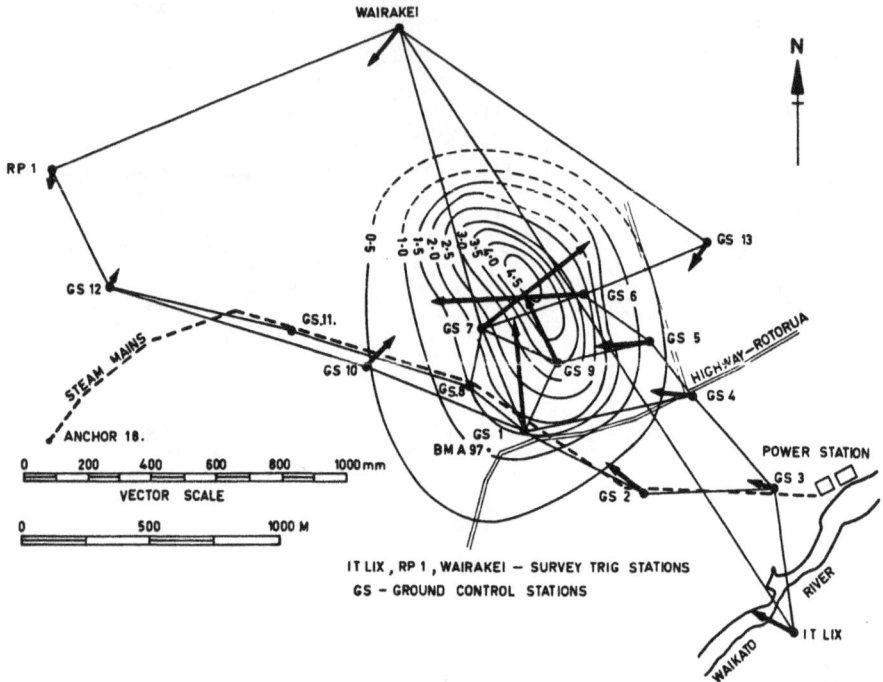

Figure 8.21—Vector diagram of horizontal ground movement at Wairakei from 1966 to 1974 [Stilwell et al., 1975].

in the field. All movements tend to be toward the region of maximum vertical displacement, with GS6, for example, showing a displacement of nearly 470 mm (18.5 in) in a westerly direction over the 8-year from 1966 to 1974. In light of the amount of both vertical and horizontal ground movement in the Wairakei field, it is indeed fortunate that the power-house was sited along the Waikato River instead of near the middle of the bore field.

The so-called waste heat that is rejected from the plant through residual liquid geofluid from the wells and condensate from the powerhouse in-fluences the temperature of the Waikato River. Liquid flows from the bore field at about 60°C (140°F) and 1.3 m³/s (21,000 gal/min), whereas con-densate flows at roughly 33°C (91°F) and 10 m³/s (159,000 gal/min). The maximum increase in the temperature of the Waikato would be 1.3°C (0.7°F) when the river is flowing at its mean value of 127 m³/s (2×10⁶ gal/min) [Axtmann, 1975a]. This temperature change is far less than natural temperature swings during the year. However, during periods of extreme drought when the flow rate in the river is abnormally low, e.g., 28 m³/s (444,000 gal/min) as was the case during the spring of 1974, it is possible for the power plant to produce significant temperature elevations, in excess of what is permitted under New Zealand's water standards.

Baseline (i.e., pre-plant) ecological data does not exist at Wairakei since an environmental impact statement was not required when the plant was constructed. Axtmann suggests, however, that the portion of the Waikato River lying between Wairakei Stream and the Aratiatia Dam constitutes a severely stressed ecosystem [Axtmann, 1975a]. Several reasons for this conclusion are given, among them the observation that a lack of diversity exists among the various species of plankton in this part of the river.

The judgment of the esthetics of a power plant is to some degree subjective. In the view of Axtmann, the Wairkei plant lies on a scale somewhere between a typical fossil-fuel plant ("visual abomination") and a typical nuclear plant (somewhat resembling a planetarium when imaginatively designed). His description of the general scene at Wairakei [Axtmann, 1975a] is worthy of full quotation:

> "... the Wairakei borefield ranks high in New Zealand's superb hierarchy of visual delights. If a tramper on Highway 1 were to pause at dusk 8 km north of Taupo on a moist day with a stiff breeze, he would see an eerie scene of haunting beauty. Scores of fleecy plumes arc skyward only to be seized and devoured by green demons that haunt the boughs of imperial conifers; bundles of silvery bullwhips, cracked by an invisible giant who lurks behind the western hill, are caught in stop-action as they rise and fall in unison. It is an odd amalgam of technology and nature, of the Tin Woodsman of Oz and the Sorcerer's Apprentice, gently underscored by the whispering, slightly syncopated 'whuff-whuff ... whuff ... whuff' of the wellhead silencers."

8.2.6 Economic factors

Total capital investment at the Wairakei geothermal power plant as of March 31, 1969 was $NZ43,367,000. This included $NZ1,300,000 for exploration, $NZ17,176,000 for exploitation, and $NZ24,891,000 for utilization. Based on an installed capacity of 192.6 MW, the capital cost works out to about $NZ225/kW [Smith and McKenzie, 1970]. Table 8.8 lists a detailed breakdown of capital expenditures. Average annual operating costs over the 4-year period from 1965 to 1969 amounted to $NZ5,778,000. This is about 4.8 NZ mill/kW·h of electricity generation, on the average. Table 8.9 gives average operating costs for the project, including those for the bore field and for the power station. In addition to the working expenses shown in table 8.9, charges for the use of capital and administrative costs contribute to the total O & M expenses: total working expenses, $NZ1,210,-200; capital charges, $NZ239,000; administrative charges, $NZ328,500. This amounts to total operating and maintenance charges of $NZ5,777,900.

Table 8.8—*Capital costs for the Wairakei geothermal power station* [a]

Item	Cost $1000 NZ
Land acquisition	nil
Site preparation	278
Establishment items (camps, temporary housing, workshops, offices, site leveling)	1, 688
Permanent Village (Wairakei Village)	1, 318
Power stations:	
Foundations, cooling water culverts, offices, workshops, roads and landscaping	4, 674
Cooling water pumphouse (pumps, mains, valves)	1, 881
Turbines and generators	8, 874
Steam piping and valves	967
Electrical work (400-V and 11-kV switchgear, panels, control room, transformers other than generator transformers	840
Generator transformers	576
Outdoor structure and switchgear, high-tension circuit breakers	662
General services (electrical, air, telephones, etc.)	115
Workshop tools	18
Bore field:	
Well drilling (including exploration and unproductive wells)	7, 643
Wellhead equipment	1, 638
Branch pipelines	971
Main pipelines	5, 258
Roads	462
Main drainage	1, 641
Water supply	446
Landscaping	405
General services	17
Pilot hot water scheme (later abandoned):	
Modifications to wellhead equipment	213
Hot water pipeline	732
Flash plant and controls	925
No. 10 turboalternator	591
Work for plant expansion (not completed)	293
Miscellaneous charges and adjustments	241
Total	43, 367

[a] Source: Smith and McKenzie, 1970.

8.2.7 Operating experience

The Wairakei plant has an outstanding record of reliability, with forced outages essentially negligible. During 1973/1974 the station was in service 85% of the time (availability factor) with an installed capacity factor of 69% and a station load factor of 89%. This performance was unmatched by any other power station, hydro or thermal, and was significantly superior to any thermal power plant in New Zealand. In fact, the Wairakei

Table 8.9—*Average annual operating and maintenance expenses at Wairakei, 1965–1969* [a]

Item	Cost $1000 NZ
Bore field:	
Mineral rights	nil
Land and roads	9. 4
Services	6. 9
Hot water drains	16. 8
Well servicing and modifications	84. 8
New wells (to maintain production)	295. 4
Measurements	60. 3
Main pipeline servicing and modifications	53. 7
New pipelines	30. 9
Branch pipelines	15. 0
Steam traps	7. 6
Buildings (in steamfield)	0. 5
Hot water equipment	1. 0
Mechanical equipment	2. 1
General workshop	2. 6
Salaries of operating personnel	25. 5
Supervision	8. 2
Miscellaneous	1. 0
Bore field total	621. 7
Power Station:	
Land and roads	9. 8
Services	3. 8
Buildings (power stations)	19. 1
Permanent Village	22. 6
Information Center (3 years)	6. 2
Ministry of Works buildings (3 years)	65. 0
Cooling water system	12. 6
Gas extractor system	7. 1
Steam lines	36. 5
Turboalternators	108. 0
Electrical equipment	25. 0
Miscellaneous mechanical equipment	22. 0
Operating costs (salaries plus other charges)	194. 6
Supervision	12. 6
Miscellaneous	43. 6
Power station total	588. 5
Total working expenses	1, 210. 2

[a] Source: Smith and McKenzie. 1970.

geothermal power station has maintained this excellent record since it was fully commissioned in 1964 [Ravenholt, 1977]. The generating history of the Wairakei plant is given in table 8.10. Since it has never been possible to generate sufficient geothermal steam to supply fully the installed electrical capacity of 192.6 MW, the capacity factors listed in the last column of the table have been adjusted accordingly. The so-called field-limited capacity factor is based on the maximum load during any given year, as shown in the third column.

In 1969 the Wairakei plant delivered 9.9% of the entire electrical generation in New Zealand for that year. The plant produced about 10% of the electricity requirements of North Island in 1974, although this percentage is expected to fall as the electrical generating capacity of the country as a whole increases [Bolton, 1977]. For the period 1970–1978, the capacity factor (based on peak load) at Wairakei has averaged 90%.

Table 8.10—*Electricity generation at Wairakei* [a]

| Year [b] | Net generation MW·h | Peak load MW | Capacity factors, percent | |
			Installed [c]	Field-limited [d]
1959	5, 527	[e] 5. 1	25. 9	33. 0
1960	168, 589	43. 3	48. 7	44. 3
1961	384, 088	93. 8	58. 4	46. 7
1962	491, 412	107. 1	69. 5	52. 4
1963	761, 355	147. 0	70. 1	59. 1
1964	1, 004, 909	173. 0	64. 6	66. 1
1965	1, 193, 583	173. 0	70. 7	78. 7
1966	1, 255, 424	166. 0	74. 4	86. 3
1967	1, 267, 553	170. 8	75. 1	84. 7
1968	[f] 1, 057, 622	166. 8	78. 5	72. 2
1969	1, 206, 568	165. 7	71. 5	83. 1
1970	1, 243, 057	159. 8	73. 7	88. 8
1971	1, 185, 210	153. 4	70. 2	88. 1
1972	1, 174, 364	149. 3	69. 4	89. 5
1973	1, 174, 812	147. 7	69. 6	90. 7
1974	1, 161, 892	148. 6	68. 9	89. 3
1975	1, 248, 603	158. 0	74. 0	90. 2
1976	1, 272, 137	158. 1	75. 2	91. 6
1977	1, 232, 526	152. 9	73. 0	92. 0
1978	1, 157, 949	147. 5	68. 6	89. 6

[a] Source: Smith and McKenzie, 1970; Birsic, 1978; R. S. Bolton, personal communication.
[b] For the year ended March 31.
[c] Based on installed capacity, i.e., 6.5 MW for 137 days in 1959, 39.4 MW for 1960, 75.1 MW for 1961, 80.8 MW for 1962, 123.9 MW for 1963, 177.1 MW for 1964, and 192.6 MW thereafter, except 153.4 MW for 1968.
[d] Based on a bore-field-limited effective maximum capacity, i.e., based on peak station load.
[e] Test runs.
[f] Low generation caused by cutback in geofluid flow rate for reservoir pressure recovery test over four months during which the maximum plant capacity was 75 MW.

On average 150–160 people are involved in the various phases of station operation, including administration (8), operation of the powerhouse (43), operation of the bore field (8), maintenance of the powerhouse (50), maintenance of the bore field (40), and operations related to the Permanent Village and services (12) [Smith and McKenzie, 1970].

The most serious accident associated with operation of the plant occurred in January 1960 when Bore 26 exploded, creating a violently steaming crater and several smaller steam jets and boiling mud pools. The blowing well was eventually brought under control through the drilling of another well, Bore 26A, spudded 61 m (200 ft) away from Bore 26. The new well was sunk with a deviated bore and passed within 1.2 m (4 ft) of Bore 26 at a depth of 453 m (1486 ft). Figure 8.22 shows the deviated bore in its programmed and actual form. The operation was supervised by an American drilling team. In order to secure the foundation of the new well, it was necessary to grout an area of 9.186 ha ((0.46 acres) by means of 250 holes of 30-m (100-ft) depth, into which was poured 1500 Mg (3.3×10^6 lbm) of cement. After Bore 26 had been intersected in the open hole below the production casing (November 1960), large amounts of 1762 kg/m^3 (110 lbm/ft^3) cement slurry were pumped down Bore 26A, to no avail. Finally about 7.6 m^3 (10 yd^3) of pea gravel were poured in, followed by another large batch of cement, and the violent thermal activity came to an abrupt halt. Restoration of the landscape was completed, and the two highly productive wells, Bore 26A (extended) and Bore 26B, now occupy the area [NZED, 1974; Craig, 1961].

There are at present no plans to expand the installed capacity at Wairakei [N. C. McLeod, personal communication], although exploration is continuing at the nearby geothermal field of Te Mihi [Smith and McKenzie, 1970] and at other promising sites in the thermal belt. Additional geothermal power in New Zealand will likely come from new plants at such sites as Kawerau and Ohaki (Broadlands), which are discussed in the following sections.

8.3 Kawerau

8.3.1 Geology

Multiple use is being made of the geothermal resource at Kawerau, 97 km (60 mi) northeast of Wairakei. The Tasman Pulp and Paper Company, in fact, relocated their mills in the early 1950's specifically to take advantage of the geothermal energy available at Kawerau. Steam and hot water from a number of wells are used for production of electricity, for generation of clean steam by means of heat exchangers, and for a number of process applications including timber drying, recovery boiler shatter sprays, liquor heaters, and log handling equipment. The geothermal steam field layout including the pulp and paper mill is shown in figure 8.23. The most active surface manifestations of geothermal energy lie to the west of

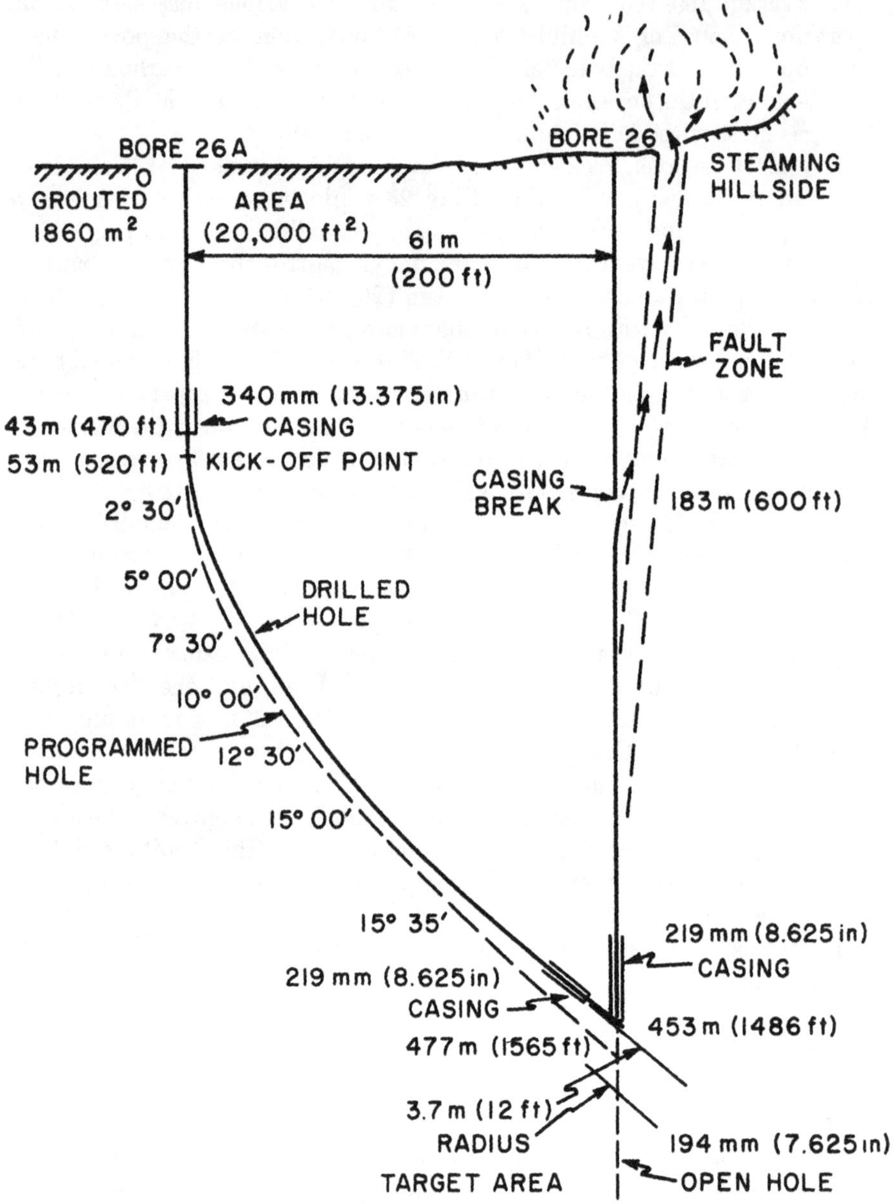

FIGURE 8.22—Drilling of deviated Bore 26A to control blowout well Bore 26 [after NZED, 1974; Craig, 1961].

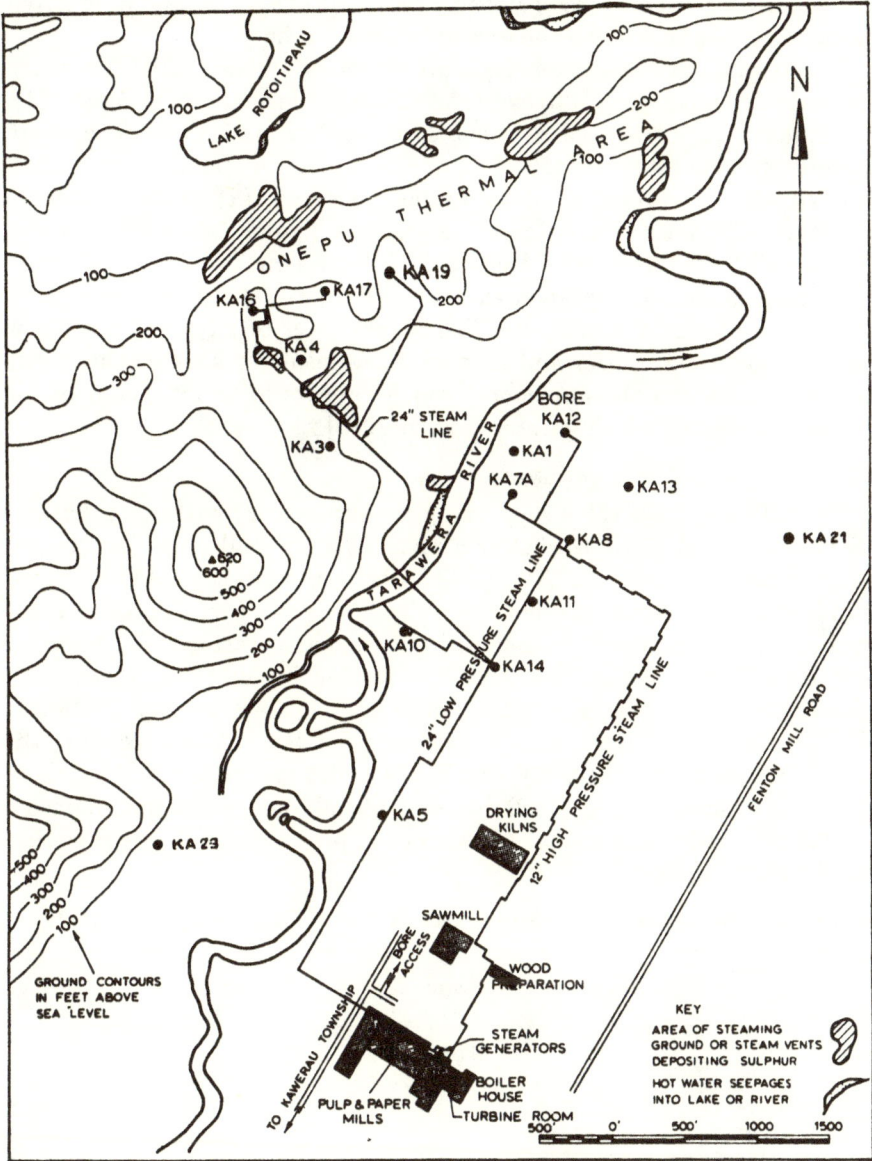

FIGURE 8.23—Steam field layout at Kawerau [after Smith, 1970; Bolton, 1977]

the Tarawera River (crosshatched areas), although highly productive wells have been drilled on the east bank close to the mill.

The region possesses a top layer of recent alluvium (which probably conceals a number of faults) with the following formations at depth: (1) a shallow layer of breccia to about 75 m (250 ft); (2) a thick layer of rhyolite to about 400 m (1300 ft); (3) an aquifer of pumice breccia imbedded with sandstones to about 730 m (2400 ft); (4) another layer of rhyolite; (5) another aquifer of extensively fissured andesite to about 820 m (2700 ft); and (6) a basement of dense ignimbrite. Several wells have been drilled to depths in excess of 915 m (3000 ft) and have penetrated the basement ignimbrite. The deeper layer of rhyolite serves as a cap for the lower aquifer and may act as an insulator between the lower, hotter aquifer and the upper one, which appears to be influenced by an influx of cooler circulating water [Smith, 1970].

8.3.2   Steam wells and gathering systems

From 1952 to 1955 three small-diameter exploratory wells were drilled: KA1, KA4, and KA5, the latter a nonproducing well owing to the low temperature encountered, 85°C (185°F) bottom-hole. Table 8.11 lists geofluid production from the other two wells during this period. Both of these wells were fitted with 102–mm (4-in) production casings; KA1 ran to 449 m (1473 ft) and KA4 to 499 m (1636 ft). During 1956/1957 seven additional wells were sunk: KA7 at 605 m (1985 ft), KA8 to 585 m (1918 ft), KA10 to 634 m (2080 ft), KA11 to 624 m (2046 ft), KA12 to 622 m (2042 ft), KA13 to 603 m (1979 ft), and KA14 to 615 m (2018 ft). All of these had a 311-mm-diameter (12.25-in) drilled hole and a 219-mm-diameter (8.625-in) production casing. All were good producers, with the exception of KA13.

Table 8.11—*Production characteristics of wells KA1 and KA4 at Kawerau, 1952–1955* [a]

Wellhead pressure lbf/in²	Steam flow 10³ lbm/h	Liquid flow 10³ lbm/h	Total flow 10³ lbm/h	Quality percent
	Well KA1			
64. 7	18. 0	20. 0	38. 0	47. 3
74. 7	17. 0	20. 9	37. 9	44. 8
94. 7	14. 7	22. 2	36. 9	39. 9
114. 7	12. 4	22. 6	35. 0	35. 4
134. 7	10. 2	21. 3	31. 5	32. 4
154. 7	8. 0	18. 7	26. 7	30. 0
174. 7	5. 8	14. 5	20. 3	28. 6
	Well KA4			
62. 7	8. 0	17. 0	25. 0	32. 0
74. 7	8. 0	18. 0	26. 0	30. 8
84. 7	7. 0	17. 0	24. 0	29. 2
94. 7	6. 0	16. 0	22. 0	27. 3
104. 7	5. 0	15. 0	20. 0	25. 0

[a] Source: Smith, 1970.

From 1958 to 1960 these wells deteriorated considerably. Several (KA11, 12 and 14) developed calcite deposits and required reaming, after which the flow rates were still poor. Significant declines in temperature were observed, ranging from 6° to 20°C (11° to 36°F), caused by an invasion of cooler water in the production zones. Another problem was caused by probable clogging of the slots in the production casing from calciting, together with the fact that the slotted liners were installed as a single string cemented to the surface above the slots, thus making it impossible to remove them for cleaning.

During the period 1961–1969 several wells were deepened, increasing their depths by 221–374 m (726–1227 ft), and three new wells were sunk; KA3 to 1092 m (3584 ft), KA16 to 972 m (3189 ft), and KA17 to 1033 m (3388 ft). Each of these consists of a 219-mm (8.625-in) production casing, a 194-mm-diameter (7.265-in) open hole, and a 168-mm-diameter (6.625-in) slotted liner. Each slotted liner has 52 slots/m (16/ft), with each slot being 19×51 mm (0.75×2 in). A few of the wells have become nonproductive either from calciting or low aquifer permeability. As of late 1969 the following four wells were producing steam at a pressure of 791 kPa (114.7 lbf/in²) : KA7A—13.6 Mg/h (30 klbm/h), KA14—10.9 Mg/h (24 klbm/h), KA16—56.7 Mg/h (125 klbm/h), and KA17—13.6 Mg/h (30 klbm/h). Well KA8 delivered 59.0 Mg/h (130 klbm/h) of steam at a pressure of 1480 kPa (214.7 lbf/in²) ; an additional 13.6 Mg/h (30 klbm/h) of steam was flashed from the liquid of KA8 at a pressure of 791 kPa (114.7 lbf/in²). With the abrupt escalation of world petroleum prices in 1973/ 1974, renewed interest in the development of the Kawerau field led to the drilling of two more wells, KA21 and 22.

Noncondensable gases constitute about 2.5% (by weight) of the geothermal steam; about 91% is carbon dioxide, with the rest mainly hydrogen sulfide. The gathering system includes branch pipes from the individual wellheads to the main steam lines; these branch lines are 203, 305, and 406 mm (8, 12, and 16 in) in diameter. The low-pressure wells feed the plant through a 610-mm-diameter (24-in) steam main capable of handling 145 Mg/h (320 klbm/h) of steam at 791 kPa (114.7 lbf/in²). The high-pressure well (KA8) delivers through a 305-mm (12-in) supply line, at a maximum flow rate of 36 Mg/h (80 klbm/h) and a pressure of 1480 kPa (214.7 lbf/in²). New steam lines are insulated with 51 mm (2 in) of resin-bonded fiberglass; older pipes have 51 mm (2 in) of asbestos and 85% magnesia. All are protected with a covering of aluminum sheet [Smith, 1970].

8.3.3 Energy conversion system

Details about the nonelectrical use of the geothermal fluid may be found elsewhere [Bolton, 1975, 1977; Lindal, 1973; Smith, 1970; Stilwell et al., 1975]. It is sufficient to note that geothermal energy contributes about 21% of the energy required for process applications, the rest being supplied by the burning of waste products and fuel oil. The plant purchases 80% of its electricity from the NZED grid and produces the other 20% in-house. A bank of turboalternators operate in parallel, fed by boiler steam and geothermal steam. The latter supplies one 10-MW noncondensing unit; table 8.12 gives its technical specifications. Since the steam that supplies this unit is excess geothermal steam beyond the process needs of the plant, this unit is part-loaded most of the time. Nevertheless, it is capable of operating at full load in the event of failure of the other turboalternator units. At full output it has a specific steam consumption of about 14.5 kg/kW · h (32 lbm/kW · h). This corresponds to a geofluid utilization efficiency for electrical production of about 24%, assuming a wellhead quality of 30% and taking the available sink temperature as 27°C (80°F).

Table 8.12—*Turbine specifications for Kawerau geothermal plant* [a]

Turbine data....	Single-cylinder, single-flow, noncondensing
Installed capacity................ ...	10 MW
Speed.....................	3000 rev/min
Steam inlet pressure.......	114.7 lbf/in²
Steam inlet temperature........... .	337.9°F
Noncondensable gas content..........	<2.5% wt of steam
Exhaust pressure........	~30 in Hg
Steam flow rate [b]...................	319×10³ lbm/h

[a] Year of start-up: 1961
[b] Up to 100×10³ lbm/h of exhaust steam is condensed in a black liquid pre-evaporator.

8.3.4   Environmental considerations and future plans

No detailed environmental study exists for this plant. A level network to keep track of subsidence was established in 1970, and checks made in 1972 revealed a subsidence rate of 15 mm/yr (0.6 in/yr) in the area of maximum fluid drawoff. However, since the mill is located near the center of the field, the effects of subsidence will unfortunately be largest at the mill. The rate of subsidence there is 28 mm/yr (1.1 in/yr). Differential settling throughout the plant may lead to trouble with the operating equipment and may necessarily limit the ultimate exploitation of the field by restricting the rate of fluid withdrawal [Stilwell et al., 1975]. Reinjection of waste fluid has not been reported; excess liquid is presumably dumped in the adjacent Tarawera River, which empties into the Bay of Plenty about 20 km (12 mi) northeast of the plant.

Although the main thrust of the plan for geothermal energy utilization at Kawerau has been aimed at process heating and other industrial applications, serious consideration will likely be paid to expansion of the facility for generation of electricity. Encouragement comes from the fact that one of the newest wells, KA21 (see figure 8.23), by itself appears capable of supporting a 20–30-MW generator. This is four to six times larger than the potential of an average geothermal well. It is expected that a separate generating station, one able to supply all the electrical needs of the mill, will be constructed at the site as soon as present investigations justify the additional investment [Ravenholt, 1977].

8.4 Ohaki (Broadlands)

The New Zealand Electricity Department is proposing to construct a 150-MW geothermal power plant near Broadlands, about 28 km (17 mi) northeast of Taupo and Wairakei. The plant will be situated adjacent to the Waikato River and will be supplied with geothermal steam from the liquid-dominated reservoirs at Ohaki and Broadlands. It is expected that the plant will be built in three phases, with each phase culminating in the installation of a 55-MW (gross)/50-MW (net) turboalternator unit. As of early 1979 statutory clearances were being obtained and preliminary design work was underway. Authorization to proceed with the first 100 MW was expected to be granted by the end of 1979 or early 1980. The first 50-MW unit may be in operation by late 1983, while the second may be commissioned during 1984. The third unit is expected to come on-line in the mid-1980's. In the following sections we shall discuss the geology of the Ohaki-Broadlands thermal area, the drilling program and productivity of the wells, the proposed energy conversion scheme, and some of the anticipated environmental problems and suggested solutions.

8.4.1 Geology

The Broadlands geothermal field has been studied intensively and several reports on the geology of the site are available [Browne, 1970; Grindley, 1970; Grindley and Browne, 1975; Hochstein and Hunt, 1970; Macdonald, 1975]. The drilling of exploration wells began in 1965 with well BR1 and has continued since that time. Figure 8.24 [after Bauer et al., 1977] shows the general layout of the field including the existing wells, both productive and nonproductive, the proposed steam pipelines, and a possible position for the powerhouse and cooling towers. The cross-section line, A-A, is also indicated. The geologic cross-section through the field along this line is drawn in figure 8.25, in which some of the wells are also shown [Grindley, 1970].

There are three surface areas where the ground temperature is at least 5°C (9°F) above the ambient at a depth of 1 m (3.3 ft). These are: (1) an area of roughly 365-m (1200-ft) radius centered on well BR7 (the so-called Broadlands thermal anomaly) ; (2) an area of roughly 550-m (1800-ft) radius centered on well BR9 (the so-called Ohaki thermal anomaly) ; and (3) an elongated area lying between wells BR6 and BR13 and extending

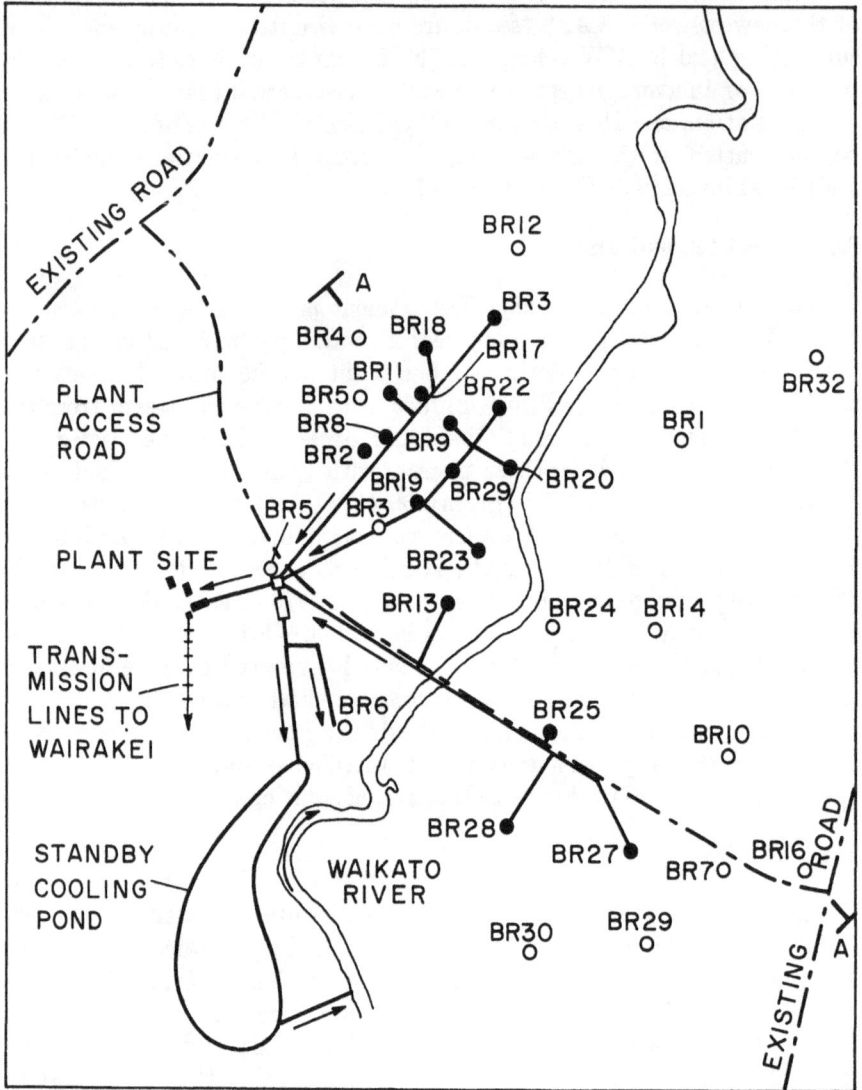

FIGURE 8.24—Preliminary layout of the Broadlands geothermal power plant, steam field, and gathering system. Key: ●=producing well; ○=nonproducing well [Bauer et al., 1977].

about 1220 m (4000 ft) in a north-south direction [Smith, 1970]. The first two thermal areas coincide with the Broadlands and Ohaki faults, respectively (see figure 8.25). Of the two, the Ohaki area is by far the more productive.

The complex geologic formations shown in Figure 8.25 may be described as follows, with main emphasis on the hydrological functions of

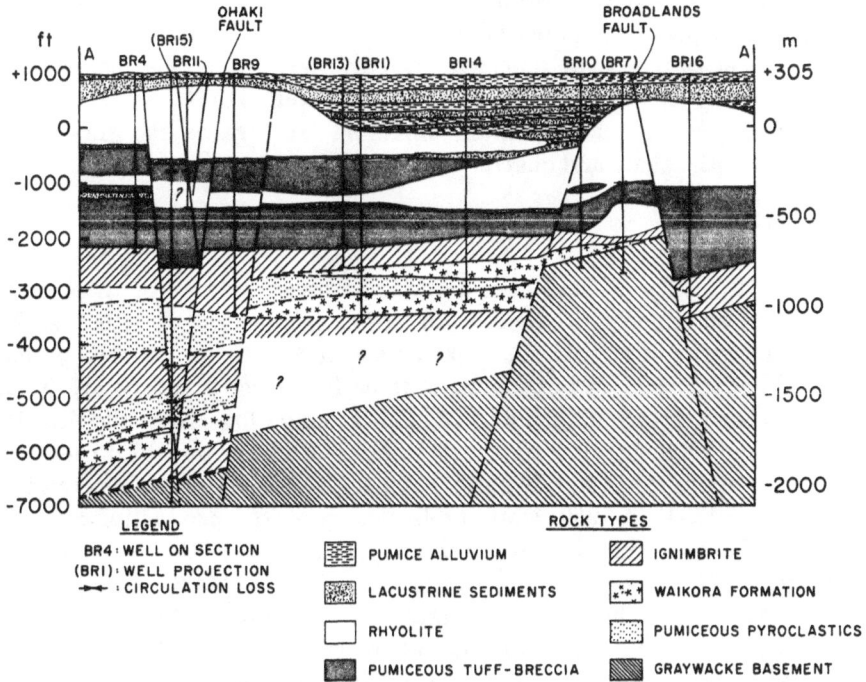

FIGURE 8.25—Geologic cross-section through the Broadlands geothermal field with Ohaki thermal area at left and Broadlands thermal area at right [after Browne, 1970; Grindley, 1970].

the several layers. The surface layers consist basically of recent pumice alluvium. The Huka Falls is made up of lacustrine sediments, tuffs, and grits, and serves as a cap rock for the reservoir. The Ohaki rhyolite consists of pumiceous and spherulitic rhyolite with underlying mudstone layer to form a partial cap. The Waiora formation is an aquifer of pumiceous tuff-breccia, the shallowest useful aquifer in the Broadlands geothermal region. The thickness of the permeable layer is about 60–200 m (200–660 ft), much thinner than the corresponding area at Wairakei. It should be noted also (see figure 8.25) that this layer becomes vanishingly thin toward the southeast section of the field, at the Broadlands thermal anomaly.

The Broadlands rhyolite forms the main cap rock in the Broadlands thermal area, ranging in thickness up to 460 m (1500 ft). The Rautawiri formation of vitric-crystal-lithic tuff and tuff-breccia constitutes the next aquifer, spaning the entire thermal field. The thickness of this zone ranges from 180 to 335 m (600 to 1100 ft). The lower layer of Broadlands rhyolite at one time was thought to be a source of essentially dry steam for well BR7; however, this well is no longer a producer and the value of this formation has diminished. The Rangitaiki ignimbrite forms a dense cap layer. The Waikora formation appears fairly impermeable, as does the Ohakuri group, a potential aquifer consisting of pumiceous pyroclastics. The graywacke basement is essentially impermeable, except possibly along the dipping fault line between wells BR14 and BR10 (see figure 8.25). Mainly, however, all joints and fractures in the graywacke are sealed with calcite and quartz.

8.4.2 Well programs and productivity

As of 1977 32 wells had been drilled; however, only 16 of these were considered sufficiently productive for power production. It will take 20 producing wells to supply the required steam flow for a 150-MW power plant. The wells currently believed suitable for power plant use are BR2, 3, 8, 9, 11, 13, 17–23, 25, 27, and 28. The geologic logs from the first 16 wells drilled [Smith,1970] are shown in figure 8.26; technical information on these wells is given in table 8.13. Figure 8.27 is a log of well operations for the period 1966–1969, while table 8.14 gives the productivity of selected wells as a function of wellhead pressure. Note that the average quality is a fairly high 42% at a pressure of 0.79 MPa (114.6 lbf/in²) which will be required at the primary separators for the power station.

The Ohaki area behaves as a connected reservoir, whereas the Broadlands area, with considerably lower permeability, contains essentially isolated

Table 8.13—*Technical information on wells BR1–16 at Broadlands* [a]

Well No.	Date drilled	Wellhead elevation, m	Depth, m drilled	cased	Pressure [b] MPa	Bottomhole temperature, °C [c]
BR1..........	12/65	293	1936	607	3. 37	275
BR2..........	7/66	301	1034	417	5. 30	278
BR3.	2/67	299	912	469	5. 21	275
BR4..	8/67	313	1019	517	5. 74	263
BR5..........	9/67	317	1268	842	[d] 0. 1	243
BR6..........	9/67	293	1082	653	1. 55	143
BR7..........	12/67	306	1119	538	7. 20	277
BR8......... .	12/67	302	776	444	4. 94	270
BR9..........	3/68	308	1368	500	5. 72	289
BR10.........	4/68	302	1087	496	5. 40	278
BR11.........	5/68	311	760	484	4. 79	270
BR12.........	10/68	294	1369	653	6. 16	275
BR13.........	7/68	292	1080	813	0. 79	254
BR14.........	12/68	297	1282	587	3. 99	293
BR15.........	9/69	304	2418	1801	4. 63	297
BR16.........	6/69	303	1404	626	5. 84	273

[a] Source: Smith, 1970.
[b] Maximum pressure at the wellhead.
[c] At the bottom of the well.
[d] Nonflowing well.

Table 8.14—*Mass flow rates from selected wells at Broadlands* [a]

	Wellhead Pressure, MPa					
	0. 44	0. 79	1. 13	1. 48	1. 82	2. 17
Well No.	*Steam Flow Rate, Mg/h*					
BR2...................	102	98	93	88	80	72
BR3...................	65	59	55	47	38	25
BR4...................	30	28	24	19	15	11
BR7...................	36	35	33	30	28	24
BR8...................	93	89	86	82	78	73
BR9...................	25	24	20	17	14	11
BR11..................	103	96	90	86	77	66
BR13..................	40	36	32	27	22	17
Total steam..........	494	465	433	396	352	299
Well No:	*Liquid Flow Rate, Mg/h*					
BR2...................	105	110	114	115	114	109
BR3...................	112	118	120	114	113	88
BR4...................	26	27	27	26	24	21
BR7...................	10	10	10	10	10	10
BR8...................	86	90	93	87	80	74
BR9...................	40	42	44	45	46	46
BR11..................	151	158	163	168	163	146
BR13..................	94	98	98	99	99	99
Total liquid..........	624	653	669	664	649	593
Total.............	1118	1118	1102	1060	1001	892
Avg. quality, percent .	44	42	39	37	35	34

[a] Source: Smith, 1970.

individual wells. Pressure effects induced by well production communicate, however, across the Ohaki reservoir in about one year, effectively isolating even the Ohaki wells on a time scale of, say, a few months. This behavior should be contrasted with that of the Wairakei field, in which all wells appear to behave in unison [Grant, 1977]. Furthermore, in Wairakei the hydrothermal reservoir is probably a single-phase fluid (i.e., a pressurized liquid) with boiling occurring during draw-down. In the case of the Broadlands field, two-phase (i.e., liquid and vapor) conditions exist down to depths of about 2 km (6600 ft) owing to the pressure of a large amount of noncondensable gas, i.e., 6% CO_2 in the deep reservoir at 300°C (572°F). In the production zone—the Waiora formation—the temperature is 260°C (500°F), the gas content is 0.6% (by weight) in the liquid phase and 23% (by weight) in the vapor phase, the pure steam saturation pressure is 4.7 MPa (681.5 lbf/in²), and the CO_2 partial pressure is 1.5 MPa (217.5 lbf/in²) [Grant, 1977].

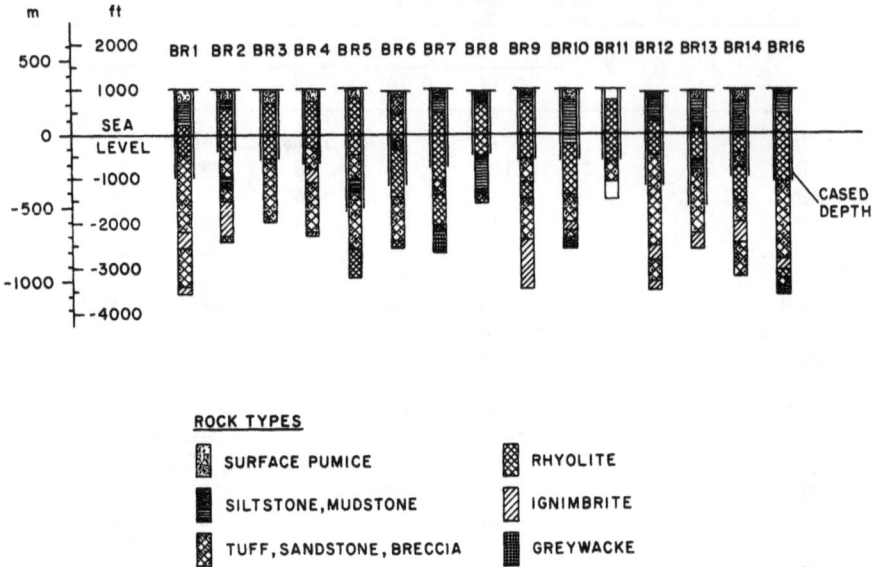

FIGURE 8.26—Geologic logs for wells BR1–16 at Broadlands [after Smith, 1970].

The amounts of carbon dioxide (CO_2), hydrogen sulfide (H_2S), and ammonia (NH_3) in the geothermal bore steam are shown in table 8.15 for several wells tested in 1976. These nine wells produced 1490 Mg/h (3.28×16^6 lbm/h) of total mass flow; the total flow of CO_2 was 12.4 Mg/h

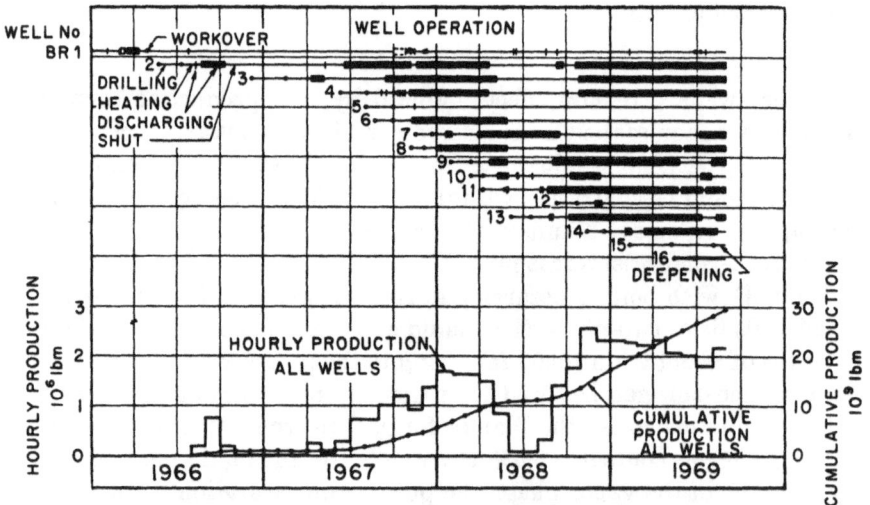

FIGURE 8.27—Log of well operations for wells BR1–16 for the period 1966–1969 [after Smith, 1970].

Table 8.15—*Composition of steam from selected wells at Broadlands* [a]

Well No.	Date	Pressure [b] MPa	Constituents in steam		
			CO₂ [c] % (wt)	H₂S [c] % (wt)	NH₃ ppm
BR8..........	10/76	0. 95	3. 4	0. 037	33. 5
BR11.........	10/76	2. 06	1. 3	0. 031	29. 5
BR17. ...	9/76	0. 99	1. 2	0. 022	22. 8
BR18.........	9/76	0. 79	6. 8	0. 077	49. 0
BR22.........	9/76	3. 14	2. 3	0. 039	43. 1
BR23.........	9/76	0. 79	2. 4	0. 035	41. 5
BR25.........	10/76	2. 74	4. 1	0. 045	26. 0
BR27.........	10/76	3. 89	10. 8	0. 071	41. 5
BR28.........	10/76	1. 03	4. 4	0. 053	28. 0

[a] Source: Bauer et al., 1977.
[b] Maximum pressure at the wellhead.
[c] For a wellhead pressure of 0.79 MPa (114. 6 lbf/in²).

$(27 \times 10^3$ lbm/h) or 0.83% (by weight), and that of H_2S was 0.159 Mg/h (351 lbm/h) or 0.011% (by weight). This amounts to about 2.4% (wt) of CO_2 and 0.031% (wt) of H_2S of the steam flow (based on an assumed quality of 35%) [Bauer et al., 1977]. The composition of the liquid separated from the wellbore (at atmospheric pressure) is given in table 8.16. The principal impurities in the water are chloride, sodium, silica, bicarbonate, and potassium. Boron and arsenic, both potentially dangerous

Table 8.16—*Composition of separated bore liquid at Broadlands* [a]

Constituent	Concentration, ppm
Chloride, Cl........	1488
Sodium, Na.....	997
Silica, SiO₂.........	771
Bircarbonate, HCO₃......	175
Potassium, K............	175
Boron, B....................... .	43. 7
Sulfate, SO₄.....	26. 8
Lithium, Li......	10. 7
Arsenic, As....... .;	3. 3
Calcium, Ca....	1. 68
Cesium. Cs...	1. 63
Rubidium, Rb......	1. 62
Hydrogen sulfide, H₂S....	1. 0
Aliminum, Al..	0. 8
Antimony, Sb.	0. 5
Tungsten, W......	0. 24
Total others....	0. 212–0. 218

[a] Source: Bauer et al.. 1977.

elements, are also present in non-negligible amounts. We shall return to a discussion of these impurities in section 8.4.4.

Pressure recovery and recharge studies were conducted during the 3-year period 1971 to 1974, during which time the wells were essentially closed in. Pressure recovery rate averaged about 170 kPa/yr (25 lbf/in$^2\cdot$yr) in the main producing area; amounting to a recovery of about 50% of the pressure loss sustained during flow tests up to 1971. The mass recharge was estimated to be 360 Mg/h (790 klbm/h), or the equivalent of one good producing well [Hitchcock and Bixley, 1975]. It is not known whether the recharge will increase or decrease when the field is being exploited. A gravity survey [Hunt and Hicks, 1975], on the other hand, showed no net mass loss from the field through 1974.

8.4.3 Energy conversion system

Present plans call for the construction of a 150-MW geothermal power station in three phases of 50 MW each. The energy conversion system will be a separated-steam/hot-water-flash (or so called double-flash) system. Although certain details about plant design are undecided at this time, basic plant layout will follow the generalized schematic shown earlier in figure 1.6. It remains to be determined whether the primary separators will be located at each wellhead or at the powerhouse (with two-phase geofluid transmission from wellheads to powerhouse). The condensers may be either direct-contact (as shown in figure 1.6) or shell-and-tube type, depending on the type of control used for the H_2S emissions.

Table 8.17 contains technical specifications for the plant as now proposed. On the basis of these figures, the plant would have a geothermal energy resource utilization efficiency of about 43%, relative to the thermodynamic available work of the geofluid at the wellhead (with a calculated quality of 25%, assuming an isenthalpic process between the wellhead and the primary separator) and a sink at a temperature of 27°C (80°F). This efficiency is somewhat low for a double-flash plant, corresponding more nearly to that for a single-flash (or separated-steam) plant. Compare the proposed Broadlands plant and a similar plant at Hatchobaru, Japan (see section 6.6). The resource at Hatchobaru is of slightly better quality, although the temperatures are essentially the same. The significant difference appears to be in the choice of flash temperature (or equivalently, pressure). At Hatchobaru the flash point has been chosen more nearly to agree with the accepted rule for optimum performance (i.e., equal temperature differences between geofluid, flash vessel, and condenser), whereas at Broadlands the flash point seems to be on the high side, perhaps to avoid the potential problems associated with a flash pressure near atmospheric. As a result, the Broadlands plant will have utilization efficiency of 43% while the value for Hatchobaru is 52%.

Table 8.17—*Preliminary specifications for Ohaki power plant at Broadlands* [a]

Turbine data:

Type..........................	Single-cylinder, double-flow, pass-in
Installed capacity...............	50 MW
Speed......................	3000 rev/min
Main steam pressure............	94.3 lbf/in^2
Main steam temperature..	323.6°F
Secondary steam pressure	29.0 lbf/in^2
Secondary steam temperature.....	248.0°F
Noncondensable gas content......	~2.4% (by wt) of main steam
Exhaust pressure................	[b]
Main steam flow rate............	913×10^3 lbm/h
Secondary steam flow rate........	206×10^3 lbm/h

Condenser data:

Type.........................	[b]
Cooling water temperature........	86°F
Outlet water temperature.........	109.4°F
Cooling water flow rate..........	130.5×10^6 lbm/h

Gas extraction system data:

Type.........................	Steam jet ejector or radial blower

Heat rejection system data:

Type....................... ...	Mechanically induced-draft cooling tower
Number of cells.................	~5
Water flow rate.................	133.7×10^6 lbm/h

Auxiliary power requirements......... 5000 kW

[a] Year of start-up: 1983. Unit 2 is expected to begin operating in 1984 and will be essentially identical to Unit 1; Unit 3, also identical, will follow.
[b] Not available.

8.4.4 Anticipated environmental effects

The potential damage to the environment from the Broadlands geothermal power development project has been documented in detail in the NZED Environmental Impact Report [Bauer et al., 1977], source for most of the material in this section. Our attention will be focused on the following topics: land usage, liquid and gaseous effluents, thermal discharge, and subsidence.

The total area encompassed by the Ohaki and Broadlands thermal region amounts to about 1380 ha (3410 acres), with roughly 565 ha (1396 acres) lying west of the Waikato River where the main geothermal field is located and where the power plant will be built. Most of this land is owned by the Maori. Land on which wells are drilled has been leased from the Maori owners, and it has been made clear that geothermal development will not encroach upon sacred lands or artifacts of special tribal importance. The powerhouse is expected to be 80 m long, 35 m wide, and 30 m high (263×115×98 ft). About 4 ha (10 acres) will be reserved for the switchyard, allowing for future growth.

Although four options were given consideration for the disposal of waste bore liquid, reinjection has been recommended. Nevertheless, a standby cooling pond is being included in the plant design because long-term operational characteristics of reinjection remain uncertain. Should it become necessary to dump 100% of the waste water into the 40-ha (100-acre) pond at full output, it would take 31 days to reach maximum pond level. Even so, the fluid discharge would be at a temperature only 5°C (9°F) above ambient river temperature, causing a mere 0.1°C (0.18°F) rise in river temperature upon complete mixing. Figure 8.28 shows the proposed location of the cooling pond, two powerhouse site options, and the steam/hot water mains.

The most serious chemical elements carried by the geothermal liquid (in terms of their effects on the ecology of the Waikato River) are ammonia (NH_3), arsenic (As), boron (B), lithium (Li), and mercury (Hg), in order of decreasing effect. Should it become necessary to discharge the full amount of geothermal liquid into the Waikato (say, in the event of a total failure of the reinjection system), then the *increment* of the composition of these elements in the river water would be in the following ranges: NH_3, 0.066–0.360 ppm; As, 0.016–0.080 ppm; B, 0.210–1.000 ppm; Li, 0.052–0.260 ppm; and Hg, $(22.3–101.0) \times 10^{-6}$ ppm. The range of values corresponds to average and low river flows, respectively. The NH_3 values include the amount that would enter the cooling water from the vapor phase of the geofluid. This would enter the cooling water either in the condenser (if it is of the direct-contact type) or in the wet cooling tower. About 2% of the amount shown is carried in the separated liquid stream. The same general situation exists in the case of Hg, where about 99% of this impurity is carried in the stream and only about 1% in the separated bore liquid.

Three gaseous effluents are of primary concern regarding impact on the environment: carbon dioxide (CO_2), hydrogen sulfide (H_2S), and radon (Rn–222). Under conditions of full power output at 150 MW, with no emissions controls on the plant, the following amounts of these would be released to the atmosphere: CO_2, 35 Mg/h (55,000 lbm/h); H_2S, 0.45 Mg/h (992 lbm/h); Rn–222, 11 nCi/h (7.5×10^{-17} kg/h or 1.7×10^{-16} lbm/h). This quantity of CO_2 is believed to be insignificant, certainly much less than is emitted by a typical thermal power station. For H_2S, this level of emission would mean a specific emission rate of 3000 g/MW · h (6.6 lbm/MW · h), and may be compared with the proposed standard of 200 g/MW · h (0.44 lbm/MW · h) for geothermal power plants in the United States [Hartley, 1978]. Furthermore, Broadlands would vent about 13 times as much H_2S into the atmosphere as does the plant at Wairakei. The concentration of Rn–222 in the exhaust stack gas would be about 0.3 pCi/l, even without dilution in the atmosphere.

Table 8.18 gives the concentrations of the main gaseous elements in the gas exhauster stack, assuming no emissions controls upstream of the power plant and a shell-and-tube condenser that concentrates all noncondensables

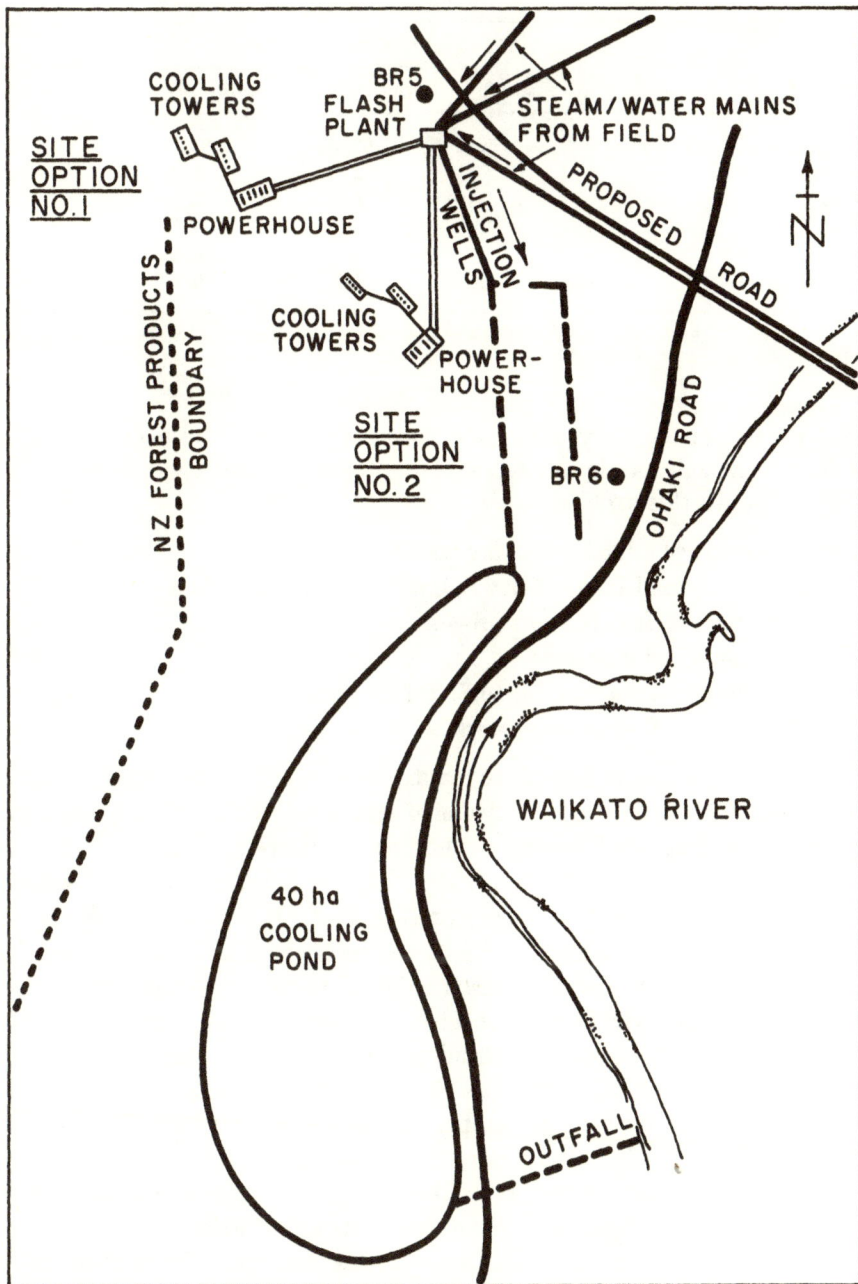

FIGURE 8.28—Location of cooling pond for Broadlands geothermal power plant
[Bauer et al., 1977].

Table 8.18—*Composition and quantity of exhaust stack gas discharge at Ohaki* [a]

Gas	Concentration		Flow rate lbm/h
	ppm (wt)	ppm (vol)	
CO_2................................	92. 5	84. 6	77, 000
H_2S.....	1. 2	1. 4	992
CH_4	0. 8	2. 1	661
N_2......	0. 7	1. 1	661
H_2................................	0. 0	0. 2	0
H_2O (vapor).....................	4. 8	10. 6	3, 970

[a] Source: Bauer et al., 1977; estimates assume no emissions controls upstream of plant and condenser of the shell-and-tube type.

into the exhauster stack. Hydrogen sulfide appears to be the only gaseous effluent that will require abatement equipment. Several techniques can be used to handle this problem:

    Iron catalyst added to cooling water
    Incineration of H_2S to form SO_2 which is then scrubbed
    Stretford sulfur recovery process
    Modified Clause process
    Takahax process (similar to Stretford process)
    Giammarco-Vetrocoke process.

In any case, it appears feasible to reduce the H_2S emissions from 0.45 Mg/h to about 0.07 Mg/h or less by means of treatment of the exhaust stack gas alone. By adding shell-and-tube condensers and preventing the H_2S from dissolving in the cooling water, this could be reduced a full order of magnitude to 0.007 Mg/h (15 lbm/h) or 47 g/MW · h (0.10 lbm/MW · h).

Land subsidence at Broadlands could result in serious inundation by the waters of the Waikato River if subsidence trends during exploration continue. Figure 8.29 shows subsidence contours during the period May 1968–March 1974. Maximum subsidence was about 190 mm (7.5 in) for a total mass drawoff of about 35×10^9 kg (77×10^9 lbm), very nearly the required *annual* drawoff to supply the 150-MW(e) plant. Because the center of the subsidence area is within about 500 m (1640 ft) of the west bank of the Waikato River, it is likely that some type of retaining wall may be necessary to prevent inundation of the field after, say, 20 years of operation. Furthermore, as the wells on the east side of the river begin producing in larger quantities, the center of the subsidence may shift even closer to the river, thereby causing a more immediate problem. The powerhouse will be located west of the area affected by subsidence and presumably will not be subjected to problems of differential ground movement (cf. figures 8.24 and 8.29).

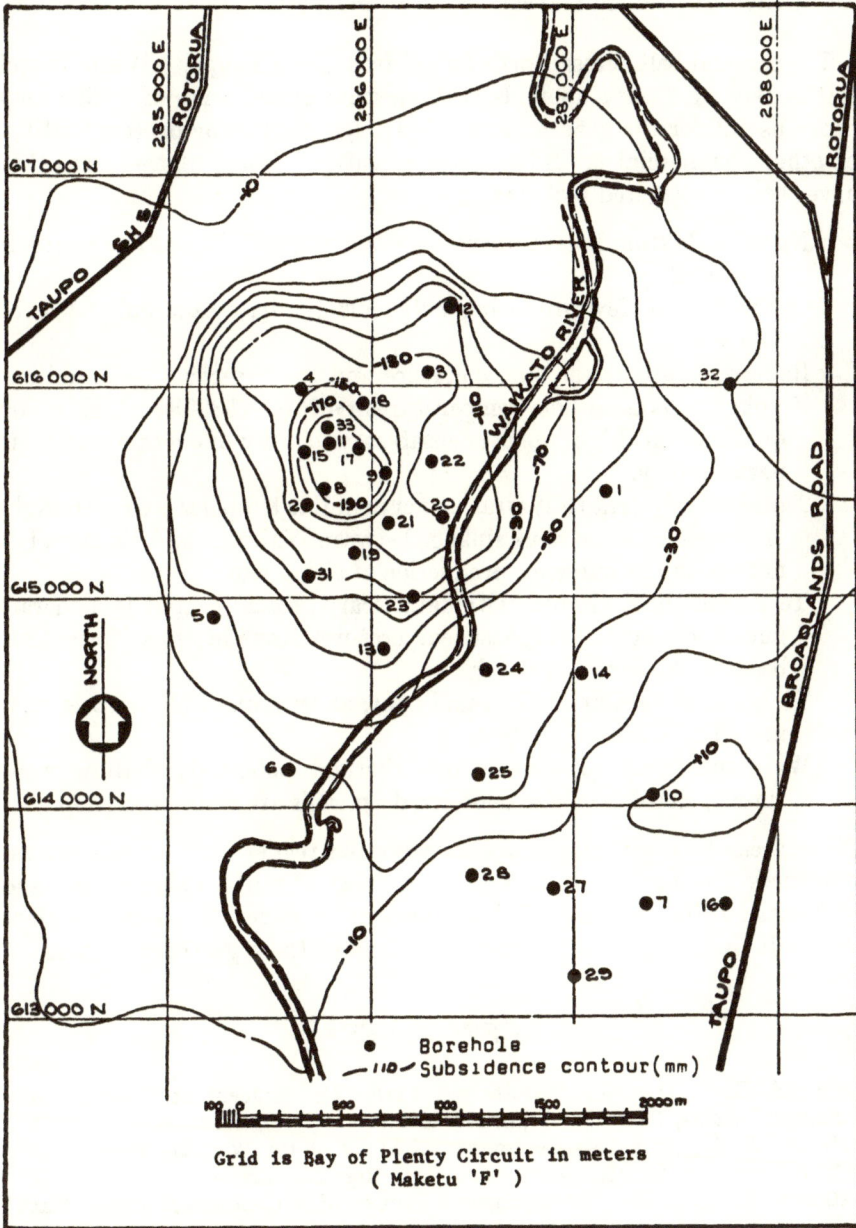

FIGURE 8.29—Subsidence contours at Broadlands for the period May 1968–March 1974 [after Bauer et al., 1977].

8.5   Other geothermal areas in New Zealand

The thermal belt across North Island from Lake Taupo to White Island in the Bay of Plenty abounds with thermal areas, some of which may prove useful for the generation of electricity, district or process heating, or other commercial or industrial applications. Some of the areas that have been investigated include:

Ngawha: bottom-hole temperature=236°C (457°F), but low reservoir permeability.

Orakeikorako: few producing wells, low-quality steam, infiltration of cold water.

Reporoa: unimpressive temperature and low reservoir permeability.

Rotokawa: bottom-hole temperatures ≈306°C (583°F); high steam quality, but high noncondensables and only moderate reservoir permeability.

Tauhara: adjacent to Wairakei; very similar temperatures with higher pressures, some weak linkage between Wairakei and Tauhara but not enough to influence production at either site.

Te Kopia: field aligned with fault scarp; steam output is moderate but of low quality, highest temperatures occur in upper formation, become indifferent at depth.

Te Mihi: extension of Wairakei field; at least one well has been connected to Wairakei system.

Waiotapu: area of considerable thermal potential; shallow wells rapidly develop calcite deposits, deep wells are more promising.

Other areas hold promise, and the interested reader may consult several references for further details [Dench, 1961; Smith, 1970; Smith and McKenzie, 1970; Bolton, 1977]. At present, however, there are no plans to install electric generating stations at ony of these geothermal areas.

REFERENCES [a]

Armstead, H. C. H., 1961. "Geothermal Power Development at Wairakei, New Zealand," *Rome, 1961,* Vol. 3, pp. 274–283.

Axtmann, R. C., 1974. "An Environmental Study of the Wairakei Power Plant," N. Z. Dept. Sci. Ind. Res., Phys. Eng. Lab. Rep. No. 445.

Axtmann, R. C., 1975a. "Environmental Impact of a Geothermal Power Plant," *Science,* Vol. 187, pp. 795–803.

Axtmann, R. C., 1975b. "Emission Control of Gas Effluents from Geothermal Power Plants," *Environ. Letters,* Vol. 8, No. 2, p. 135.

Bauer, H. E., Jones, E. M. E., Stirling, J. J., Willis, D.J., and Wu, P., 1977. "Environmental Impact Report for the Broadlands Geothermal Power Development," New Zealand Electricity Department, Power Development Section, Wellington.

Birsic, R. J., 1978. "Geothermal Steam Paths of New Zealand," *Geothermal Energy Magazine,* Vol. 6, No. 11, pp. 29–32.

[a] See note on p. 24.

Bolton, R. S., 1970. "The Behaviour of the Wairakei Geothermal Field During Exploitation," *Pisa, 1970*, Vol. 2, pp. 1426–1439.

Bolton, R. S., 1975. "Recent Developments and Future Prospects for Geothermal Energy in New Zealand," *San Francisco, 1975*, Vol. 1, pp. 37–42.

Bolton, R. S., 1977. "Geothermal Energy in New Zealand," in *Energy Technology Handbook*, D. M. Considine, ed.-in-chief, McGraw-Hill, New York, pp. 7.14–7.33.

Browne, P. R. L., 1970. "Hydrothermal Alteration as an Aid in Investigating Geothermal Fields," *Pisa, 1970*, Vol. 2, pp. 564–570.

Craig, S. B., 1961. "Geothermal Drilling Practices at Wairakei, New Zealand," *Rome, 1961*, Vol. 3, pp. 121–133.

Dench, N. D., 1961. "Investigations for Geothermal Power at Waiotapu, New Zealand," *Rome, 1961*, Vol. 2, pp. 179–191.

Fisher, W. M., 1961. "Drilling Equipment Used at Wairakei Geothermal Power Project, New Zealand," *Rome, 1961*, Vol. 3, pp. 154–168.

Fooks, A. C. L, 1961 "The Development of Casings for Geothermal Boreholes at Wairakei, New Zealand," *Rome, 1961*, Vol. 3, pp. 170–183.

Glover, R. B., 1970. "Interpretation of Gas Compositions from the Wariakei Field over 10 Years," *Pisa, 1970*, Vol. 2, pp. 1355–1366.

Grant, M. A., 1977. "Broadlands—A Gas-Dominated Geothermal Field," *Geothermics*, Vol. 6, pp. 9–29.

Grindley, G. W., 1961. "Geology of New Zealand Geothermal Steam Fields," *Rome, 1961*, Vol. 2, pp. 237–245.

Grindley, G. W., 1970. "Subsurface Structures and Relations to Steam Production in the Broadlands Geothermal Field, New Zealand," *Pisa, 1970*, Vol. 2, pp. 248–261.

Grindley, G. W. and Browne, P. R. L., 1975. "Structural and Hydrological Factors Controlling the Permeabilities of Some Hot-Water Geothermal Fields," *San Francisco, 1975*, Vol. 1, pp. 377–386.

Haldane, T. G. N. and Armstead, H. C. H., 1962. "The Geothermal Power Development at Wairakei, New Zealand, *Proc. Inst. Mech. Engrs.*, Vol. 176, No. 23, pp. 603–649.

Hartley, R. P., 1978. "Pollution Control Guidance for Geothermal Energy Development," EPA Rep. No. 600/7-78-101, Ind. Env. Res. Lab., Cincinnati.

Hitchcock, G. W. and Bixley, P. F., 1975. "Observations of the Effect of a Three-Year Shutdown at Broadlands Geothermal Field, New Zealand," *San Francisco, 1975*, Vol. 3, pp. 1657–1661.

Hochstein, M. P. and Hunt, T. M., 1970. "Seismic, Gravity and Magnetic Studies, Broadlands Geothermal Field, New Zealand," *Pisa, 1970*, Vol. 2, pp. 333–346.

Hunt, A. M., 1961. "The Measurement of Borehole Discharges, Downhole Temperatures and Pressures, and Surface Heat Flows at Wairakei," *Rome, 1961*, Vol. 3, pp. 196–207.

Hunt, T. M., 1977. "Recharge of Water in Wairakei Geothermal Field Determined from Repeat Gravity Measurements," *N.Z.J. Geol. Geophys.*, Vol. 20, No. 2, pp. 303–317.

Hunt, T. M. and Hicks, S. R., 1975. "Repeat Gravity Measurements at Broadlands Geothermal Field, 1967–1974," Geophys. Div. Rep. No. 113, Geophysics Division, DSIR, Wellington.

Kestin, J., 1978. "Available Work in Geothermal Energy," Brown Univ. Rep. No. CATMEC/20, DOE No. COO/4051-25, Providence, RI.

Lindal, B., 1973. "Industrial and Other Applications of Geothermal Energy," *Geothermal Energy: Review of Research and Development*, H. C. H. Armstead, ed., UNESCO, Paris, pp. 135–148.

Macdonald, W. J. P., 1975. "The Useful Heat Contained in the Broadlands Geothermal Field," *San Francisco, 1975*, Vol. 2, pp. 1113–1119.

McNitt, J. R., 1965. "Review of Geothermal Resources," *Terrestrial Heat Flow*, W. H. K. Lee, ed., Geophysical Monograph Series No. 8, Amer. Geophy. Union of NAS–NRC, Pub. No. 1288.

Milora, S. L. and Tester, J. W., 1976. *Geothermal Energy as a Source of Electric Power*, MIT Press, Cambridge.

Miner, S., 1969. "Preliminary Air Pollution Survey of Hydrogen Sulfide: A Literature Review," Nat. Air Poll. Cont. Adm., Raleigh, NC.

NZED, 1974. "Wairakei—Power from the Earth," New Zealand Electricity Department, Pamphlet No. 50392H, A. R. Shearer, Gov. Printer, Wellington.

NZED, 1975. "Wairakei Power Station," New Zealand Electricity Department, Pamphlet No. 60657H, A. R. Shearer, Gov. Printer, Wellington.

Ravenholt, A., 1977. "Wairakei and New Zealand's Geothermal Power," *Fieldstaff Reports*, Southeast Asia Series, Vol. 25, No. 6.

Reay, P. F., 1972. "The Accumulation of Arsenic from Arsenic-Rich Natural Waters by Aquatic Plants," *J. Appl. Ecol.*, Vol. 9, pp. 557–565.

Smith, J. H., 1961a. "Casing Failures in Geothermal Bores at Wairakei," *Rome, 1961*, Vol. 3, pp. 254–261.

Smith, J. H., 1961b. "The Organization for and Cost of Drilling Geothermal Steam Bores," *Rome, 1961*, Vol. 3, pp. 264–268.

Smith, J. H., 1970. "Geothermal Development in New Zealand," *Pisa, 1970*, Vol. 2, pp. 232–247.

Smith, J. H. and McKenzie, G. R., 1970. "Wairakei Power Station New Zealand—Economic Factors of Development and Operation," *Pisa, 1970*, Vol. 2, pp. 1717–1723.

Stilwell, W. B., 1970 "Drilling Practices and Equipment in Use at Wairakei," *Pisa, 1970*, Vol. 2, pp. 714–724.

Stilwell, W. B., Hall, W. K., and Tawhai, J., 1975. "Ground Movement in New Zealand Geothermal Fields," *San Francisco, 1975*, Vol. 2, pp. 1427–1434.

Usui, T. and Aikawa, K., 1970. "Engineering and Design Features of the Otake Geothermal Power Plant," *Pisa, 1970*, Vol. 2, pp. 1533–1545.

Wahl, E. F., 1977. *Geothermal Energy Utilization*, John Wiley & Sons, New York.

Weissberg, G. B. and Zobel, M. G. R., 1973. *Bull. Environ. Contam. Toxicology*, Vol. 9, p. 148.

Wigley, D. M., 1970. "Recovery of Flash Steam from Hot Bore Water," *Pisa, 1970*, Vol. 2, pp. 1588–1591.

Woods, D. I., 1961. "Drilling Mud in Geothermal Drilling," *Rome, 1961*, Vol. 3, pp. 270–273.

CHAPTER 9

PHILIPPINES

9.1 Outlook

Geothermal energy presently accounts for 58 MW of electricity in the Philippines. According to optimistic projections this will rise to 1320 MW by 1985, when geothermal energy would be supplying nearly one-quarter of the total electric generating capacity of the country. This chapter will cover the main geothermal areas in some detail and briefly discuss the ambitious geothermal development program of the Philippines.

9.2 Tiwi

9.2.1 Geology and exploration

Tiwi is the site of one of the principal geothermal fields in the Philippines. One of the most popular hot springs on the island of Luzon in Albay Province, it is located at the far southeastern tip of Luzon, about 300 km (185 mi) from Manila [Muffler, 1975]. The Tiwi field has been investigated with a number of techniques, including Wenner and dipole-dipole resistivity surveys, geologic, heat flow, and geochemical methods. Exploratory work began in 1964 through the Philippine Commission on Volcanology, supported with financial assistance from the National Science Development Board.

As a result of the surveys an area of 2300 ha (5680 acres) was outlined as a potential reservoir at drillable depths. Fourteen wells sunk inside the resistivity low confirmed the indications of the surveys. The wells produced a mixture of liquid and vapor at high flow rates, revealing the nature of the reservoir. The Tiwi system is a liquid-dominated field in a reservoir of Quaternary andesites and subsidiary decites. A system of microfractures is believed to lend permeability to the reservoir. These findings were reported by A. P. Alcaraz in 1976 and quoted by Ravenholt [1977].

A total of 20 wells have now been drilled; 19 of these are producers extending to depths of between 760 and 2130 m (2500 and 7000 ft). The wells were drilled by Philippine Geothermal, Inc., a wholly owned subsidiary of Union Oil Company, which subcontracted the actual drilling operations to Richter Drilling International Pty., Ltd., of Australia. Two rigs are kept active with three work shifts.

Table 9.1—*Technical specifications for Philippine geothermal power stations* [a]

	Tiwi Units 1–4 1979*	Los Baños [b] Wellhead Unit 1977*	Los Baños [b] Units 1–4 1979*	Leyte Wellhead Unit 1977*
Turbine data:				
Type	Single-cylinder, double-flow, dual-admission, 6×2	Single-cylinder, one Curtis stage, noncondensing, geared	Single-cylinder, double-flow, dual-admission, 5×2	Single-cylinder, one Curtis stage, noncondensing, geared
Rated capacity, MW	55	1.2	55	3.0
Speed, rpm	3600	7129/1800	3600	7554/1800
Main steam pressure, lbf/in²	101.4	95.5	94.8	114.2
Main steam temperature, °F	329.0	324.1	324.1	338.0
Secondary steam pressure, lbf/in²	26.8	—	24.8	—
Secondary steam temperature, °F	244.4	—	240.1	—
Exhaust pressure, in Hg	4.0	30.4	4.0	37.6
Main steam flow rate, 10³ lbm/h	(ᶜ)	49.2	776.0	117.0
Secondary steam flow rate, 10³ lbm/h	(ᶜ)	—	276.0	—
Condenser data:				
Type	Barometric, spray jet	—	Barometric, spray jet	—
Cooling water temperature, °F	87.1	—	87.1	—
Outlet water temperature, °F	120.0	—	120.0	—
Cooling water flow rate, 10⁶ lbm/h	27.8	—	29.1	—

Gas extractor data:

Type	(c)	3-stage blower and steam jet ejector
Number	(c)	1 set of each
Suction pressure, in Hg	(c)	3.75
Capacity, each set, ft³/min	(c)	16,275
Power consumption	(c)	430 kW (blower), 19.33 Mg/h (ejector)

Heat rejection system data:

Type	Crossflow, induced-draft cooling tower	Crossflow, induced-draft cooling tower
Number of cells	(c)	6
Design wet-bulb temperature, °F	(c)	80
Water flow rate, 10^5 lbm/h	(c)	30.6

*Year of startup.
a Sources: Toshiba, 1977; Aikawa et al., 1978.
b Now called Makiling Banahaw.
c Not available.

9.2.2   Energy conversion system

The preliminary design, equipment procurement, specifications, and contract documents for the first two units at Tiwi were carried out by Rogers International. The Tokyo Shibaura Electric Company of Japan (Toshiba) holds orders for the turbogenerators for the first four units to be installed at Tiwi. Each of these are identical 55-MW, single-cylinder, double-flow, 6×2 stage machines of the dual-admission type. Technical specifications for these units are given in table 9.1 [Toshiba, 1977]. The generators for the units will be rated at 69,000 kVA at 13.8 kV and 60 Hz with 0.8 power factor. The stator will be conventionally hydrogen cooled; the rotor, direct hydrogen cooled.

The energy conversion scheme will be a separated-steam/hot-water-flash ("double-flash") system (see figure 1.6). Since the flow rate of steam required for plant operation is not known at present, we cannot compute precisely the geothermal resource utilization efficiency for the plant. However, it is anticipated that 10 producing wells will be needed for each 55-MW unit, and using as a guide Units 9 and 10 at The Geysers, California [DiPippo, 1978], which are both 55-MW condensing units manufactured by Toshiba, each unit will require roughly 454 Mg/h (1.0 x 10⁶ lbm/h of geothermal steam. For an average wellhead quality $\overline{x}=25\%$, the resource utilization efficiency, η_u, would be approximately 41%; for $\overline{x}=30\%$, $\eta_u \approx 49\%$; for $\overline{x}=35\%$; $\eta_u \approx 57\%$. These estimates are all based on the available work of the geofluid at the wellhead.

Each unit of the power plant will produce about 1247 Mg/h (2.75 x 10⁶ lbm/h) of waste liquid which will be disposed of through reinjection wells. It was originally planned to have Units 1 and 2 on-line during 1978, but this date was not met. The first unit began operating in January 1979, the second is now expected on-line during the fall of 1979, and the other two units should be operating by 1981.

9.3   Makiling Banahaw (Los Baños)

9.3.1   Steam wells

Los Baños lies about 70 km (43 mi) southeast of Manila on the island of Luzon. Part of a huge area with a geothermal potential of around 720 MW, the area extends over 153,000 ha (378,000 acres) in the Makiling-Banahaw volcanic region. At least 14 wells have been drilled at the Los Baños thermal area. Bottom-hole temperatures are in the range 280°–310°C (540°–590°F), and geofluid qualities as high as 36% at the wellhead have been reported [Ravenholt, 1977]. Most of the wells are located about 450 m (1475 ft) above sea level near Mount Bulalo.

9.3.2   Energy conversion systems

A small wellhead auxiliary geothermal power unit has been operating at Los Baños (now called Makiling Banahaw) since early 1977. It was supplied by Mitsubishi Heavy Industries, Ltd. (MHI); technical particulars for this machine may be found in table 9.1. The main power units for Makiling Banahaw will consist of four identical units, each of 55-MW capacity, of the single-cylinder, double-flow, mixed pressure, impulse-reaction design. The process flow diagram will be similar to that shown earlier in figure 1.6, i.e., a separated-steam/hot-water-flash ("double-flash") system. Table 9.1 contains the technical specifications for each of the first four units. According to these particulars, the geothermal resource utilization efficiency, η_u, will be as follows, depending on the actual quality, $\bar{x}$, of the geofluid at the wellhead: $\eta_u = 54\%$ for $\bar{x} = 25\%$; $\eta_u = 57\%$ for $\bar{x} = 30\%$; $\eta_u = 60\%$ for $\bar{x} = 35\%$.

9.4   Leyte (Tongonan)

A 3-MW portable geothermal unit is operating at Tongonan on the island of Leyte. The unit consists of a noncondensing turbine (single Curtis stage) connected to a generator through a helical reduction gear. The entire unit is mounted on a platform to facilitate transfer from one site to another. It was manufactured by Mitsubishi Heavy Industries, Ltd.; technical specifications are given in table 9.1. Figure 9.1 shows the powerhouse; figure 9.2 shows the assembled power unit being tested in the MHI shop [Ravenholt, 1977; MHI, 1977].

9.5   Other Philippine areas with geothermal potential

The potential for geothermal development in the Philippines is significant, estimated at 1320 MW by 1985. Table 9.2 gives a breakdown of the

Table 9.2—*Potential geothermal power generation in the Philippines* [a]

Geothermal Field	1978	1979	1980	1981	1982	1983	1984	1985	Est. Max. Capacity
Tiwi.............	[c]110	—	55	55	—	55	55	—	560
Los Baños [b].......	[c]55	55	—	55	55	—	55	55	720
Tongonan........	—	—	55	55	—	55	55	—	([d])
S. Negros.........	—	—	—	55	55	—	55	55	425
Manat-Masara..	—	—	—	55	55	—	55	55	500
Total Annual..	165	55	110	275	165	110	275	165	....
Cumulative.......	165	220	330	605	770	880	1155	1320	....

[a] Source: Ravenholt, 1977. All values in megawatts; wellhead units not included.
[b] Now called Makiling Banahaw.
[c] Now expected in 1979.
[d] Not available.

FIGURE 9.1—Portable geothermal power unit at Tongonan, Leyte [MHI, 1977].

FIGURE 9.2—Tongonan turbogenerator and auxiliaries being tested in the Mitsubishi facilities at Nagasaki [MHI, 1977].

distribution of this expected generating capacity. In addition to the sites listed, initial investigations are underway at Mambucal in northern Negros. Resistivity surveys indicate that the potential of the fields included in the table exceeds 2200 MW.

Some idea of the scope of the effort that will be needed to achieve an installed capacity of 1320 MW in 1985 can be derived from the fact that about 10 producing wells are needed for each 55-MW unit. This means that 240 producing wells must be drilled; allowing 3 out of every 4 wells drilled to be producers, a total of 320 wells must be sunk. Assuming that each well costs $1,000,000 (current costs are $750,000), this will necessitate a capital investment, for wells alone, of $320,000,000. Taking into account both the time required to prove a field and the construction lead time for a power plant, it is difficult to imagine how such an enormous project can be accomplished within 7 years. Simply to drill the required wells within that time would mean that roughly 4 wells would have to be drilled each month.

Small wellhead units in the 1–10 MW range are expected to find application in cases where large units are either unnecessary or impractical, particularly on the smaller islands of the Philippines such as the Visayan group of Leyte, Cebu, Bohol, Negros, and Panay. The electricity produced from them is cheaper than that generated by diesel engines. In any case, the use of wellhead units provides a source of revenue and local power during the early stages of development of a geothermal field. The revenue obtained helps alleviate the cash flow problem faced by field developers.

The opportunity exists for the Philippines to supply a significant percentage of its electrical needs from indigenous geothermal energy. The high cost of imported petroleum products provides a great deal of motivation for development of the geothermal resources of the country.

REFERENCES [a]

Aikawa, K., Fukuda, S., and Tahara, M., 1978. "MHI's Recent Achievements in Field of Geothermal Power Generation," *MHI Technical Review*, Vol. 15, No. 3, Ser. 43, pp. 195–207, Mitsubishi Heavy Industries, Ltd., Tokyo.

DiPippo, R., 1978. "Geothermal Power Plants of the United States—A Technical Survey of Existing and Planned Installations," Brown Univ. Rep. No. CATMEC/14, DOE No. COO/4051-20, Providence, RI.

MHI, 1977. "Portable Geothermal Power Plant—3,000 kW Leyte Geothermal Power Plant in Philippines," Mitsubishi Heavy Industries, Ltd., Nagasaki Shipyard & Engine Works, Nagasaki.

Muffler, L. J. P., 1975. "Summary of Section I—Present Status of Resources Development," *San Francisco, 1975*, Vol. 1, pp. xxxiii–xliv.

Ravenholt, A., 1977. "Energy from Heat in the Earth—Philippines Taps Immense Resources of Pacific Fire Belt," *Fieldstaff Reports*, Southeast Asia Series, Vol. 25, No. 5.

Toshiba, 1977. "Toshiba Experience List on Geothermal Power Plant," Tokyo Shibaura Electric Company, Ltd., Heavy Apparatus Division, Tokyo.

[a] See note on p. 24.

CHAPTER 10

TURKEY

10.1 Introduction

The focus of geothermal energy development in Turkey is at the Kızıl-dere field in the Menderes River valley, western Anatolia. A small wellhead power-generating unit is in operation at this site, and there are plans to expand the installed capacity to about 12 MW in the near future. Turkey is situated on an active tectonic zone and possesses great potential for geothermal energy, as evidenced by more than 600 hot springs, some with temperatures as high as 102°C (216°F), and numerous areas exhibiting hydrothermal alteration. Since 1962 the Mineral Research and Exploration Institute of Turkey (MTA) has been conducting surveys of the geothermal resources of the country by means of geologic, geophysical, geochemical, and drilling studies. Fourteen promising areas have been identified, the best of which is located at Kızıldere, near Sarayköy in the Denizli Province [Alpan, 1975].

10.2 Kızıldere

The geothermal field at Kızıldere consists of two producing reservoirs, one lying between 300 and 800 m (984 and 2625 ft), and one between 400 and 1100 m (1312 and 3609 ft). The deeper reservoir is considered the main producer and has a temperature of 200°C (392°F), whereas the upper zone is at 170°C (338°F). The chemical composition of the fluids from the two zones are similar [Tezcan, 1975a]. Portions of the field may consist of isolated dry steam caps [Tezcan, 1975b].

From 1966 to 1975, 14 wells were drilled in the area, with 12 of these being producers. The wells range in depth from 370 to 1241 m (1214 to 4072 ft). Half the producing wells terminate in the upper reservoir; half reach the deep reservoir. In general, the produced fluid may be characterized as follows [Alpan, 1975]: maximum temperature, 207.4°C (405.3°F); maximum wellhead pressure, 2.16 MPa (314 lbf/in²); maximum total flow rate (single well), 1003.5 t/h (2.2×10⁶ lbm/h; maximum vapor flow rate (single well), 67.6 t/h (149×10³ lbm/h); fluid dryness fraction (at wellhead), 2%–12%, 10% average.

Table 10.1—*Geofluid characteristics at Kızıldere field, Turkey* [a]

Substance	Concentration, ppm	
	Well KD-XIII	Avg., 12 Wells
Bicarbonate, HCO_3..........................	2116	2247
Sodium, Na.......................................	1174	1240
Sulfate, SO_4	641	811
Silica, SiO_2.................................	327	288
Potassium, K...................................	131	128
Chloride, Cl....................................	115	107
Boron, B...	24. 5	24. 5
Fluoride, F......................................	18. 2	18. 15
Ammonium, NH_4	5. 8	3. 95
Calcium, Ca.....................................	4. 1	3. 2
Magnesium, Mg.................................	1. 5	0. 95
Arsenic, As.....................................	0. 51	0. 17
pH..	8. 9	8. 8

[a] Source: Alpan, 1975.

A 0.5-MW power unit has been installed on well KD–XIII by MTA. The composition of the geofluid produced by this well is shown in table 10.1, along with the average composition for all 12 producing wells [Alpan, 1975]. Well KD–III is 760 m (2494 ft) deep and produces from the lower reservoir, which it enters at a depth of 590 m (1936 ft). The maximum temperature is 197°C (386.6°F), the production pressure is 1.08 MPa (157 lbf/in²), and the flow rates of liquid and vapor are 552 and 20 t/h (1.15×10⁶ and 44×10³ lbm/h), respectively [Alpan, 1975]. Characteristics of the wellhead turbine are given in table 10.2. The plant has a specific geofluid consumption rate of 79.8 kg/kW · h (176 lbm/kW · h).

Table 10.2—*Characteristics of wellhead power generator at KD–XIII, Kızıldere, Turkey* [a]

Turbine type........................	Single-cylinder, one Curtis stage, noncondensing
Rated capacity.....................	0.5 MW
Speed..............................	4500 rpm
Inlet steam pressure...............	~70.5 lbf/in²
Inlet steam temperature	~302°F
Noncondensable gas content........	17% (by wt of steam)
Exhaust steam pressure..	~34.0 in Hg
Maximum allowable pressure..	~114 lbf/in²
Turbine steam flow rate [b]	~3255 lbm/h
Last-stage blade height . .	3 in

[a] Year of startup: 1975.
[b] Total geofluid flow rate is ~88×10³ lbm/h; dryness fraction is 3.7%.

A realistic appraisal of the ultimate potential of the Kızıldere field is difficult owing to serious problems of well plugging. Out of a total of 12 producing wells only 6 have been judged suitable for production. However, only 3 of these can be relied on at any given time because of the necessity for periodic reaming of clogged wells. With 3 wells in operation and 1086 t/h (2.4×10^6 lbm/h) of geofluid being produced, it has been estimated that 11,430 kW (gross) or 10,550 kW (net) could be generated. Under the best conditions, if 6 wells could be used simultaneously to produce 1640 t/h (3.6×10^6 lbm/h), then the system could support 32 MW (gross) or 28 MW (net). To achieve this level of output, a reliable and effective system of reinjection of waste water would need to be implemented [Alpan, 1975].

10.3 Exploration activities

Exploration has reached the drilling phase in a number of areas, including Ankara (Ayaş, Çubuk, Kızilcahamam, and Mürtet), Afyon, Izmir (Agamemnum and Seferihisar-Doğanbey), Çanakkale (Tuzla-Kestanbol), and Söke (Germencik). Preliminary investigations are being carried out at several other sites, including Bergama-Dikili, Çan-Gönen, Eskişehir, Gediz, Nevşehir-Kozaklı, Salihli-Turgutlu, Sındırğı-Hisaralin, and Tatvan-Nemrut. The geothermal resources at these sites may be applied for generation of electricity, heating of buildings and greenhouses, and general improvement of hot springs for the tourist trade [Alpan, 1975].

REFERENCES [a]

Alpan, S., 1975 "Geothermal Energy Exploration in Turkey," *San Francisco, 1975*, Vol. 1, pp. 25–28.
Tezcan, A. K., 1975a. "Geophysical Studies in Sarayköy-Kızıldere Geothermal Field, Turkey," *San Francisco, 1975*, Vol. 2, pp. 1231–1240.
Tezcan, A. K., 1975b. "Dry Steam Possibilities in Sarayköy-Kızıldere Geothermal Field, Turkey," *San Francisco, 1975*, Vol. 3, pp. 1805–1813.

---

[a] See note on p. 24.

CHAPTER 11

UNION OF SOVIET SOCIALIST REPUBLICS

11.1 Overview

The Soviet Union has a huge potential in moderate-temperature geothermal waters that may someday be exploited for space or process heating. The only known sites, however, that are being used or contemplated for electric power production are located on the Kamchatka Peninsula, well removed from main population centers. Whereas the potential for direct heating from geothermal resources is estimated at 48,000 MW (thermal) [Tikhonov and Dvorov, 1970], a realistic estimate for the geothermal electric power capacity may be several hundreds of megawatts.

An extensive survey of potential geothermal applications in the Soviet Union concentrated on nonelectrical uses, but several areas in the Kurile Islands and on the Kamchatka Peninsula where electric power plants could be installed were identified [Makarenko et al., 1970; Tikhonov and Dvorov, 1970]: Pauzhetka, Uzono-Semyachik, Mutnovo-Zhirovo, Bolshoye-Bannoye, and Goryachy Plyazh (Yuzhno-Kurilsk). All of these sites are in regions of recent vulcanism characterized by dramatic surface thermal manifestations.

The locations of several major hydrothermal areas on the Kamchatka Peninsula are shown in figure 11.1. The major population center of the area, Petropavlovsk-Kamchatskiy, is within 75 km (47 mi) of several of these geothermal prospects. By and large the resources are of low-to-moderate temperature and are situated in relatively shallow reservoirs (<1 km). Some typical temperature profiles are shown in figure 11.2. The highest bottom-hole temperatures have been found at the Pauzhetka site, about 200°C (392°F) at 400 m (1312 ft).

At least two geothermal powerplants have been in operation in Russia and several others have been mentioned as either under construction or in planning. These include the flash steam plant at Pauzhetka, the binary plant at Paratunka, the multiple-flash steam plant under construction at Bolshoye-Bannoye, and the steam plants proposed for Makhachkala and Yuzhno-Kurilsk. Each will be discussed in the following sections.

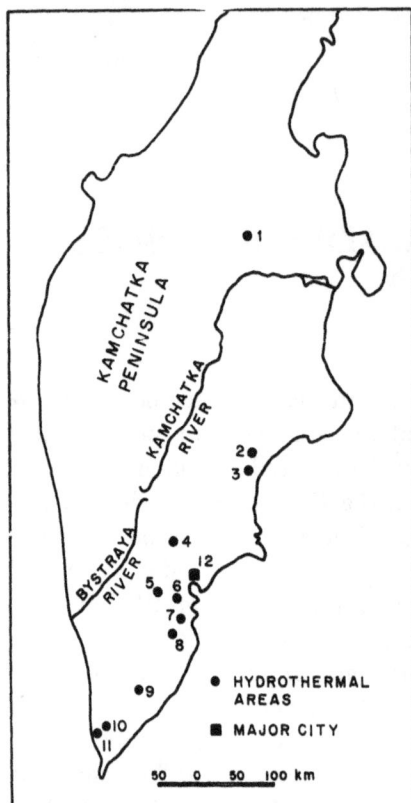

FIGURE 11.1—Location of hydrothermal areas on Kamchatka Peninsula, USSR. Key: 1. Kireunsky; 2. Uzon-Geyserny; 3. Semyachinsky; 4. Nalychevsky; 5. Bolshoye-Bannoye; 6. Paratunka; 7. Zhirovshy; 8. Severo, Mutnovsky; 9. Khodutkinsky; 10. Pauzhetka; 11. Koshelevsky; 12. Petropavlovsk-Kamchatskiy [after Vakin et al., 1970].

11.2 Pauzhetka

11.2.1 Geofluid characteristics

Approximately 20–25 wells have been drilled at the Pauzhetka geothermal area, each producing 36 t/h (79,000 lbm/h) of steam and liquid. The geofluid is a mixture of liquid and vapor with a dryness fraction of about 9%. Wellhead pressure lies between 196 and 392 kPa (28 and 57 lbf/in²). The fluid carries 1000–3400 ppm of total dissolved solids, of which about 250 ppm is silica. Noncondensable gases amount to slightly more than 0.05% (by weight) of the geofluid mixture (or 0.6% of the steam), with the bulk of the gases being CO_2 (92%), and the rest mainly H_2S (4%) and NH_3 (3%) [Tikhonov and Dvorov, 1970]. Composition of the geothermal liquid from a typical well at Pauzhetka (well No. 4) is given in table 11.1 [Vakin et al, 1970].

11.2.2 Energy conversion system

At present, a separated-steam plant of 5-MW capacity is in operation at Pauzhetka. The plant, owned and operated by the Kamchatka Electricity

Production and Distribution Administration, began production in 1967. About nine wells are required to supply the station. The steam piping from the wells to the powerhouse ranges in diameter from 210 to 370 mm (8.25 to 14.5 in) and totals 1.3 km (4300 ft) in length. The steam pipes are made of carbon steel.

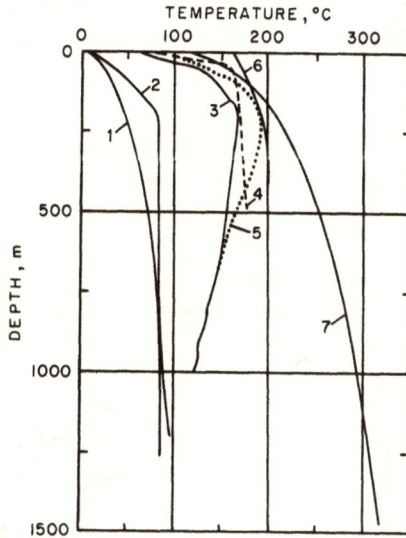

FIGURE 11.2—Temperature profiles in wells in Kamchatka region. Key: 1, 2. Paratunka wells GK-11, GK-3; 3, 4. Bolshoye-Bannoye wells GK-2, GK-38; 5, 6. Pauzhetka wells R-1, R-13; 7. Hydrostatic boiling curve [after Vakin et al., 1970].

Table 11.1—*Composition of fluid from well No. 4 at Pauzhetka* [a, b]

Component	Concentration, ppm
Chloride, Cl	1633
Sodium, Na	986
Silica, H_2SiO_3	216
Potassium, K	105
Sulfate, SO_4	78
Calcium, Ca	52
Carbonate, CO_3	22
Bicarbonate, HCO_3	4
Magnesium, Mg	3. 5
Ammonium, NH_4	0. 6
Total dissolved solids	3448

[a] Source: Vakin et al., 1970.
[b] Bottom-hole temperature = 195°C (383°F); pH = 8.2.

The design of the plant is straightforward; a flow diagram is given in figure 11.3. Cyclone separators with mist eliminators in the upper part yield steam of approximately 0.995 dryness fraction. The 5-MW output is obtained by means of two 2.5-MW turbines arranged in tandem. The machines were manufactured by the Thermal Turbine Machine Construction Plant at Kaluga, and are situated in a turbine room that has a flood area of $33 \times 9 = 297$ m² ($108 \times 30 = 3200$ ft²). Condensers are of the direct-contact type, made of stainless steel, with 11 m³ (389 ft³) of steam volume, and are fitted with 118 nozzles through which the cooling water is sprayed.

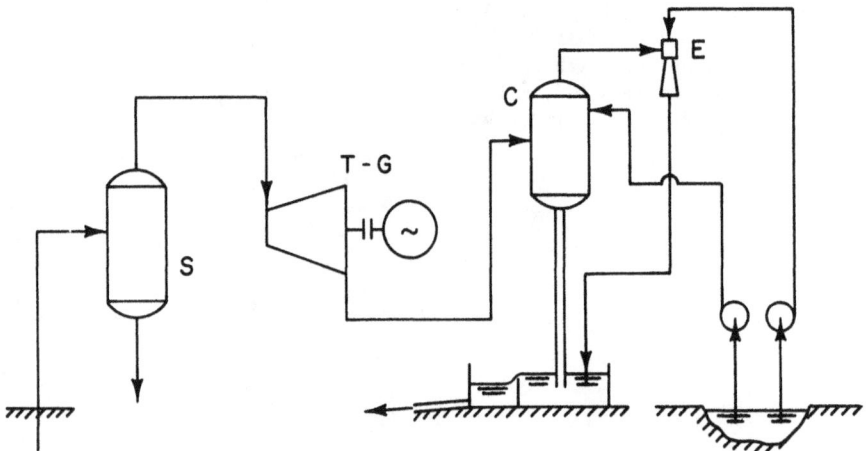

FIGURE 11.3—Flow diagram for Pauzhetka flash-steam geothermal power plant. Key: S = separator; T–G = turbogenerator; C = condenser; E=water jet ejector [after ARPA, 1970].

A summary of the technical particulars for the unit may be found in table 11.2 [Naymanov, 1970]. On the basis of the data in table 11.2 and assuming a geofluid wellhead dryness fraction of 9%, the plant would have a resource utilization efficiency of 54% and would consume 56 kg/kW · h (123 lbm/kW · h). However, values as low as 5% for the geofluid dryness fraction have been reported [Tikhonov and Dvorov, 1970], and this would reduce utilization efficiency to 28%. The actual value probably lies somewhere between these limits.

Power from the plant is transmitted to the town of Pauzhetka, the Ozernovsk fishing combine, and the collective farm at Krasnyy Truzhenik.

Table 11.2—*Technical specifications for Pauzhetka geothermal power plant* [a]

Turbine data:
Type..........................	Tandem-compound, single-flow
Rated capacity..................	2×2.5 MW
Steam inlet pressure.............	28.4 lbf/in^2
Steam inlet temperature..........	260.6°F
Noncondensable gas content.....	0.6% by weight of steam
Exhaust pressure................	0.87–2.32 in Hg
Steam flow rate.................	59.5×10^3 lbm/h

Condenser data:
Type..........	Direct-contact, barometric

Gas extractor data:
Type.........................	Water jet ejector
Water impeller power......... ..	170 kW

Heat rejection system:
Type..	Once-through, Pauzhetka River

[a] Year of startup: 1967.

The power line carries electricity at 35 kV and is 30 km (19 mi) long [ARPA, 1972]. The geothermal liquid that is separated at the wellheads is discharged into the Pauzhetka River at a temperature of 110°C (230°F) and at a rate of 100 kg/s (220 lbm/s). There were plans to make use of this hot fluid for the heating of greenhouses, although it is not known whether this scheme has been implemented [Tikhonov and Dvorov, 1970]. The same authors also reported that the cost of electricity from the Pauzhetka geothermoelectric station is less by a factor of 10 to 15 than electricity generated by diesel power plants on the Kamchatka Peninsula.

Although there were plans to expand the capacity of the plant to 12.5 MW, and eventually to 20 MW, these intentions remain unfulfilled at present [ARPA, 1972]. The ultimate potential of the Pauzhetka reservoir has been estimated at between 50 and 70 MW of electrical power [Tikhonov and Dvorov, 1970].

11.3 Paratunka

11.3.1 Objectives

The Paratunka geothermal power plant was an ambitious attempt at providing a form of total-energy system, albeit on a rather limited scale. The power plant was a binary fluid cycle employing Refrigerant-12 as the working fluid in conjunction with geothermal waters at temperatures as low as 81°C (178°F). The power from the plant served a small village and several Soviet state farms. In addition, the geothermal water, after leaving the powerhouse and having been cooled to 45°C (113°F) in the plant's heat exchangers, was used to heat the soil in a series of greenhouses. Finally, the cooling water leaving the condensers of the power plant was used to water the plants in the greenhouses. Direct use of water from the Paratunka River is not feasible because of its low temperature of 5°–7°C (41°–45°F).

It is generally acknowledged that the Paratunka plant was the first binary geothermal pilot plant to generate electricity, having begun operations in 1967. Although the plant apparently operated successfully for several years, recent reports indicate that the power station has been closed and dismantled because of difficulties with leaks in the Refrigerant-12 piping [Smith, 1978]. Although the properties of Refrigerant-12 are not ideally suited for geothermal applications [Naymanov, 1970], it is instructive to examine the details of the Paratunka plant.

11.3.2 Geofluid characteristics

The geothermal hot water is obtained from a number of shallow wells located about 1.5 km (0.9 mi) from the plant site. The wells range in depth from 302 to 604 m (991 to 1982 ft), and in diameter from 127 to 200 mm (5 to 7.875 in) [ARPA, 1972]. Eight wells were completed in 1964 before construction got underway on the plant; six wells were used to supply the plant with about 280 t/h (617×10^3 lbm/h) of hot water [Moskvicheva and Popov, 1970] with one well kept on reserve. Composition of the geothermal water from a typical well (Stredne-Paratunka well No. 2) is given in table 11.3 [Vakin et al., 1970].

Table 11.3—*Composition of fluid from well No. 2 at Stredne-Paratunka* [a, b]

Component	Concentration, ppm
Sulfate, SO_4 . .	556
Sodium, Na	206
Calcium, Ca.	71
Chloride, Cl . . .	39
Silica, H_2SiO_3 . . .	32
Carbonate, CO_3	19
Potassium, K	10
Fluoride, F.	2
Total dissolved solids	935

[a] Source: Vakin et al., 1970.
[b] Bottom-hole temperature = 90°C (194°F); pH = 9.0.

11.3.3 Energy conversion system

A simplified flow diagram of the power plant is given in figure 11.4; technical specifications are listed in table 11.4. Values given in the table are the actual values achieved during the tests reported by Moskvicheva and

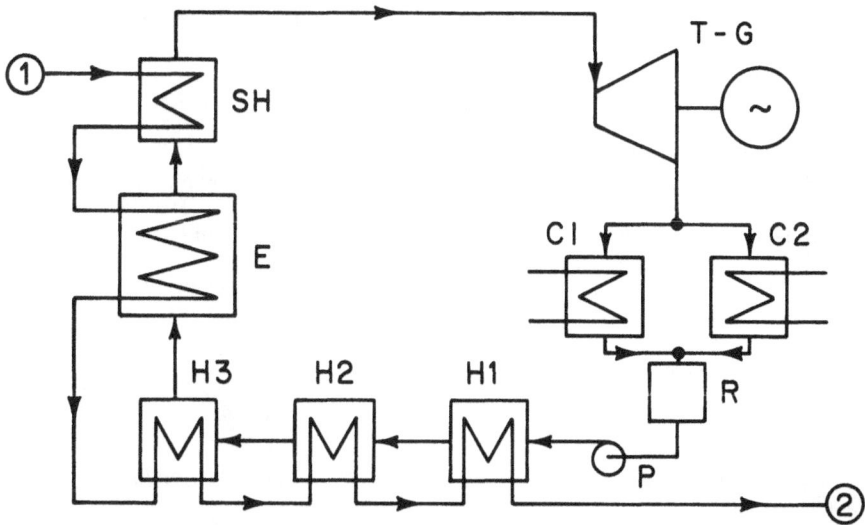

FIGURE 11.4—Flow diagram for Paratunka binary geothermal power plant. Key: 1-2 = hot water inlet-outlet; H1, H2, H3 = heaters; E = evaporator; SH = superheater; T–G = tubogenerator; C1, C2 = condensers; R = receiver; P = pump.

Popov [1970]. In certain respects, these differ from the design values. For example, the hot water inlet temperature should have been 90° C (194°F), the cooling water from the Paratunka River should have been 5°C (41°F), and the condensation temperature of the Refrigerant-12 should have been 15°C (59°F), instead of 81.5°C (178.7°F), 6°–8°C (42.8°–46.4°F), and 32°C (89.6°F), respectively. Note also that the performance data quoted in the above reference leads to a negative pinch-point temperature difference in the geofluid/Refrigerant-12 heat exchanger. This result is prohibited by the laws of thermodynamics and casts doubt on the validity of the data as reported.

Nevertheless, the efficiency of the energy conversion system can be determined in terms of the amount of hot water required for a given output. Figure 11.5 shows hot water flow rate as a function of gross power output. The actual fluid consumption was roughly 65% higher than the design value at 680 kW because of the 8.5°–9°C shortfall in geofluid temperature. In fact, the actual fluid requirement exceeded the calculated value by about 9% at the actual temperature. The specific hot water consumption at maximum load (680 kW) was about 412 kg/kW·h (908 lbm/kW·h). This converts to a geothermal resource utilization efficiency of 23% (gross), or 15% (net) when 110 kW for the cooling water pump and 130 kW for the two Refrigerant-12 circulating pumps is subtracted from the gross output. The turboexpander reached an isentropic efficiency of 82% at full load, two percentage points below the design value.

The turbine-generator, three preheaters, the boiler/superheater, two condensers, and associated auxiliary equipment were located in a machine

Table 11.4—*Technical specifications for Paratunka geothermal power plant* [a]

Turbine data:

Type.......................................	Radial outflow
Rated capacity.............................	680 KW
Maximum capacity.........................	750 KW
Secondary working fluid...................	Dichlorodifluoromethane, CCl_2F_2 (Ref-12)
Ref-12 inlet pressure.......................	202.7 lbf/in²
Ref-12 inlet temperature...................	149°F
Ref-12 exhaust pressure....................	113.8 lbf/in²
Ref-12 exhaust temperature................	~105°F
Ref-12 mass flow rate......................	~640 × 10³ lbm/h

Geothermal fluid data:

Inlet pressure.............................	42.6 lbf/in²
Inlet temperature..........................	178.7°F
Outlet temperature........................	~113°F
Hot water flow rate........................	617 × 10³ lbm/h

Condenser data:

Type.......................................	Surface type, shell-and-tube
Cooling water inlet temperature............	43°–46°F
Cooling water outlet temperature...........	55°–58°F
Cooling water flow rate....................	3.307 × 10⁶ lbm/h

Heat rejection system:

Type.......................................	Once-through, Paratunka River
Water pump power.........................	110 kW

[a] Year of startup: 1967 (now dismantled).

hall 12 m wide, 24 m long, and 8 m high (39×79×26 ft). A photograph of the turbine-generator is shown in figure 11.6. The specific installed cost of the plant has been reported to be four times that of the other Soviet geothermal power plant located at Pauzhetka. Paratunka's high cost was attributed to the small size of the unit, the costs associated with development of the unique halocarbon turboexpander, and with installation of the piping system to supply the adjacent greenhouse facilities [ARPA, 1972].

11.4 Bolshoye-Bannoye

A 1965 report stated that a sophisticated, multiple-flash geothermal power plant was under construction at Bolshoye-Bannoye. [ARPA, 1972]. Only 20 wells, however, had been completed and the rate of construction was slow. Composition of the geothermal liquid from a typical well in the Bolshoye-Bannoye region (well No. 35) is given in table 11.5 [Vakin et al., 1970].

Whether the plant was completed or has ever been operated is unknown. It was to have a rated output of 8 MW, with two 2.5-MW low-pressure turbines and four 750-kW very low pressure turbines. A flow diagram for the plant is shown in figure 11.7. A mixture of geothermal steam and hot

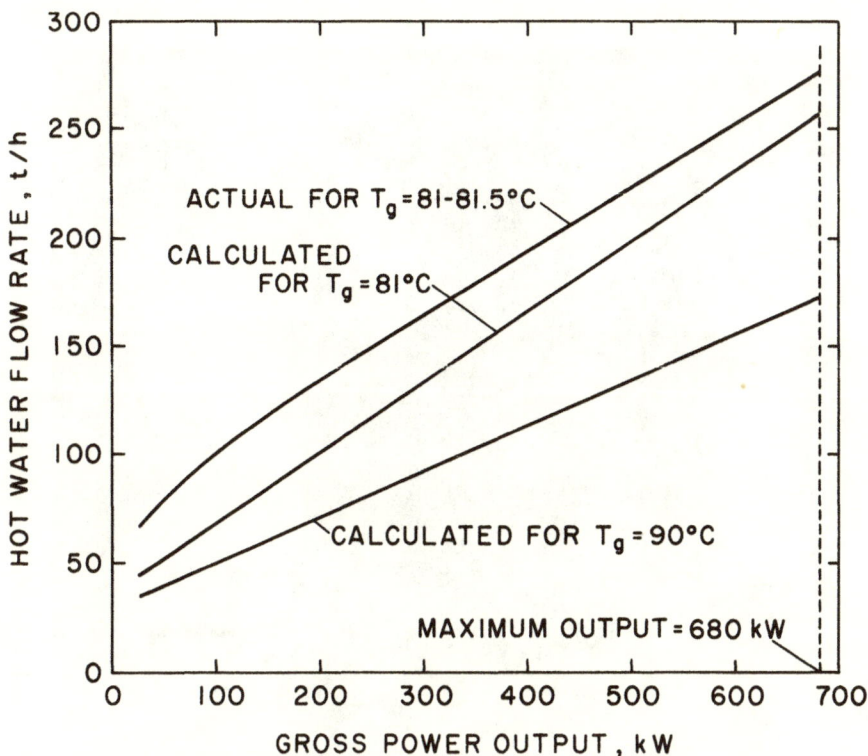

FIGURE 11.5—Performance characteristics of Paratunka binary plant
[after Moskvicheva and Popov, 1970].

Table 11.5—*Composition of fluid from well No. 35 at Bolshoye-Bannoye* [a, b]

Component	Concentration, ppm
Sulfate, SO_4...	532
Sodium, Na...	324
Silica, H_2SiO_3...	290
Chloride, Cl........	127
Bicarbonate, HCO_3..	42. 7
Carbonate, CO_3.....	37. 2
Potassium, K....	25
Calcium, Ca..	23
Arsenic, As...	0. 12
Ammonium, NH_4..	0. 1
Total dissolved solids....	1400

[a] Source: Vakin et al., 1970.
[b] pH=8.7.

FIGURE 11.6—Paratunka turbogenerator unit [Moskvicheva and Popov, 1970].

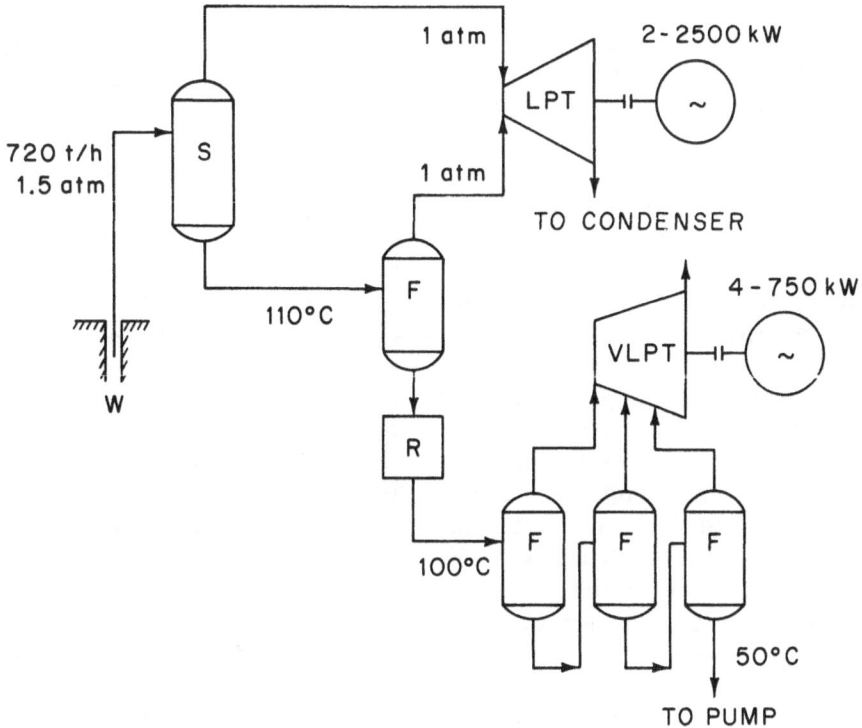

FIGURE 11.7—Flow diagram for proposed multiple-flash plant at Bolshoye-Bannoye [after ARPA, 1972].

water from a number of wells is fed to a series of separators at a pressure of 152 kPa (22 lbf/in²). By the time the separated steam reaches the low-pressure turbines, the pressure has fallen to 101 kPa (14.7 lbf/in²). An intermediate flash tank generates additional steam at one atmosphere for the low-pressure turbines from the hot water separated at the wellheads. The remaining hot water is first collected in a receiver and then flashed successively to produce three streams of subatmospheric steam for use in a set of four multiple-admission turbines. The liquid effluent from the final flasher must be pumped back to atmospheric pressure for disposal.

Based upon the exergy of the geofluid at the wellhead and the indicated geofluid flow rate and power output, the plant would have a gross geothermal resource utilization efficiency of 35%, or a specific geofluid consumption of 90 kg/kW·h (198 lbm/kW·h). Quality of the geofluid mixture at the wellhead would be about 7%. The cost of electricity from the plant was estimated to be about one-sixth the cost of electrictiy from conventional sources serving the city of Petropavlovsk-Kamchatskiy [ARPA, 1972].

11.5   Potential Soviet geothermal power plants

A limited amount of information is available on Soviet geothermal power plant developments. The sites described below seem to be the best prospects for new plants in the Soviet Union.

Makhachkala. A 12-MW flash-steam geothermal plant has been proposed to satisfy the electrical and heating requirements of the town of Makhachkala in the Dagestan ASSR. Estimates are that a total geofluid discharge of about 100 t/h (220×10^3 lbm/h) will be required to supply the plant. Very deep wells of the order of 4–4.5 km (2.5–2.8 mi) are needed to tap waters at 160°C (320°F). Water at 120°C (248°F) may be obtained from wells 2.5 3 km (1.6–1.9 mi) in depth. No other technical details on this plant have been made available.

Yuzhno-Kurilsk. A geothermal power plant of about 5 to 6 MW capacity has been proposed for the Goryachy Playazh geothermal area on Kumashir Island of the Kurile Island group, about 8 km (5 mi) from the town of Yuzhno-Kurilsk. The plant will be designed to use geofluid at 130°C (266°F). Numerous surface thermal manifestations exist in the Guryachy Playazh region, with several steam vents having temperatures of 100°–130°C (212°–266°F).

Nizhne-Koleshevskaya. A recent report indicated that a plant of 50 to 70 MW capacity will soon be constructed at Nizhne-Koleshevskaya [ECPE, 1977], but no additional data were included.

Avachinski Volcano. The same source also reported that plans are underway to tap the Avachinski Volcano on the Kamchatka Peninsula at a depth of 3.5 km (11,500 ft) in the hope of establishing a resource that might supply a 5000-MW geothermal plant for 500 years.

11.6   Future use of geothermal energy in the Soviet Union

Beneath the surface of Siberia lies the largest known reservoir of geothermal hot water. This area is sparsely populated, and the temperature of the water is not high enough at currently drillable depths to permit efficient production of electricity. This resource will likely be developed for domestic heating and agricultural purposes where it is feasible to do so.

With regard to geothermal power plants in the Soviet Union, information generally is difficult to obtain, and once obtained is difficult to confirm. The bulk of the information on which this report is based appeared in papers included in the 1970 United Nations Symposium on the Development and Utilization of Geothermal Resources in Pisa, Italy. It is known, however, that the Soviet research and development effort in geothermal energy has included the design, construction, and testing of a so-called total-flow turbine to make use of the total geofluid mixture as it emerges from the wellhead [Naymanov, 1970]. It seems that the Soviet Union intends to exploit its geothermal resources by providing an efficient match between the characteristics of its relatively low-temperature hydrothermal fluids and the requirements of selected applications such as domestic space heating, agricultural soil heating, or electric power production.

REFERENCES [a]

ARPA, 1972. "Soviet Geothermal Electric Power Engineering," Report 2, ARPA Order No. 1622–3, Informatics, Inc., Rockville, MD.

ECPE, 1977. "5000 MW Geothermal Power Plant?", *Energy in Countries with Planned Economies*, Vol. 1, No. 4, Dec. 14.

Kutateladze, S. S., Moskvicheva, V. N., and Petin, Yu. M., 1975. "Applications of Low-Temperature Heat Carriers in Geothermal Energetics for Use as a Secondary Energy Source in Industry," *San Francisco, 1975*, Vol. 3, pp. 2031–2035.

Makarenko, F. A., Mavritsky, B. F., Lokshin, B. A., and Konov, V. I., 1970. "Geothermal Resources of the USSR and Prospects for Their Practical Use," *Pisa, 1970*, Vol. 2, pp. 1086–1091.

Moskvicheva, V. N. and Popov, A. E., 1970. "Geothermal Power Plant on the Paratunka River," *Pisa, 1970*, Vol. 2, pp. 1567–1571.

Naymanov, O. S., 1970. "A Pilot Geothermoelectric Power Station in Pauzhetka, Kamchatka," *Pisa, 1970*, Vol 2, pp. 1560–1566.

Smith, R. A., 1978. "Soviet Failure," *Geothermal Report*, Vol. 7, No. 24, Dec. 15, p. 3.

Tikhonov, A. N. and Dvorov, I. M., 1970. "Development of Research and Utilization of Geothermal Resources in the USSR," *Pisa, 1970*, Vol. 2, pp. 1072–1078.

Vakin, E. A., Polak, B. G., Sugrobov, V. M., Erlikh, E. N., Belousov, V. I., and Pilipenko, G. F., 1970. "Recent Hydrothermal Systems of Kamchatka," *Pisa, 1970*, Vol. 2, pp. 1116–1133.

---

[a] See note on p. 24.

CHAPTER 12

UNITED STATES

12.1 Historical background

"The most important factors having been worked out—a successful method of drilling into high pressure natural steam bearing areas and a successful method established for the control of steam thus produced—and with a solution of the minor factors now assured, it would appear that the successful commercial use of this natural resource is on the eve of accomplishment at The Geysers."

These words were written by H. N. Siegfried, an engineer for the Southern Sierras Power Company, in a 1925 report to A. B. West, president and general manager of the same company [Siegfried, 1925]. Thirty-five years later the first commercial geothermal power plant came on-line at The Geysers. At the time of Siegfried's comment, six steam wells had been drilled (all of them producers), flow tests had been carried out, and estimates of power output had been made. His report contained a layout of a simple power plant with three 10,000-kW turbines. The Geysers Unit 1 produced 11,000 kW from a single turbine starting on September 25, 1960.

The history of geothermal energy in the United States dates back more than 130 years. Explorer-surveyor William Bell Elliott is credited with discovering The Geysers natural steam field while hunting bears in April 1847 between Cloverdale and Calistoga [Lengquist and Hirschfeld, 1976]. The awesome sight of clouds of water vapor shooting high into the air accompanied by the roar of escaping steam and the odor of sulfur fumes led Elliott to believe he had discovered the very gates to the Inferno.

The region was exploited at first as a tourist attraction boasting alleged therapeutic qualities for the hot fluids and advertised as "The Eighth Wonder of the World," "Nature's Gift to California," and "The Tourists' Paradise." The register of the resort hotel contained the names of Theodore Roosevelt, William McKinley, Ulysses Grant, and Mark Twain. Following the passing of the resort as a popular attraction, an attempt was made in the early 1920's to develop its potential for electric power production. John D. Grant, a rock, gravel, and cement contractor from Healdsburg, not far from The Geysers, deserves the credit for initiating the development of The Geysers [Siegfried, 1925]. Grant secured the services of J. D. Galloway, a San Francisco consulting engineer, who supervised the technical aspects of the work. The drilling was done by Fred Stone, part owner of the Doheny-Stone Drilling Company, who used his invention called the "Hydrill," a hydraulically operated drilling head that could control the flow of well fluid through the casing annulus from the wellhead [Anderson and Hall, 1973].

Large quantities of underground steam were tapped with relatively shallow wells. Eight wells were drilled to depths of between 47 and 194 m (154 to 363 ft.) The steam was used to power a 250-kW generator driven by a reciprocating, noncondensing engine. Further power development at that time, however, was not carried out. Although the major technical obstacles to the use of geothermal energy seemed to have been overcome, geopower plants would have to compete with the hydroelectric plants that were appearing in the same areas where geothermal energy was present. Furthermore, the 1920's, 1930's, and 1940's were years of very rapid growth of the fossil-fuel industry, and oil and gas were both plentiful and relatively inexpensive.

And so the potential of The Geysers lay unexploited until B. C. McCabe, a Los Angeles lumber merchant with no engineering education and no experience in the power industry, decided to invest in the site in the early 1950's. He leased 14.6 km² (3620 acres) from The Geysers Development Company and established the Magma Power Company. In 1955 he drilled his first well, Magma No. 1; it was 249 m (817 ft) deep and produced 68 Mg/h (150,000 lbm/h) of dry steam at a wellhead pressure of 790 kPa (114 lbf/in²).

McCabe joined forces with Dan A. McMillan, Jr., of Thermal Power Company and by 1957 they had completed six wells ranging in depth from 161 to 431 m (527 to 1414 ft). They signed a contract on October 30, 1958, with Pacific Gas and Electric Company (PG&E) that obligated Magma-Thermal to supply steam at a flow rate of 107 Mg/h (235,000 lbm/h) and at a pressure of 790 kPa (114 lbf/in²) to the strainer inlet of a 12,500-kW turbine-generator [Lengquist and Hirschfeld, 1976]. In 1967 Union Oil Company of California joined Magma-Thermal in development of the geothermal field. A total of about 61 km² (15,000 acres) is under lease. The U.S. Department of the Interior has classified an additional 668 km² (165,000 acres) surrounding The Geysers as a Known Geothermal Resource Area (KGRA).

In the early days of geothermal energy discovery in the United States, several areas besides The Geysers were explored. Most of these appeared promising because of surface manifestations such as steam vents, hot springs, and boiling mud pots. Potential sites in the Imperial Valley of California such at Niland and Mullet Island (Salton Sea) were studied and drilled as early as 1927. Fred Stone, who carried out the drilling, reported that the wells at the Salton Sea encountered gas (probably carbon dioxide), considerable steam, water, and a large quantity of "slush" [Anderson and Hall, 1973]. The wells were good steam producers, but the slush caused a number of problems. Although Stone believed these difficulties would clear up in time, they persist to the present. The geothermal brines of the Salton Sea contain 250,000–300,000 ppm of total dissolved solids [Austin et al., 1977], and utilization of these fluids remains one of the major unsolved problems in geothermal energy in the United States.

A problem with nearly all geothermal regions in the early days was their remoteness. Expensive transmission lines would have to be constructed to transmit the power from geothermal sites to populated areas. Although the population of the western United States has grown considerably during the past 50 years, this economic factor still plays a role in development of the geothermal resources of the country.

12.2 The Geysers at Sonoma and Lake Counties, California

The largest geothermal electric power complex in the world is located at The Geysers in Sonoma County of northern California. Pacific Gas and Electric Company produces more than 660 MW at the site; 245 MW of additional capacity are under construction, and PG&E intends to install an additional 220 MW by 1982. The present proved capacity of The Geysers exceeds 2000 MW.

12.2.1 Geology and exploration

The Geysers is one of several areas of hot springs and fumaroles that occur along a section of a long fault zone in the Mayacmas Mountains in northern California. The geothermal reservoir is of the vapor-dominated type and extends over an area 21.5×8.6 km (13.3×5.3 mi), bounded by the Mercuryville fault zone on the southwest and the Collayomi fault zone on the northeast [Donnelly et al., 1976]. The drilled area covers more than 50 km² (12,350 acres) in an 11.2×4.5 km (7×2.5 mi) strip lying roughly between Big Sulphur Creek in Sonoma County and the border between Sonoma and Lake counties. An adjacent, and as yet untapped, liquid-dominated reservoir is believed to lie to the northeast.

The Geysers-Clear Lake geothermal region lies in a zone of transform faulting between the Pacific and North American plates. The region is characterized by very young faults, and the geology is quite complex. The Mayacmas Mountains, a post-Pliocene uplift, are formed of an internal structure of differentially uplifted fault blocks tilted 30°–40° to the northeast. The majority of the faults dip steeply to the southwest, with vertical displacements along individual blocks ranging from 1.5 to 3 km (0.9 to 1.9 mi). Surface rocks are from the Jurassic to Cretaceous periods and are of marine sediments from both shallow and deep depositional environments [Reed and Campbell, 1975].

The precise nature of the hydrothermal reservoir and the exact mechanism for production of steam in The Geysers field remain controversial. A large negative Bouguer gravity anomaly ($\sim$25 mgal) exists in the center of the field, accompanied by an electrical resistivity low. Large P-wave delays ($\sim$1 sec) have been observed, as well. Although no clear and consistent interpretation of these observations has been formulated, we can conclude that the source of the thermal anomaly is a deep-seated magmatic intrusion that lies embedded in the crust at a depth of about 10 km (6 mi).

The steam-producing areas are highly fractured regions with nearly vertical orientation. The fractures occur in hard, dense graywacke (a sandstone), and steam is found in two depth ranges: a shallow zone at 300–600 m (984–1968 ft) and a deep zone at 1500–3000 m (4920–9840 ft) [Reed and Campbell, 1975]. The vapor-dominated reservoir is characterized by a pressure far less than the hydrostatic pressure and extends to depths of at least 3 km (9840 ft). The pressure anomaly is so drastic that at a depth of about 3 km (9840 ft) the reservoir pressure is only 3.5 MPa (500 lbf/in^2), whereas at a horizontal distance of merely 0.8 km (2600 ft) away from the reservoir it is 30 MPa (4300 lbf/in^2). A very efficient self-sealing mechanism is clearly present here to provide a tight cap rock, otherwise the reservoir would quickly be inundated by near-surface ground water. In addition, there is a linear temperature profile through the cap rock signifying that heat transfer in that layer is purely by conduction. The solidarity of the cap rock is all the more remarkable in light of the high tectonic activity in the region.

In regard to the origin and nature of the vapor-dominated reservoir, there appear to be at least two schools of thought. It has been proposed that the presence of the vapor-dominated zone is the result of the boiling of a deep water table and that the pressure anomaly is preserved by a seal of minerals at the boundary caused by deposition of solids from the geothermal fluid, in particular calcite ($CaCO_3$) and anhydrite ($CaSO_4$) [White et al., 1971; Truesdell and White, 1973]. Opponents of this view argue that the water table would need to fall by at least 2 km (6560 ft) to account for the size of the vapor-dominated region [Weres et al., 1977]. Furthermore, it has been shown that there is insufficient volume in the reservoir to contain only steam in the quantities required to explain the observed production from the field [Witherspoon, 1977]. An alternate model, is one in which the pores or fractures in the reservoir contain liquid as well as vapor. In this model the liquid may even be at relatively shallow depths, sealed in pores that have been closed by movements of elements of the fault zone. Witherspoon [1977] speculates that the liquid water is of meteoric origin (rain water) at least 50 years old.

In summary, the existence and exploitability of The Geysers dry-steam field may be attributed to three causes. First, a network of passageways allows meteoric water to percolate into the reservoir, serving to recharge but not drown the reservoir. Second, favorable local structural traps exist to accumulate steam in faults and fractures or in crests of structural highs. Third, the magma intrusion at depth serves as a strong heat source.

12.2.2   Steam wells and gathering system

More than 200 wells have been drilled at The Geysers; 75 of these actively deliver steam to the first 11 units, which have a total installed capacity of 502 MW. Approximately 15 wells are needed to support a typical 110-MW unit. A steam well will produce 34 to 159 Mg/h (75,000 to 350,000 lbm/h) at a wellhead pressure of 960 kPa (140 lbf/in²). A steam flow rate of 91 Mg/h (200,000 lbm/h) may be taken as typical. The closed-in pressure is about 3.1 MPa (190 lbf/in²), and the corresponding temperature is 240°C (465°F), with a specific enthalpy of 2800 kJ/kg (1204 Btu/lbm).

A combination of mud and air drilling is used, with mud used for the larger portions of the wells where diameter exceeds 317.5 mm (12.5 in). Although air drilling is faster, since the cuttings are removed more quickly from beneath the drill bit, it can be used only in the smaller-diameter sections where seepage of liquid into the hole is not a serious problem. A typical casing program is shown in figure 12.1. Mud is used down to about 600 m (1968 ft), mud or air (depending on the presence or lack of ground-

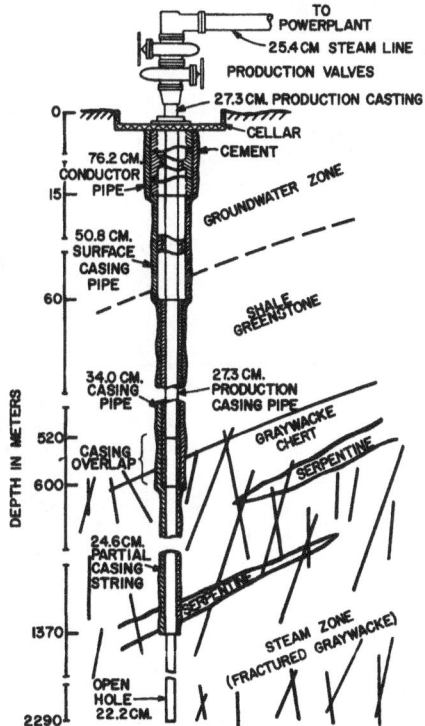

FIGURE 12.1—Typical well casing program, The Geysers [Reed and Campbell, 1975].

water) from there to 1370 m (4494 ft), and air from the top of the fractured graywacke to the bottom of the well. Table 12.1 lists characteristics of the two methods of drilling wells; table 12.2 gives the composition of the drilling mud used by Union-Magma-Thermal.

Table 12.1—*Characteristics of mud and air drilling methods* [a]

Characteristic	Mud	Air
Hole diameter.......................	>444.5 mm (17.5 in).	<317.5 mm (12.5 in)
Drilling rate.......................	1 (reference)........	4
Drill bit life.......................	1 (reference)........	2.3–4.6
Time in drilling typical well [b]..........	7 days.............	14–21 days
Wear item.......................	Drill bit............	Drill pipe
Recovery of lost drill string..........	Possible............	Impossible
Steam pressure countered by..........	Mud weight........	[c] Wellhead BOPE
Directional control.................	Fair..............	Poor
Directional devices.................	Bent; kick sub.....	Whipstock
Ground-water seepage..............	>several Mg/h......	<a few Mg/h
Pipe cooling.......................	Good.............	Fair
Noise (unabated)..................	Moderate..........	Extreme
Dust blown from well..............	None.............	Great
Water use.......................	Some.............	For control of dust and noise
Water pollution....................	Possible............	Unlikely
Air pollution.......................	Diesel exhaust......	Diesel exhaust; some steam and rock dust

[a] Source: Weres et al., 1977.
[b] Total drilling time is the sum of the figures shown.
[c] BOPE=blowout prevention equipment.

Table 12.2—*Composition of drilling mud as used by Union-Magma-Thermal at The Geysers* [a]

Component	Formula	Percentage (volume)
Water.......................	H_2O	93. 09
Bentonite....................	$Na_{0.33}Al_{2.67}Si_{3.6}O_{20}(OH)_2$	5. 39
Quebracho....................	Organic, wood extract	0. 45
Caustic soda.................	NaOH	0. 32
Lignin (Tannathin)............	$C_{212}H_{171}O_{41}N_3S$	0. 12
Sodium bicarbonate............	$NaNCO_3$	0. 09
Cottonseed hulls [b].............	Organic	
Walnut shells [b]................	Organic	
Mica [b].......................	$KAl_3Si_3O_{10}(OH)_2$	

[a] Source: Reed and Campbell, 1975.
[b] Materials added to control loss of circulation by plugging the fissures causing the loss.

The spacing between wells is determined by economics and the physical setting in the field. Many drill pads are built on filled land owing to the rugged terrain. The density of wells differs from company to company; Union-Magma-Thermal drills one well per 8 ha (20 acres) and Aminoil drills one well per 16 ha (40 acres) [Weres et al., 1977]. Directional drilling is used to drill under areas that would otherwise be inaccessible, to intersect nearly vertical fissures, to intercept a blown-out well in order to control the blowout, to reduce the number of pads required to drill a number of wells, and to simplify the steam-gathering system necessary to transmit the steam to the powerhouse. Directional drilling with air is accomplished by means of a wedge placed down-hole to deflect the bit off to the side. The wedge is called a whipstock and is capable of producing a deflection of $6.6°–8.2°/100$ m ($2°–2.5°/100$ ft). It is relatively easy to achieve a lateral displacement of 400 m (1310 ft) in a well of 1500 m (4920 ft) nominal depth. Since drilling a directional well may take about one-third longer than drilling a vertical well of the same depth, directional drilling is more costly than vertical drilling. These added drilling costs are partially offset by savings resulting from consolidation of wellhead equipment at a single site; only a small additional area is needed to accommodate up to six wellheads, whereas a single wellhead requires between 0.4 and 1.2 ha (1 to 3 acres).

A typical gathering system for a 55-MW unit consists of a network of carbon steel pipes, starting with 254-mm (10-in) O.D. pipes at the wellheads and ending with 914-mm (36-in) O.D. pipes of 9.5-mm (0.375-in) wall thickness at the powerhouse. Usually seven wells must be connected to the system to supply the required 450 t/h (10^6 lbm/h) of steam. A centrifugal axial separator is situated on the steam line at each well to remove particulate matter that can cause erosion of the steam pipes and turbine blades [Matthew, 1975]. A map of The Geysers area in figure 12.2 shows the locations of the first 15 units. The steam pipelines are no longer than about 2 km (6560 ft) to control loss of availability of the steam from the wellhead to the turbine.

12.2.3 Energy conversion systems

The power units at The Geysers have evolved from relatively small units with barometric, external condensers and no emissions controls to units of 110-MW capacity with low-level, surface-type condensers and Stretford-type H_2S removal systems. Table 12.3 lists the geothermal units at The Geysers and provides technical information on each of them. Table 12.4 summarizes technical specifications for power units in operation at The Geysers as of early 1979. In the sections that follow, each unit will be described in some detail.

Unit 1. The first two units of The Geysers project are housed in a single powerhouse located adjacent to Big Sulphur Creek, not far from the site

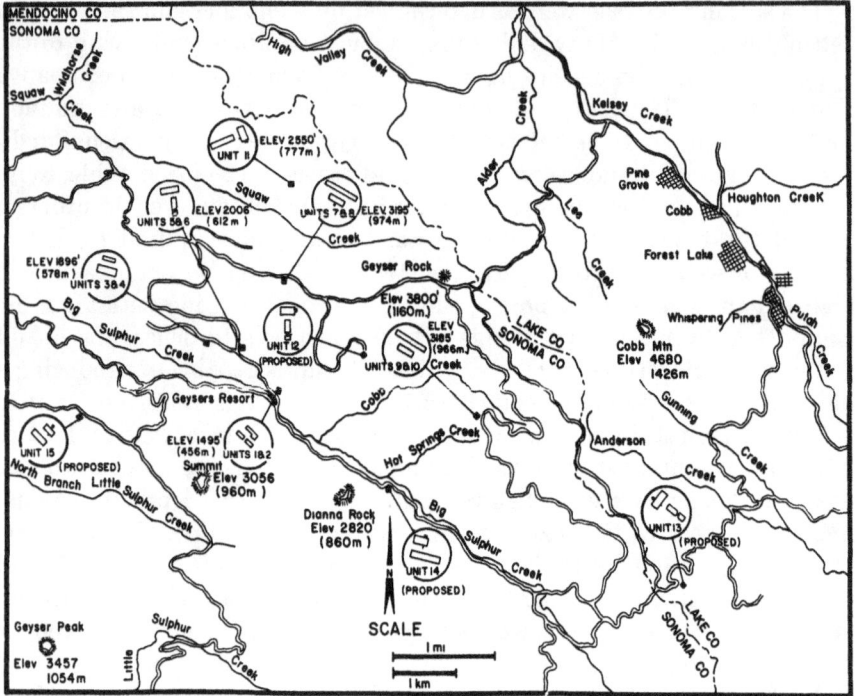

FIGURE 12.2—Map of The Geysers area showing location of Units 1–15 [Dan et al., 1975].

of the original Geysers Resort. An aerial photograph of the plant site is shown in figure 12.3; an elevation drawing of the unit including the cooling tower is given in figure 12.4; a heat balance diagram for Unit 1 is shown in figure 12.5. Table 12.4 contains technical specifications for the major mechanical elements of the unit. Although the maximum capacity of the unit is 12.5 MW, it is usually operated at 11 MW to allow for flexibility in load factor. The steam turbine for Unit 1 is a reconditioned General Electric machine of 1924 vintage originally in service at a PG&E plant in Sacramento. The original version had nine stages and was designed for steam at 1.31 MPa (190 lbf/in²) and 248°C (478°F) with exhaust at 5.1 kPa (1.5 in Hg). The first three stages were removed, the control valves were replaced by two butterfly valves of 152 mm (6 in) and 254 mm (10 in), and the last two stages of buckets were rebuilt of 11–13% chrome steel. Other modifications were made to allow the turbine to accommodate the geothermal steam conditions and the higher back-pressure [Bruce, 1961].

Unit 2. Unit 2 came on-line in 1963, 3 years after the first unit, and shares the same powerhouse with Unit 1. The arrangement of the unit is similar to the first, although the turbine is a newly designed unit by the Elliott Company, rated at 13 MW [Barton, 1970; Bruce, 1970]. Turbine inlet

Table 12.3—*Geothermal power plant development at The Geysers natural steam field, California*

Utility/Unit [a]	Year of startup	Ca-pacity MW	Turbine Manufacturer	Type	Con-denser type [b]	Steam supplier [c]
PG&E, No. 1..	1960	11	G.E.	6×1	DCEB	U-M-T
PG&E, No. 2...	1963	13	Elliott	5×1	DCEB	U-M-T
PG&E, No. 3..	1967	27	Elliott	7×1	DCEB	U-M-T
PG&E, No. 4. .	1968	27	Elliott	7×1	DCEB	U-M-T
PG&E, No. 5. .	1971	53	Toshiba	6×2	DCLL	U-M-T
PG&E, No. 6..	1971	53	Toshiba	6×2	DCLL	U-M-T
PG&E, No. 7...	1972	53	Toshiba	6×2	DCLL	U-M-T
PG&E, No. 8 ..	1972	53	Toshiba	6×2	DCLL	U-M-T
PG&E, No. 9...	1973	53	Toshiba	6×2	DCLL	U-M-T
PG&E, No. 10.	1973	53	Toshiba	6×2	DCLL	U-M-T
PG&E, No. 11	1976	106	Toshiba	6×4	DCLL	U-M-T
PG&E, No. 12.	1979	106	Toshiba	6×4	DCLL	U-M-T
PG&E, No. 13..	1980	135	G.E.	6×4	STST	Aminoil
PG&E, No. 14	1980	110	Toshiba	6×4	STST	U-M-T
PG&E, No. 15.	1979	55	Toshiba	5×2	STST	Thermogenics
PG&E, No. 16.	1983	110	Toshiba	6×4	STST	Aminoil
PG&E, No. 17	1982	110	Toshiba	6×4	STST	U-M-T
PG&E, No. 18	1982	110	([d])	([d])	STST	U-M-T
PG&E, No. 19..	1982	110	([d])	([d])	STST	Aminoil
PG&E, No. 20..	1983	110	([d])	([d])	STST	U-M-T
PG&E, No. 21..	1983	110	([d])	· ([d])	STST	U-M-T
NCPA, No. 1. .	1981	110	Fuji	([d])	STST	Shell Oil

[a] PG&E=Pacific Gas & Electric Company; NCPA=Northern California Power Agency.
[b] DCEB=Direct-contact, external, barometric type; DCLL=Direct-contact, low-level type; STST=Shell-and-tube, surface type.
[c] U-M-T=Union Oil-Magma Power-Thermal Power.
[d] Not available.

pressure is about 550 kPa (80 lbf/in²), somewhat lower than that of the first unit. Table 12.4 gives the unit's technical specifications.

Units 3 and 4. Units 3 and 4 are essentially identical. Each has about twice the capacity of either Unit 1 or 2 [Barton, 1970; Bruce, 1970]. Steam flow to the turbine of Unit 4 is about 1.2% greater than that for Unit 3 because Unit 4's generator is air cooled rather than hydrogen cooled. The photograph in figure 12.6 shows the site of these units, including a portion of the steam-gathering system. The rugged nature of the terrain is evident from the photograph. Figure 12.7 is a flow diagram for Unit 3, and table 12.4 lists technical data for the major components.

Units 5-10. Units 5-10, each 55 MW, are about double the capacity of Units 3 and 4. The increase in unit size was made possible by confidence in the ability of the geothermal reservoir to produce the required steam flow

Table 12.4—*Technical specifications for geothermal units in operation at The Geysers*

	Unit No.					
	1	2	3–4	5–10	11–12	15
Turbine data:						
Type [a]	SCSF	SCSF	SCSF	SCDF	TCFF	SCDF
Rated capacity, MW	11	13	27	53	106	55
Speed, rpm	1800	3600	3600	3600	3600	3600
Steam pressure, lbf/in²	93. 9	79. 7	78. 9	[b] 113. 0	113. 3	113. 7
Steam temperature, °F	348	348	341. 8	355	355	338
Noncondensable gas, % wt	<0. 3	<0. 3	<0. 5	<0. 5	<1. 0	~1. 0
Exhaust pressure, in Hg	4. 0	4. 0	4. 0	4. 0	4. 0	5. 5
Steam flow rate, 10³ lbm/h	240	255	[c] 510	907. 5	1808	~1100
Condenser data:						
Cooling water temperature, °F	80. 6	80. 6	80. 6	80. 0	80. 0	80. 0
Outlet water temperature, °F	120. 8	120. 0	119. 8	118. 4	118. 4	120. 4
Water flow rate, 10⁶ lbm/h	5. 5	~6. 0	6. 41	21. 3	~62. 8	~35
Gas extractor data:						
Type	All units have steam jet ejectors.					
Gas capacity, ft³ min	~350	~350	~1000	~1830	~3660	[d]
Steam consumption, 10³ lbm/h	~10	~10	~23	58. 4	120	~50
Heat rejection system data:						
Type	All units have crossflow, mechanically-induced-draft water cooling towers.					
Number of cells	3	3	6	5	10	5
Design wet-bulb temperature, °F	66. 5	65	66	65	65	65
Water pump power, kW	[d]	[d]	[d]	930	1860	~900
Fan motor power, kW	[d]	[d]	355	605	1210	~620

[a] SCSF=Single-cylinder, single-flow; SCDF=Single-cylinder, double-flow; TCFF=Tandem-compound, four-flow.
[b] 113.7 for Units 5 and 6.
[c] 516 for Unit 4.
[d] Not available.

over the anticipated plant lifetime [Finney, 1972; Matthew, 1975]. Such confidence was established through 9 to 10 years of operating experience at the site. For these units a change was made from the barometric jet condensers used on Units 1–4 to low-level, tray-type jet condensers. The turbines, supplied by the Tokyo Shibaura Electric Company, Ltd. (Toshiba), are all still of the single-cylinder type, but of the double-flow variety. On the ground floor of the powerhouse are the condensers and various auxiliaries; the turbine and operating floor are on an elevated deck. As illustrated in the heat balance diagram of figure 12.8, no cooling

FIGURE 12.3—Units 1 and 2, The Geysers [photo by PG&E News Bureau].

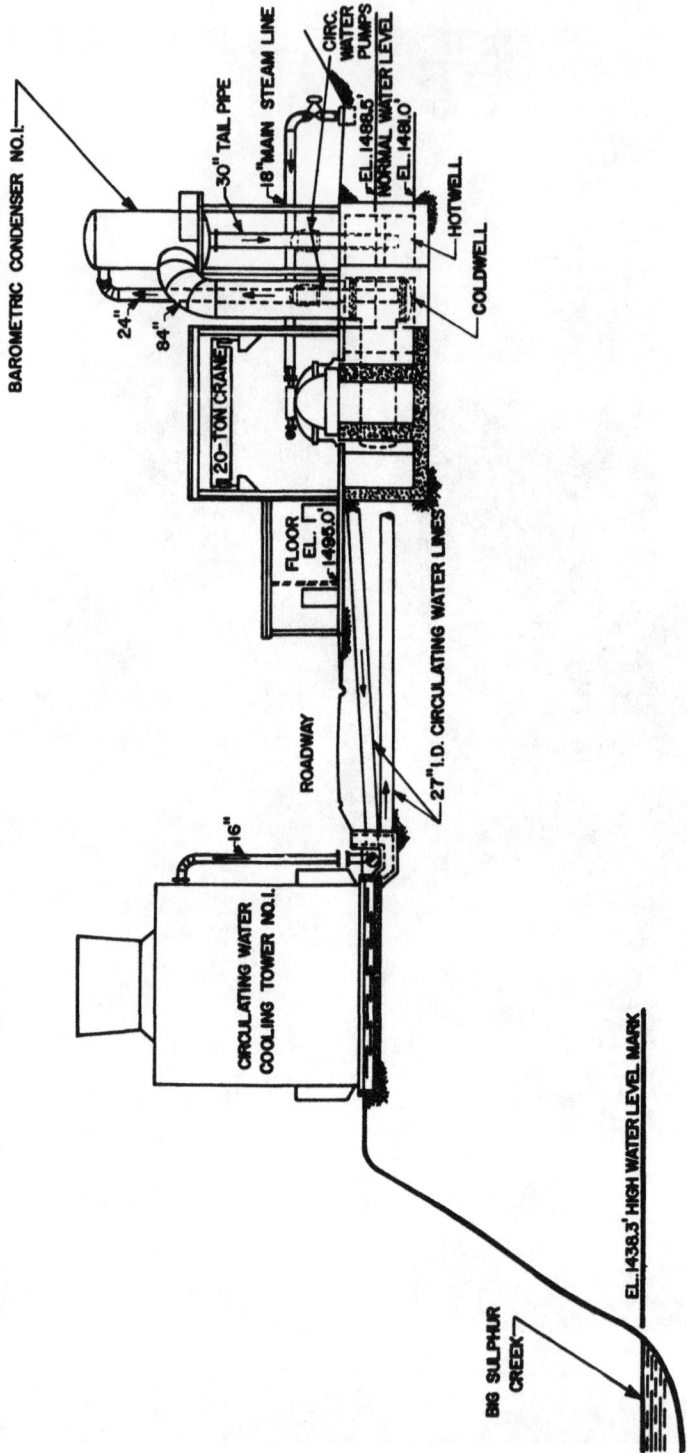

FIGURE 12.4.—Elevation view of Units 1 and 2, The Geysers [Bruce, 1961].

PERFORMANCE

THROTTLE FLOW lb/h 240,000
GENERATOR ELEC. OUTPUT 12,500 kW
AUXILIARY POWER (ELECTRICAL)
  CIRCULATING WATER PUMPS 229.0
  COOLING TOWER FANS 96.2
  EXCITER 69.0
  OTHER 7.8
         TOTAL 402 kW

NET UNIT OUTPUT 12,098 kW
HEAT INPUT 293×10⁶ Btu/h
NET HEAT RATE 24,215 Btu/kW·h
REFERRED TO 60 °F

CONDITIONS

GENERATOR POWER FACTOR = 1.0
CONDENSER BACK PRESSURE 4.0" Hg
DRY BULB AIR TEMPERATURE = 96 °F
WET BULB AIR TEMPERATURE = 66.5 °F
GAS SHOWN ENTERING COOLING
TOWER IS AIR ABSORBED BY WATER.
GASES TO AND FROM COOLING
TOWER DO NOT INCLUDE COOLING
AIR BUT ONLY THAT WHICH IS
ABSORBED AND DEGASSED.
ANY AIR ABSORBED IN COOLING
WATER IN HOT WELL IS NOT
INCLUDED.

LEGEND

- - - - - MAIN STEAM LINE
— — — WATER LINES
— · — · — STEAM LINES
· · · · · GAS LINES
S   POUNDS PER HOUR STEAM
W   POUNDS PER HOUR WATER
g   POUNDS PER HOUR GASES
F   DEGREES FAHRENHEIT
H   STEAM ENTHALPY
A   PRESSURE PSI ABSOLUTE
G   PRESSURE PSI GAGE

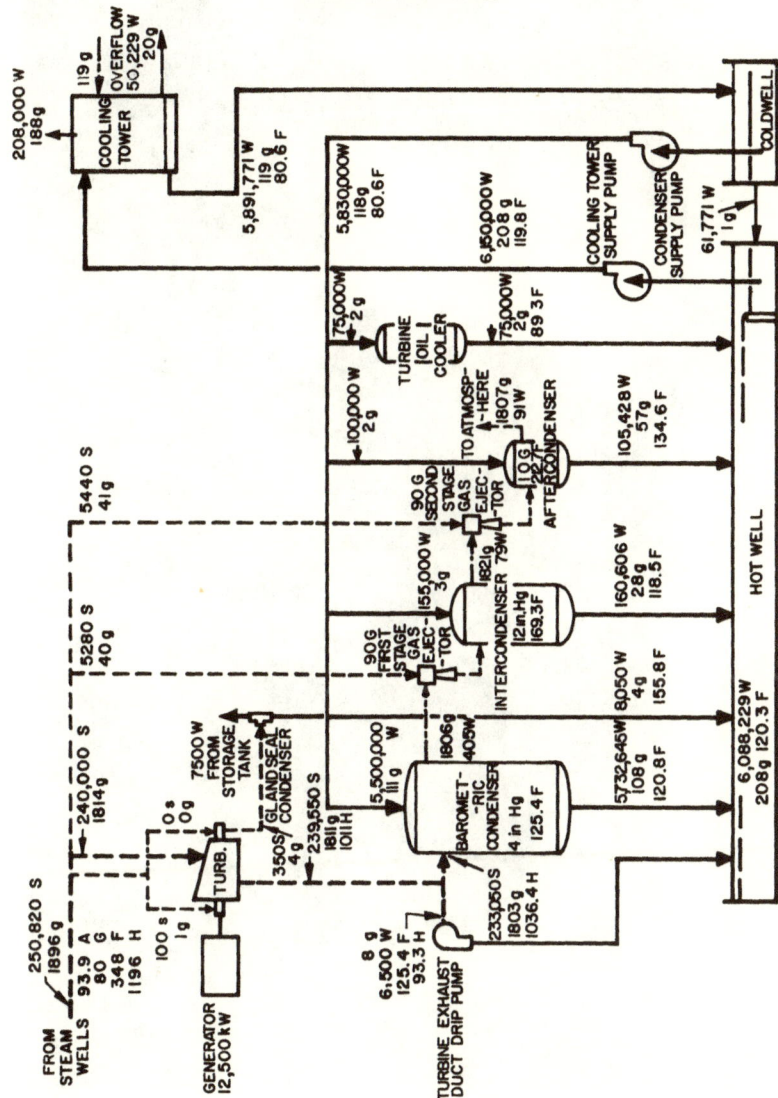

FIGURE 12.5—Heat balance diagram for Unit 1, The Geysers [Bruce, 1961].

FIGURE 12.6.—Units 3 and 4, The Geysers [photo by PG&E News Bureau].

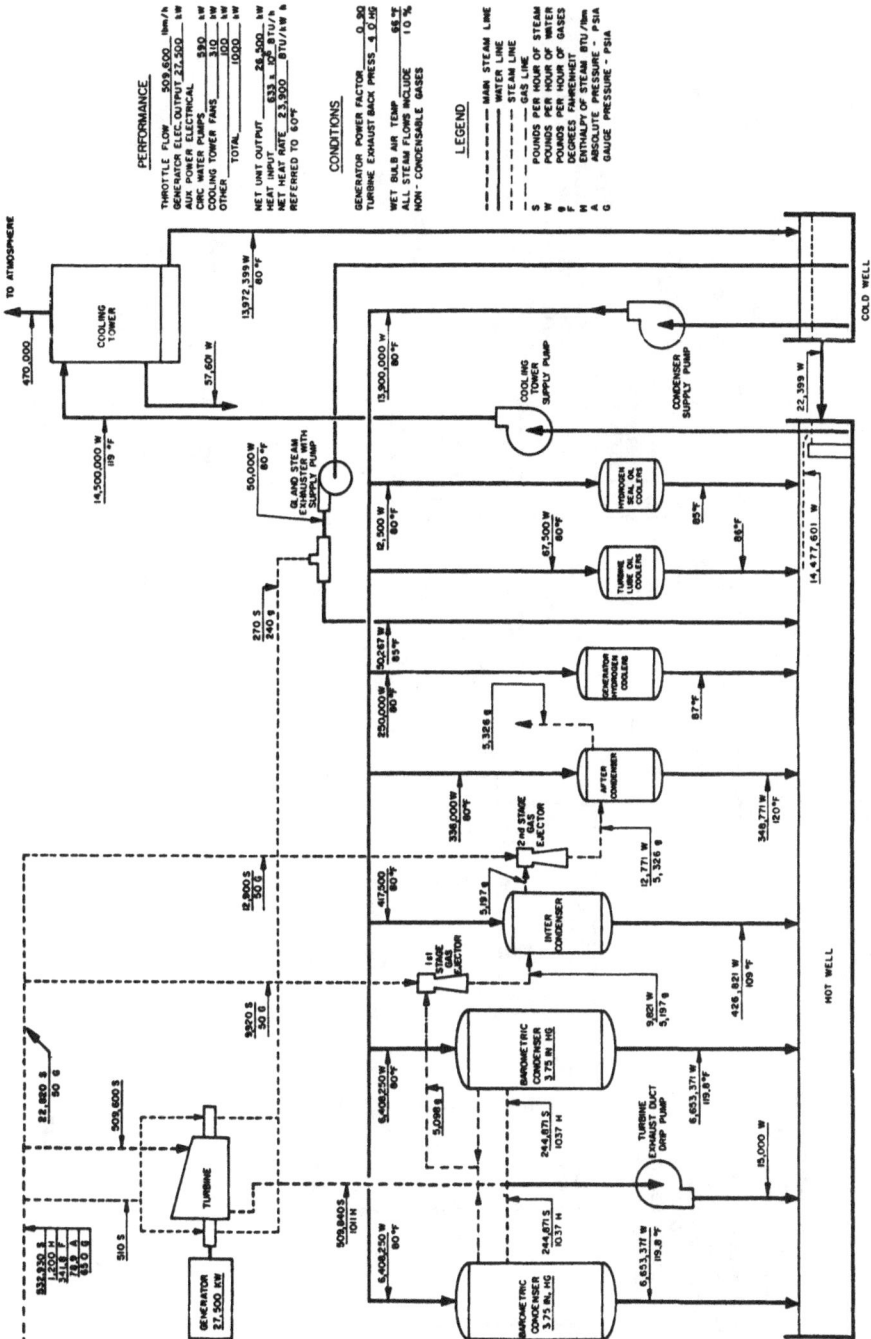

FIGURE 12.7.—Heat balance diagram for Unit 3, The Geysers [Lengquist and Hirschfeld, 1976].

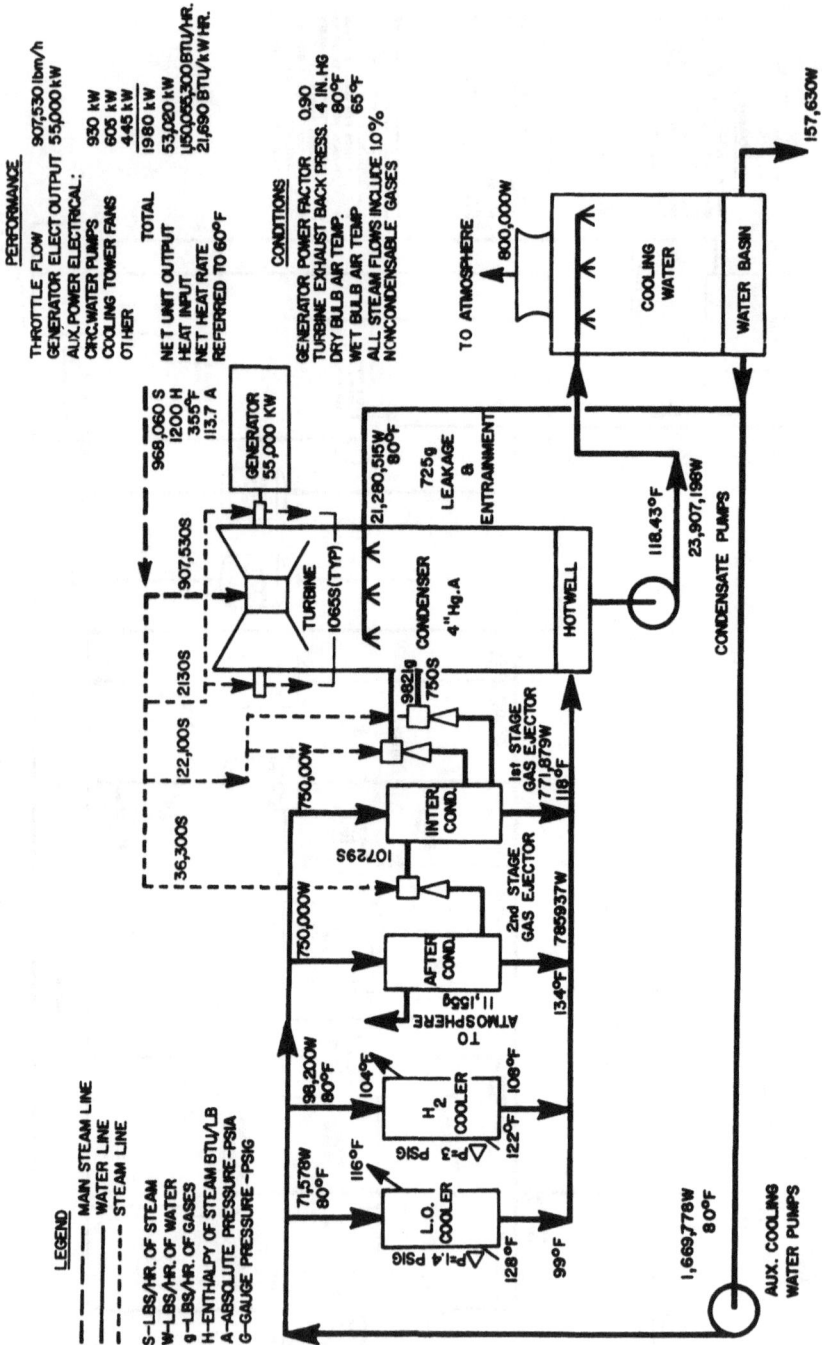

FIGURE 12.8—Heat balance diagram of Units 5 and 6, The Geysers [Matthew, 1975].

water pump is needed between the cooling tower water basin and the low-level jet condenser owing to the condenser vacuum. Auxiliary cooling water pumps are needed, however, to handle the requirements of the inter- and after-condensers, the cooler for the hydrogen used to cool the generator, and the turbine lubricating oil coolers. About 4% of the gross electrical output is required for plant auxiliaries such as circulating water pumps and cooling tower fans. Figures 12.9 and 12.10 show the site of Units 5 and 6 and the site of Units 7 and 8, respectively. Table 12.4 contains technical data on the major components.

Units 11 and 12. Although Units 11 and 12 do not share a common power-house (see figure 12.2), they are essentially twin units and represent another step upward in unit capacity. Each is rated at 106 MW and uses a tandem-compound, four-flow turbine manufactured by Toshiba. Technical data for these units are given in table 12.4. These units represent about the maximum feasible capacity (per unit) in The Geysers; with the exception of Units 13 and 15, all future units are planned to have 110-MW capacities.

Unit 13. Of the two units now under construction, Unit 13, which is expected on-line in February of 1980, is unique in several respects. It is the first unit to be built in Lake County rather than Sonoma County, and the first to be supplied with steam from a producer other than Union-Magma-Thermal; in this case, the supplier is Aminoil (formerly Signal Oil and Gas Company). It will be the largest single geothermal unit in the world, with a rated capacity of 135 MW. Furthermore, the unit will be fitted with a turbine manufactured in the United States, the first American turbine installed at The Geysers in more than a decade. This unit will be located in the lower southwest section of Lake County, just over the boundary from Sonoma County (see figure 12.2). It will be sited on the eastern side of the Mayacmas Mountains in a sheltered valley known as Hidden Valley. Lying about 4 km (2.5 mi) southeast of Units 9 and 10, closest of the other units at The Geysers, it will feed power to PG&E's Fulton substation by means of a new 230-kV transmission line which will run 3.4 km (2.1 mi) to join the one from Units 9 and 10 [GEM, 1977b].

Although the unit was initially designed with a direct-contact condenser of the low-level jet type, it will be built with a surface condenser of the shell-and-tube type in order to assist the hydrogen sulfide abatement system to be installed on the unit. This will be the first unit to have a surface condenser at The Geysers, and the first unit of any dry steam geothermal plant in the world so equipped.[a] Furthermore, this and all future units will be fitted with a means of controlling hydrogen sulfide emissions from the plant. A Stretford system will be employed for units expected on-line in the near future. Because of these late design changes, technical specifications for this unit are not yet known reliably; table 12.5 lists the specifica-

[a] Unit 15, p. 314, which came on-line in June 1979, has taken this particular honor from Unit 13. (*Note added in proof.*)

FIGURE 12.9—Units 5 and 6, The Geysers [photo by PG&E News Bureau].

FIGURE 12.10—Units 7 and 8, The Geysers [photo by PG&E News Bureau].

Table 12.5—*Preliminary specifications for units under construction at The Geysers*

	Unit No.	
	13	14
Turbine data:		
Type [a]...................................	TCFF	TCFF
Rated capacity, MW......................	135	110
Speed, rev/min...........................	3600	3600
Steam pressure, lbf/in²	114	113. 7
Steam temperature, °F.....................	338	355
Noncondensable gas, % wt.................	~1. 0	~1. 0
Exhaust pressure, in Hg....................	5. 5	3. 0
Steam flow rate, 10³ lbm/h.................	~2640	1940
Condenser data:		
Cooling water temperature, °F..............	80. 0	80. 0
Outlet water temperature, °F...............	118. 4	118. 4
Water flow rate, 10⁶ lbm/h.................	([b])	70. 6
Gas extractor data:		
Type.....................................	Both units will have steam jet ejectors.	
Steam consumption, 10³ lbm/h..............	([b])	60
Heat rejection system data:		
Type.....................................	Both units will have cross-flow, mechanically-induced-draft water cooling towers.	
Number of cells...........................	10	10
Design wet-bulb temperature, °F.............	65	65
Water pump power, kW.....................	([b])	~1860
Fan motor power, kW......................	([b])	~1210

[a] TCFF=Tandem-compound, four-flow.
[b] Not available.

tions as they are currently understood, and figure 12.11 is a simplified schematic of the preliminary heat balance diagram.

Unit 14. The design of Unit 14 is similar to that of Units 11 and 12, having a capacity of 110 MW, with Toshiba turbines and steam furnished from Union-Magma-Thermal. The plant will be situated near Big Sulphur Creek, about 2 km (1.2 mi) southeast of Units 1 and 2 (see figure 12.2), and should start up in June of 1980. However, unlike Units 11 and 12, Unit 14 will be equipped with a Stretford hydrogen sulfide removal plant, and a shell-and-tube surface condenser. Compare the plant specifications for the two cases: (1) no H_2S abatement (figure 12.12) and (2) with Stretford abatement system (figure 12.13). In the latter case it may be possible to produce about 3% more output from the same input by lowering the turbine exhaust pressure from 13.5 kPa (4 in Hg) to 10.2 kPa (3 in Hg). In addition, the steam flow to the turbine will increase by about 1% owing to the reduced steam requirements of the gas ejection system when the Stretford plant is used. In this configuration, the flow of cooling water through

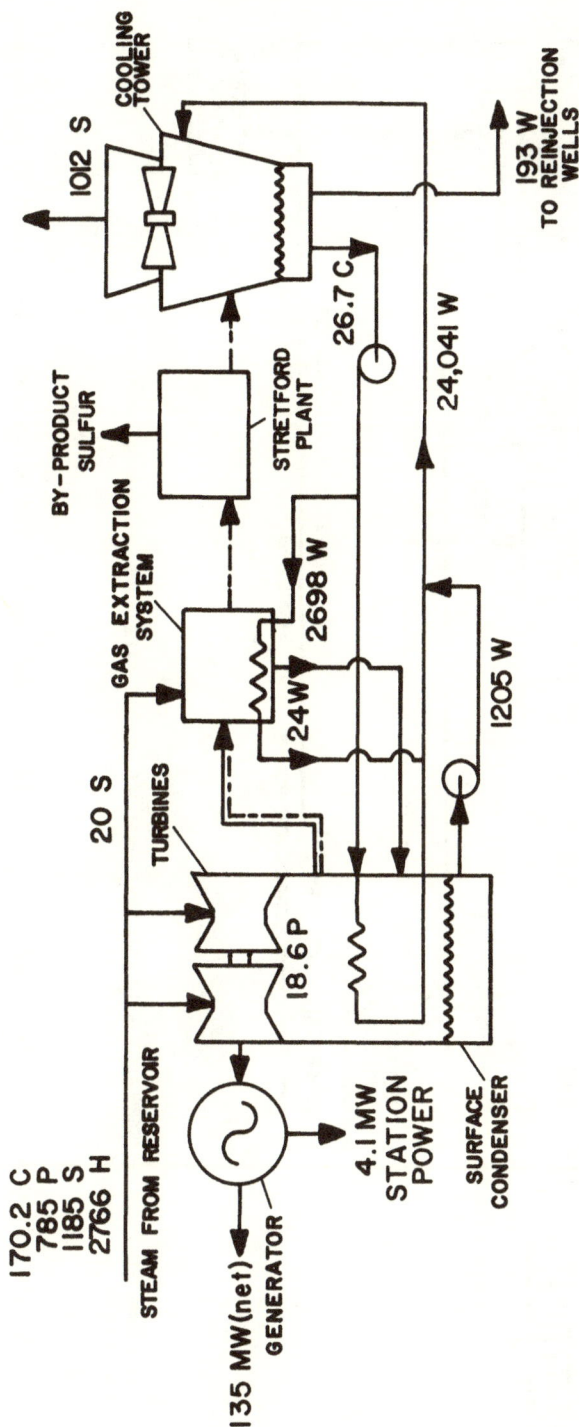

170.2 C
785 P
1185 S
2766 H

STEAM FROM RESERVOIR

135 MW(net)
GENERATOR

4.1 MW
STATION POWER

TURBINES

20 S

18.6 P

SURFACE CONDENSER

GAS EXTRACTION SYSTEM

BY-PRODUCT SULFUR

STRETFORD PLANT

24 W

2698 W

1205 W

COOLING TOWER

1012 S

26.7 C

24,041 W

193 W
TO REINJECTION WELLS

LEGEND

S = STEAM, Mg/h   W = WATER, Mg/h
H = ENTHALPY, kJ/kg   C = TEMPERATURE, °C
P = PRESSURE, kPa

FIGURE 12.11—Heat balance diagram for Unit 13, The Geysers [PG&E, 1975].

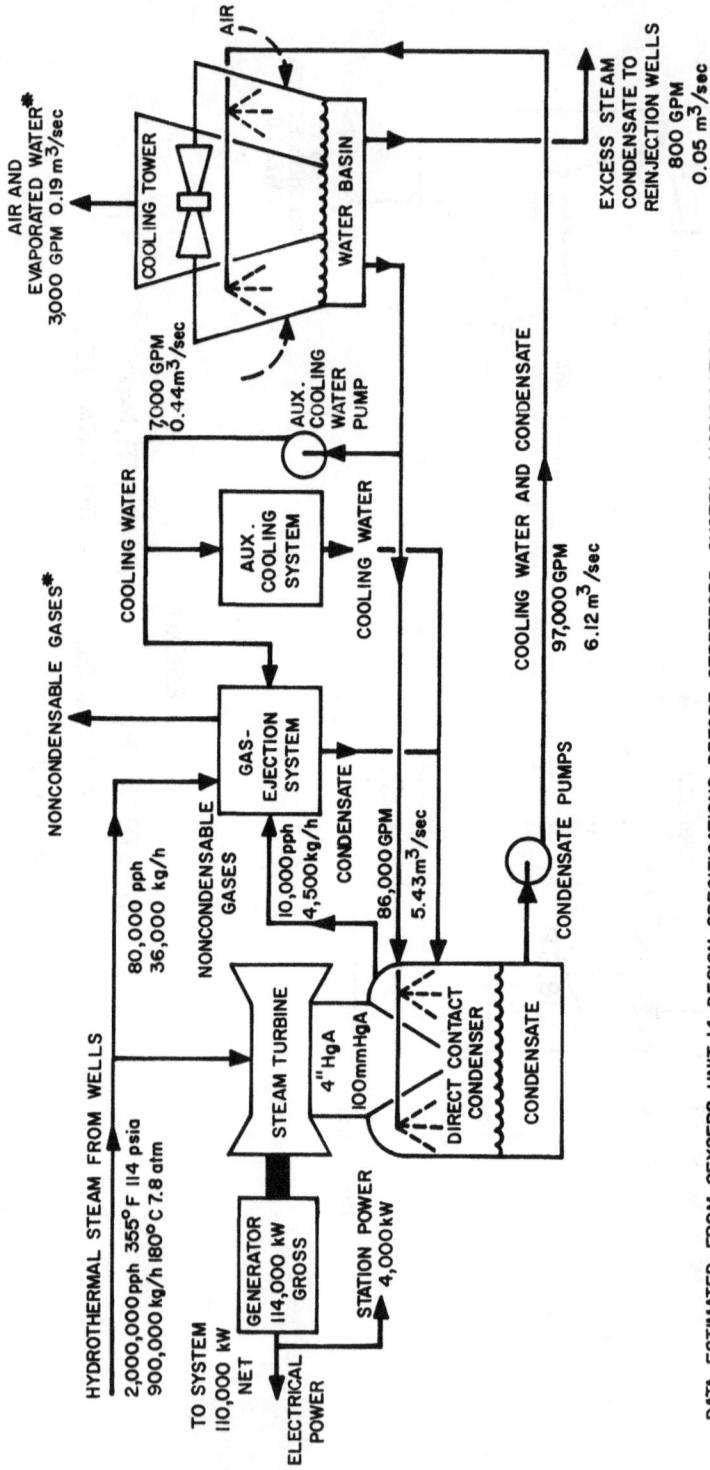

DATA ESTIMATED FROM GEYSERS UNIT 14 DESIGN SPECIFICATIONS BEFORE STRETFORD SYSTEM MODIFICATION.

* SOURCE OF H₂S EMISSION.

FIGURE 12.12—Schematic flow diagram for Unit 14 before H₂S abatement modification, The Geysers [Ramachandran, 1977].

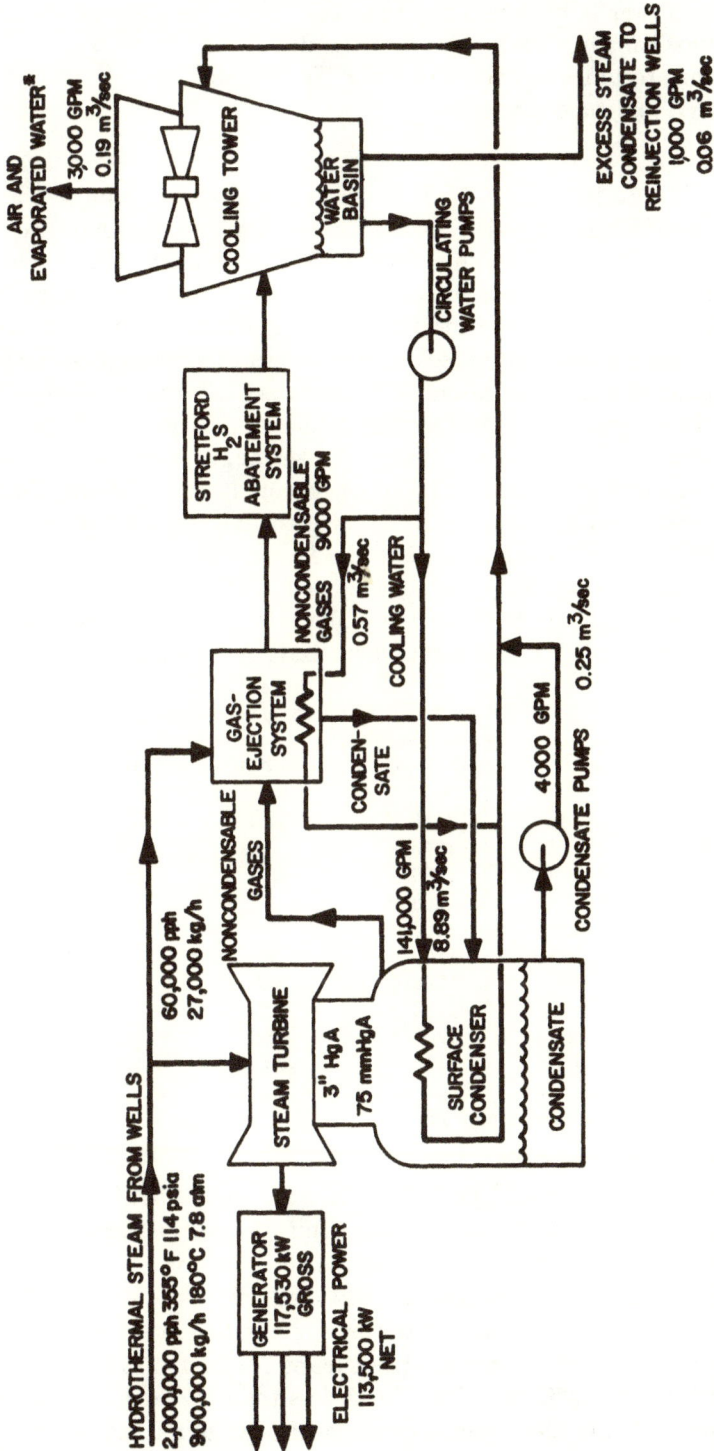

*SOURCE OF H_2S EMISSION.

FIGURE 12.13—Schematic flow diagram for Unit 14 with Stretford H_2S abatement system, The Geysers [Ramachandran, 1977].

the cooling tower is increased significantly (59%) because the shell-and-tube heat exchanger is used instead of the direct-contact condenser. Table 12.5 contains technical specifications for Unit 14 with the Stretford plant and the shell-and-tube condenser.

Unit 15. Unit 15 is rated at 55 MW, has a double-flow, impulse-reaction turbine manufactured by the General Electric Company, is supplied with steam from Thermogenics, and is fitted with H_2S abatement equipment. It is located in the western portion of the geothermal field (see figure 12.2). Table 12.5 lists technical specifications for the modified unit; many of the values shown have been estimated and are subject to refinement as the plant goes through its initial operations. The unit was started up in June of 1979.

Table 12.6 summarizes geothermal resource utilization efficiency for the first 15 power units at The Geysers. Values range from 50%–56% based on the thermodynamic availability of the steam at the wellhead. Units 13 and 15 are expected to operate with steam at the lowest temperature of any unit, 170°C (338°F), and may have the lowest efficiency of all units at The Geysers. Although they are not included in the table, Units 16 and 17 will be similar to Unit 14 in design and performance; i.e., they may be expected to operate at a 56% resource utilization efficiency.

Table 12.6—*Resource utilization efficiency of The Geysers units* [a]

Unit No.	$T_1/°F$	$h_1/(Btu/lbm)$	$w_o/(Btu/lbm)$	$w/(Btu/lbm)$	$\eta_u/\%$
1	348	1200. 5	324. 9	164. 6	51
2	348	1203. 2	316. 2	166. 9	53
3	341. 8	1200. 2	314. 8	169. 7	54
4	341. 8	1200. 2	314. 8	167. 7	53
5–6	355	1200. 4	335. 6	187. 0	56
7–10	355	1200. 6	335. 3	187. 0	56
11–12	355	1200. 5	335. 5	187. 6	56
13	338	1190. 3	332. 8	165. 3	50
14	355	1200. 4	335. 8	187. 7	56
15	338	1190. 6	332. 4	164. 8	50

[a] Based on a sink temperature of 80°F.

Future units. Units 16 and 17, both to be rated at 110 MW, are expected to be commissioned by PG&E in 1981. Unit 16 will be located in Lake County and will be supplied with steam from Aminoil; Unit 17 will be located in Sonoma County and be supplied with steam from Union-Magma-Thermal. Toshiba will manufacture the turbines for both units, which will operate with steam at a pressure of 784 kPa (113.7 lbf/in²) and a temperature of 169°C (336.2°F), with less than 1% of noncondensable gas by weight. Units 18–21 should come on-line two at a time, each with an installed capacity of 110 MW. Units 18, 20, and 21 will use steam from

Union-Magma-Thermal; Unit 19 will be supplied by Aminoil. Unit 18 is expected to be sited in Sonoma County, and Units 19–21 in Lake County [B. J. Cossette, personal communication].

The Northern California Power Agency (NCPA) plans to construct a 110-MW plant to serve the communities of Lodi, Roseville, Santa Clara, Alameda, Ukiah, Healdsburg, and Lompoc in conjuction with the Plumas-Sierra Rural Electric Cooperative [GEM, 1977a]. The plant will be furnished with steam from Shell Oil Company wells and will be built on one of Shell's leases in The Geysers steam field. Expected to come on-line in 1981, the unit will require about 907 Mg/h (2×10^6 lbm/h) of steam, similar to PG&E's 110-MW units at The Geysers.

The Sacramento Municipal Utility District (SMUD) has recently announced its intention to become active in the construction of geothermal power plants at The Geysers.

12.2.4 Construction materials

A primary factor in determining the materials used in manufacturing components for a geothermal powerplant is the composition of the geothermal fluid. Table 12.7 lists the concentration of the constituents of the steam found at The Geysers, using data taken from measurements on 61 wells during the period 1972–1974 by PG&E. Carbon dioxide, hydrogen sulfide, methane, and ammonia are the main constituents carried by the steam, amounting to about 95% of the total impurities. Steam from The Geysers field is relatively noncorrosive since it comes from the wells in a slightly superheated state. Carbon steel (ASTM-A106 Gr.B or equivalent) can therefore be used in the gathering system, including main-steam pipelines, valves, and strainers. For the most part, the turbines are made from manufacturer's standard materials with items of cast, forged, or fabricated steel. The casing is carbon steel, but the blading is of 13%

Table 12.7—*Constituents of steam at The Geysers* [a]

| Constituents | Concentration, mg/kg | | | Average Flow into 110-MW Unit kg/h[b] |
	Low	Average	High	
Carbon dioxide, CO_2.........	290	3260	30600	2700
Hydrogen sulfide, H_2S........	5	222	1600	180
Methane, CH_4............ ..	13	194	1447	160
Ammonia, NH_3....	9. 4	194	1060	160
Boric acid, H_3BO_3..........	12	91	233	75
Hydrogen, H_2..............	11	556	218	46
Nitrogen, N_2...............	6	52	638	43
Ethane, C_2H_6.............	3	8	19	6. 6
Arsenic, As...... .. .	0. 002	0. 019	0. 050	0. 016
Mercury, Hg..............	0. 00031	0. 005	0. 018	0. 004

[a] Source: Reed and Campbell, 1975.
[b] Based on a steam flow of 821 Mg/h (1.81×10^6 lbm/h) to the turbine.

chrome steel. Moisture removal provisions exist in the lower-pressure stages where expansion leads to higher moisture content. Such moisture traps are of standard design and are used in conventional steam turbines as well [Finney, 1972]. The quality of the steam at the turbine exhaust hood is typically about 90%.

The corrosive nature of the geofluid becomes manifest when the steam condenses, especially in the presence of air. As with all turbines that operate under vacuum conditions, some infiltration of air into the turbine through the seals is unavoidable. Under condensation the noncondensable gases become more concentrated, the hydrogen sulfide in the presence of air oxidizes to weak sulfuric acid, and the fluid becomes highly corrosive to such materials as carbon steel, cast iron, copper-based alloys, zinc, cadmium, silver, wood, and concrete. The condenser consists of a shell of carbon steel plate overlaid with 1.6 mm ($\frac{1}{16}$ in) of Type 304 (19% Cr/ 9% Ni) stainless steel and internals made of solid stainless steel. Condensate lines are fabricated from Type 304 stainless steel pipe; condensate pumps are of conventional canned, vertical design but all wetted parts, including impellers and bowls or volutes, are made of austenitic Type 304 stainless steel. The circulating water lines above ground are made of aluminum pipe of Type 3003, 3053, or 6061. Aluminum alloys with no copper content are used. Reinforced concrete was used for underground water pipes in earlier units, but this material has not been totally satisfactory. Newer units use glass-reinforced plastic for all buried piping between the powerhouse and cooling tower. In addition, good results have been obtained with cement-asbestos, epoxy-lined pipe.

The cooling towers are of the conventional mechanically-induced-draft design, but Type 304 stainless steel or aluminum is used for wetted metal parts. Redwood is used for structural support members of the water distribution headers and basins; transite forms the tower siding. Where cast iron must be used, it is covered with a thick coating of coal-tar epoxy resin or other plastic material. Wood is unsuitable as tower fill material; polystyrene, polypropylene, or polyvinyl chloride (PVC) are acceptable alternatives. The preferred material is PVC because of its strength and fire-retardant properties.

Stainless steel, aluminum, or plastic-lined material is used for all pipes carrying condensate or cooling water. Concrete surfaces exposed to these waters are coated with coal-tar epoxy compounds or synthetic rubber coatings. Heat exchangers are made from aluminum or austenitic stainless steel. Since copper alloys and silver are susceptible to corrosive attack by hydrogen sulfide, electrical equipment should not be made of these materials. Experience shows that tin alloy coatings are effective in resisting corrosion but are unsatisfactory on current-carrying contact surfaces. Aluminum, stainless steel, and some precious metals are particularly effective. Platinum inserts or plating has been used on these contacts. Outdoor equipment must be protected since the air contains corrosive elements. The fog and mist produced by evaporation from the cooling towers are quite corrosive. Epoxy paints and heavy galvanizing resist these corrosive gases.

12.2.5   Effluent and emissions handling systems

All units currently operating at The Geysers employ evaporative water cooling towers in conjunction with direct-contact jet condensers.* Since the amount of condensate produced from the geothermal steam exceeds the amount of water evaporated in the towers, proper disposal of the excess liquid is a problem. In addition, emission of gaseous impurities in the geothermal steam into the atmosphere must be considered because of the potential hazard to health, wildlife, and vegetation.

Liquid effluent. Approximately 20% of the mass of geothermal steam produced from the wells must be disposed of as excess liquid from the cooling tower basin. From 1960 to 1970 the problem was solved simply by allowing the liquid to run into nearby Big Sulphur Creek. Beginning with Units 5 and 6, however, excess water has been reinjected into the producing reservoir. For each 55-MW unit 1700 m^3/day (450,000 gal/day) must be reinjected. In 1974 four wells were employed for the reinjection of approximately 14,000 m^3/day (3.7×10^6 gal/day) of excess water [Reed and Campbell, 1975]. Since the steam-producing reservoir is of anomalously low pressure relative to hydrostatic conditions, it is not necessary to pump the liquid down the wells; pumping is required only to move the liquid from the cooling tower sites to the reinjection wells.

Gaseous emissions. The noncondensable gases in the geothermal steam are vented to the atmosphere at two places in the plant, the gas ejector and the cooling tower, as shown in figure 12.12. The composition of the gas stream leaving the gas ejector is shown in table 12.8 for nine of the early

*Unit 15 has a shell-and-tube, surface condenser.

Table 12.8—*Composition of gas streams from gas ejectors at The Geysers* [a]

Constituents	Average Concentration, % by weight	Average Flow from 110-MW Unit kg/h [b]
Carbon dioxide, CO_2	59. 9	2500
Nitrogen, N_2	[c] 25. 8	[c] 1100
Oxygen, O_2	[c] 8. 4	[c] 350
Methane, CH_4	3. 8	160
Hydrogen sulfide, H_2S	1. 2	50
Hydrogen, H_2	0. 9	38

[a] Source: Reed and Campbell, 1975; based on measurements by PG&E of nine units over the period 1967–1974.
[b] Based on a steam flow of 821 Mg/h (1.81×10^6 lbm/h) to the turbine.
[c] Nitrogen and oxygen are from air entering at the turbine.

units at The Geysers. Measurements were taken by PG&E from 1967 to 1974; the values shown are averages. The output from the stacks of the cooling tower has been estimated by PG&E, and table 12.9 lists the results for plants without emissions controls.

Table 12.9—*Estimated emissions from cooling towers on uncontrolled units at The Geysers* [a]

Constituents	Average Concentration mg/kg	Average Flow from 110-MW Unit kg/h [b]
Ammonia, NH_3............................	240	160
Carbon dioxide, CO_2.....................	210	140
Hydrogen sulfide, H_2S....................	200	130
Hydrogen, H_2............................	12	8

[a] Source: Reed and Campbell, 1975.
[b] Based on a flow rate of 658 Mg/h (1.45×10^6 lbm/h) of water vapor from the tower

The most objectionable of the gases discharged is hydrogen sulfide, H_2S. It has an unpleasant smell that can be detected by the human olfactory sense at extremely low levels of concentration. The California ambient air quality standard for H_2S is 30 parts per billion (ppb), based on an assumed odor detection threshold [Semrau, 1976]. Although no Federal standards exist for H_2S, the U.S. Environmental Protection Agency (EPA) has suggested a maximum of 200 g/MW·h of electrical production or its equivalent [Hartley, 1978]. The first 10 units at The Geysers were originally provided with no means of control of H_2S emissions. Daily operation of these 10 uncontrolled units produced 22 Mg/d or 2300 g/MW·h of H_2S [Weres, 1976]. All new units will be fitted with some type of H_2S abatement system. An iron hydroxide system of about 70% efficiency was tested on Unit 11. It discharges on the average 2 Mg/d or 800 g/MW·h into the atmosphere, including preplant emissions and vent emissions that occur during plant shutdown and are uncontrolled at this time [Weres, 1976]. Units 13, 14, 15, and future units will have surface condensers instead of jet condensers. Separate chemical processing plants operating on the Stretford process will remove the hydrogen sulfide on Units 13–15 [Semrau, 1976]; a simplified flow diagram of the Stretford process is shown in figure 12.14. Since this forms an independent facility, it will have no direct influence on power plant performance. The product of the Stretford process is pure, marketable sulfur.

The bulk of the H_2S (90%) will be confined to the noncondensable gas stream where it can be effectively controlled. The concentration at the exit of the gas extractor is guaranteed by the vendor not to exceed 10 ppm, which is equivalent to an efficiency greater than 99.9% [Semrau, 1976].

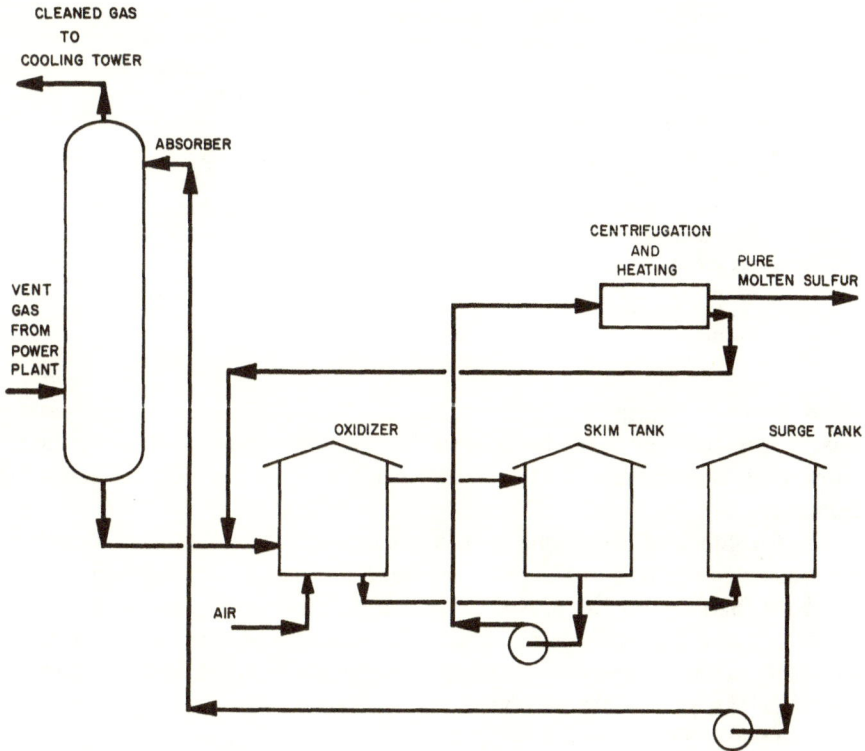

FIGURE 12.14—Typical flow diagram of Stretford H₂S removal process [Laszlo, 1976].

However, the overall plant abatement of H₂S with such a system is only expected to be between 78% and 92%, since about 10% of the H₂S will still reach the cooling tower through the condensate from the surface condenser. As illustrated in figure 12.13, about 0.25 m³/s (4000 gal/min) of condensate, containing about 10% of the H₂S originally present in the steam, is fed into the cooling water and eventually reaches the tower.

Tough, new regulations on H₂S emissions in The Geysers area will most likely require some form of emissions controls on all existing power units (with the possible exception of Units 1 and 2). Stretford or iron catalyst systems will need to be back-fitted at considerable cost to the utility.

Another potentially harmful emission from The Geysers plant is radon, a radioactive decay product of uranium found in the gaseous state in sedimentary rocks. Carried by the natural steam from the reservoir to the surface and emitted along with the other noncondensables, it is detectable in extremely minute concentrations owing to its radioactivity. The California Department of Health requires radon concentrations be less than 3 picocurie/liter (pCi/l) in uncontrolled areas (1 pCi of radon = 6.8×10^{-21} kg). Table 12.10 gives data on the emission of radon from The Geysers plants. As the table shows, the natural steam from the wells con-

Table 12.10—*Maximum measurements of radon emissions at The Geysers* [a]

Location	Concentration [b]
Steam from wells...............	8.3 picocurie/liter
Ejector gas from plant...........	5.3 microcurie/liter
Cooling tower water............	0.21 picocurie/liter
Ambient air outside plant . .	0.026 picocurie/liter

[a] Source: Reed and Campbell, 1975; measurements by PG&E.
[b] California Department of Health requires radon concentrations be less than 3 pC/l in uncontrolled areas.

tains about 2.8 times the allowable amount, and the ejector gas carries nearly two million times the allowable limit. Measurements nevertheless show that the standards are not exceeded in areas of normal access by humans.

The problem of emissions to the atmosphere is not totally solved by abatement equipment at the powerhouse. No emission control devices have yet been installed in the pre-plant elements. Thus during plant shutdown steam is vented directly to the atmosphere and constitutes a total failure, in effect, of the abatement systems. Venting is employed to avoid damage to either the producing reservoir or the steam wells, which might occur under totally closed-in conditions. A possible technique to control vent emissions of H_2S is to use a bed of iron oxide to absorb the gas. This technique would eliminate the problems of other systems requiring condensation of the steam. The projected emission levels of H_2S are shown in figure 12.15. With an overall abatement of 90%, H_2S in the year 2000 will be comparable to those at present, even though overall capacity will have grown by an order of magnitude to 5100 MW.

12.2.6   Economic data

The cost to an electric power company of producing and transmitting a unit of electricity to the customer is the most important factor influencing the choice of energy conversion system in a free marketplace. Three basic expenses contribute to the total cost of electricity: (1) capital investment, which includes the outlay for the power plant, interest incurred during construction, working capital, organization, startup expenses, and the costs of constructing transmission lines, if necessary; (2) operating and maintenance expenses; and (3) fuel cost, which in the case of geothermal power plants means the price the utility must pay to the field developer for the geofluid, be it steam or hot brine.

At The Geysers, the utility purchases dry steam "over the fence" from the supplier, according to a pricing formula that includes a surcharge levied on the utility by the supplier to cover the cost of disposing (by

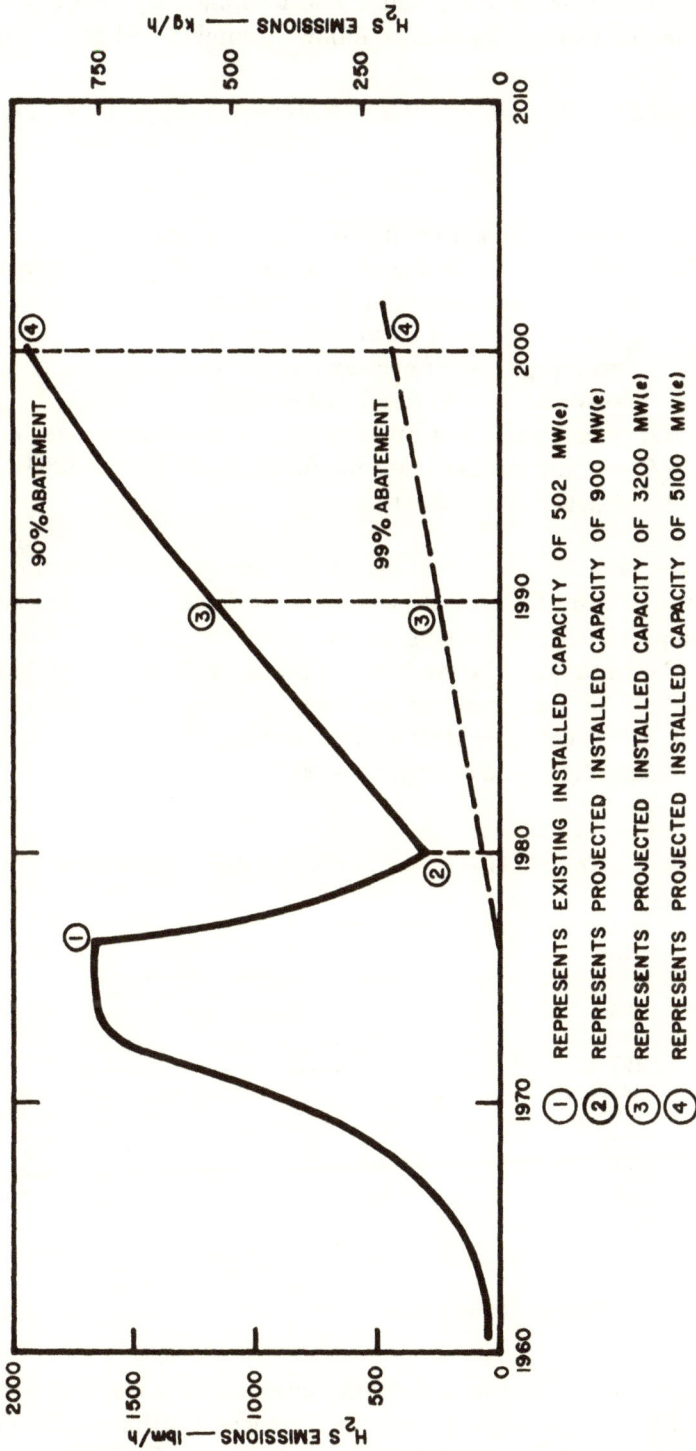

FIGURE 12.15—Estimated H_2S emissions from The Geysers power plants [Ramachandran, 1977].

reinjection) of excess condensate. The formula that PG&E uses to buy steam for its Geysers units is as follows [Dutcher and Moir, 1976]:

$$C_S = [2.11 \ E_F(\overline{C}_F/\overline{C}_F°)(MHR/MHR°) + E_N\overline{C}_N]/(E_F + E_N), \quad (12.1)$$

where

C_S = cost of steam (mill/kW · h) for year n

E_F = electricity produced from fossil fuel during year $n-1$

E_N = electricity produced from nuclear fuel during year $n-1$

$\overline{C}_F$ = average cost of fossil fuel for year $n-1$

$\overline{C}_N$ = average cost of nuclear fuel for year $n-1$

$\overline{C}_F°$ = average cost of fossil fuel in 1968

MHR = minimum heat rate for fossil plants during year $n-1$

$MHR°$ = minimum heat rate for fossil plants during 1968

2.11   = negotiable constant.

Thus the cost of geothermal steam for any year is determined by the amount and cost of electrical production by fossil and nuclear means during the previous year. Base figures are taken for the cost of fossil fuel and fossil plant heat rate during 1968. In addition there is a surcharge of 0.5 mill/kW·h for reinjection of the spent geofluid. The historical price of The Geysers steam since 1969 is given in table 12.11. These prices include the surcharge for liquid effluent disposal.

Table 12.11—*Price of The Geysers steam, 1969–1978, in mill/kW · h* [a]

Year	Price
1969	2. 65
1970	2. 64
1971	2. 74
1972	2. 90
1973	3. 15
1974	3. 73
1975	7. 39
1976	11. 35
1977	14. 10
1978	16. 05

[a] Source: Ramachandran, 1977.

The only other geothermal steam contract at The Geysers is the one between Shell Oil Company and the Northern California Power Agency (NCPA), signed June 27, 1977. The contract calls for NCPA to pay Shell according to the amount of steam delivered. The initial price, at the time of the contract, was $0.6917/1000 lbm of steam. Beginning July 1, 1977, the price will be adjusted semiannually by the Gross National Product (GNP) Implicit Price Deflator Index (IPD) published by the U.S. Department of Commerce for the preceding calendar quarter [Lindsay, 1977]. The IPD is the ratio of the GNP in current dollars to the GNP in constant 1972 dollars for the current period. The geothermal steam supplied by Shell must be dry and at a pressure no lower than 799.8 kPa (116.0 lbf/in²); when the amount of noncondensable gases exceeds 0.5% (weight), the flow rate of steam will be corrected accordingly. Uncontaminated waste liquid will be returned to Shell for disposal at a temperature not greater than 79.4°C (175°F) and at a pressure not less than 262 kPa (38 lbf/in²).

The history of capital investment by PG&E at The Geysers is given in table 12.12, where all figures are in current US dollars. A comparison between the cost of electricity generated by Units 11 and 14 is shown in table 12.13, in which cost figures are given in constant 1976 dollars. The effect of inflation is nullified in this comparison, leaving the added expense of the equipment additions and modifications needed to meet the H_2S emissions constraints as the main contributing factor causing the 1.6 mill/kW · h (9%) increase in the cost of electricity. Of course, since Unit 14 is under construction, the cost of electricity from that unit, 19.2 mill/kW · h, is

Table 12.12—*Capital investment by PG&E at The Geysers geothermal power project* [a]

Unit	Megawatts	Cumulative Megawatts	Date of Commercial Operation	PG&E Capital Investment	PG&E Cumulative Capital Investment
1	11	11	9/25/60	$4, 010, 000	$4, 010, 000
2	13	24	3/19/63		
3	27	51	4/28/67	7, 610, 000	11, 620, 000
4	27	78	3/02/68		
5	53	131	12/15/71	12, 756, 000	24, 376, 000
6	53	184	12/15/71		
7	53	237	8/18/72	10, 982, 000	35, 358, 000
8	53	290	11/23/72		
9	53	343	10/25/73	13, 520, 000	48, 878, 000
10	53	396	11/30/73		
11	106	502	11/20/76	14, 404, 000	63, 282, 000
12	106	608	° 3/01/79	° 21, 473, 000	° 84, 755, 000
b 13	135	743	° 2/01/80	° 28, 934, 000	° 113, 689, 000
b 14	110	853	° 6/01/80	° 27, 966, 000	° 141, 655, 000
15	55	908	° 6/17/79	° 17, 339, 000	° 158, 994, 000
d 16	110	1018	° 1983	° 42, 449, 000	° 201, 443, 000
d 17	110	1128	° 1982	° 41, 592, 000	° 243, 035, 000

[a] Source: Dan et al., 1975; all figures in US $.
[b] Under construction.
[c] Estimated.
[d] Permits pending.

Table 12.13—*Busbar price of electricity at The Geysers: Units 11 and 14* [a]

	Unit 11 (existing)	Unit 14 (under construction)
Installed capacity, net (MWe)...................	110	110
Capital investment, ($/kW) [b]....................	[c] 170	[d] 232
Required price of electricity: [e]		
Capital charges (mill/kW · h) [f]...............	4. 0	5. 6
Operation and maintenance (mill/k W · h)	0. 6	0. 6
Steam price (mill/kW · h)	13. 0	13. 0
Total.................................	17. 6	19. 2

[a] Source: Ramachandran, 1977.
[b] All figures are in 1976 US dollars. Capital investment includes total installed plant cost, interest during construction, working capital, organization, and startup expenses.
[c] Includes direct-contact condenser and iron catalyst treatment for H_2S abatement.
[d] Includes shell-and-tube surface condenser and Stretford H_2S abatement plant.
[e] Revenue required by the utility (PG&E) based on regulated utility economics.
[f] Conversion from $/kW to mill/kW · h is based on 17% fixed charge rate and 80% capacity factor.
[g] By existing steam pricing formula.

estimated. If it becomes mandatory to guarantee 99% abatement of H_2S, a condensate treatment system will have to be installed in addition to the noncondensable gas treatment plant. The cost of providing this extra cleanup operation is estimated to be about 0.7 to 1.8 mill/kW · h. Details are shown in table 12.14 for two types of treatment, peroxide and ozone.

Table 12.14—*Estimated cost of condensate treatment to achieve 99% H_2S abatement at The Geysers* [a]

Method of Treatment [b]	Capital Investment [c]		Operating Cost mill/kW·h	Total mill/kW·h
	$/kW	mill/kW·h		
Peroxide system.........	9–11	0. 22–0. 27	0. 43–1. 55	0. 65–1. 82
Ozone system...........	16–35	0. 4–0. 9	0. 44–0. 72	0. 84–1. 62

[a] Source: Ramachandran, 1977.
[b] Both methods assume that 10% of inlet H_2S concentration is dissoved in condensate, that it varies from 125 to 500 mg/kg of the steam, and that the unit is equipped with integrated surface condenser and Stretford system.
[c] All figures are in 1976 US dollars. Conversion from $/kW to mill/kW·h is based on 17% fixed charge rate and 80% capacity factor.

For the operating year 1976, PG&E reported that their geothermal power plants produced electricity at the lowest cost of any other type of steam plant in its system [Mahoney and Bangert, 1977]. The figures (in

1977 dollars) are as follows: geothermal, 18 mill/kW · h; nuclear, 24 mill/kW · h; coal-fired, 26 mill/kW · h; oil-fired, 36 mill/kW · h. Geothermal plants were the least expensive to construct; they were 26% cheaper than oil-fired plants, about half as expensive as coal-fired plants, and cost only 38% of a typical nuclear plant. All comparisons are on a dollars-per-kilowatt basis.

12.2.7   Operating experience

Operation of The Geysers power plants is challenging because of several factors, including the isolated location, rugged mountain terrain, and environmental and operating requirements unique to geothermal plants [Matthew, 1975]. All units are designed for automatic operation. A computer-based information and control center located at Units 5 and 6 is manned full-time. Information and alarms from each unit in the system are sent to this center for action by the operator. The units function as baseload units, since abrupt changes in well flow rates damage both the wells and the turbines owing to gross carryover of water, dust, and debris. When a unit trips off the line, the wells connected to it are allowed to vent to the atmosphere through silencers. These wells are not closed-in unless the unit must remain off-line for an extended period for maintenance or repairs.

Ambient temperature fluctuates during the year from about −7° to 46°C (20° to 115°F), necessitating forced ventilation of the electrical equipment during summer and special precautions to prevent freezing of outside equipment, particularly the cooling towers, in winter. Automobile access to the plant during winter is difficult because the narrow mountain roads are often ice covered.

Table 12.15 contains information on plant reliability. Capacity and availability factors are shown for The Geysers system over the operating history of the plant up to 1975. For comparison, figures are quoted for

Table 12.15—*PG&E power plant reliability, geothermal versus fossil* [a]

Performance Factor	The Geysers System		Fossil Plants	
	1960–1975	Avg. 1961–1974	Nonreheat [b]	Reheat [c]
Capacity factor (%) [d] . . .	70–96	75. 6	24. 8	54. 0
Availability factor (%) [e] . . .	70–96	86. 8	91. 5	87. 6
Forced outage rate (%) ..	—	6. 2	1. 3	2. 5
Scheduled outage rate (%) .	—	7. 2	7. 0	9. 8

[a] Source: Matthew, 1975.
[b] 100 MW, turbine inlet conditions: P=8.7 to 9.1 MPa (1265 to 1315 psi); T=482°C to 510°C (900°F to 950°F)
[c] 165 to 330 MW, turbine inlet conditions: P=12.5 to 16.6 MPa (1815 to 2415 psi); T=538°C to 565°C (1000°F to 1050°F)
[d] Capacity factor=generation/(unit rating×period)
[e] Availability factor=(period−outage period)/period

fossil-fired plants in the PG&E system. The Geysers has the highest capacity factor, an indication of preferential base-load usage. The availability factor is comparable to but slightly lower than that for fossil plants.

The higher forced outage rate for the geothermal plants is a result of unexpected failure of auxiliaries and components that usually function without trouble in conventional plants. These components include turbine blades and shrouds, rotating electrical apparatus (exciters and motors), small pumps, and cooling tower fan blades. A spare parts maintenance program is designed to minimize downtime. Geothermal units at The Geysers are dismantled, cleaned, and repaired every 2 to 3 years compared with 5 to 8 years for conventional units. Downtime is less, 3 to 4 weeks instead of 8 to 12 weeks, because of the smaller size and relative simplicity of the geothermal units. As experience is gained, the scheduled outage rate should continue to improve.

12.3   Magmamax dual binary plant at East Mesa, California

There are currently two practical methods for converting the energy in a liquid-dominated hydrothermal resource into electricity. The first involves either separating the vapor fraction for use in a steam turbine or combining a separation with a subsequent flashing process of the residual liquid to generate more vapor. The second involves the use of the hot geofluid as a heating medium for a secondary working fluid of a suitably low boiling point which is then used in a more or less conventional, basic Rankine cycle. The geothermal power plant being constructed at East Mesa, California by the Magma companies is of the latter type and is called the Magmamax [a] dual binary cycle. When completed in 1979, it will be the first commercial geothermal power plant of this type in the United States. The plant will have a rated capacity of 11.2 MW(e).

12.3.1   Geology and exploration

The plant is being built on the East Mesa KGRA in the Imperial Valley, about 2.4 km (1.5 mi) east of the junction of routes I–8 and 115, or about 8 km (5 mi) southeast of Holtville. The plant site is in Section 7 Township 16 South and Range 17 East (T16S–R17E) of the KGRA, a well-known thermal anomaly. The plant is located between heat flow contours of 6.0 and 7.0 heat flow units (HFU).[b] Figure 12.16 shows a heat flow contour map of the area [BuRec, 1977]. The plant is adjacent to the U.S. Bureau of Reclamation's East Mesa Test Site; Magma Power Company holds 756 ha (1868 acres) of Federal leases for which they paid $2.25 per acre [Williams et al., 1977].

The East Mesa KGRA is part of the larger Mexicali-Imperial Rift Valley, an extension of the system of mid-oceanic ridges and rifts that

---

[a] Magmamax Process, U.S. Pat. No. 3757516, Magma Energy, Inc., Los Angeles.
[b] 1 HFU=1 $\mu cal/cm^2 \cdot s$.

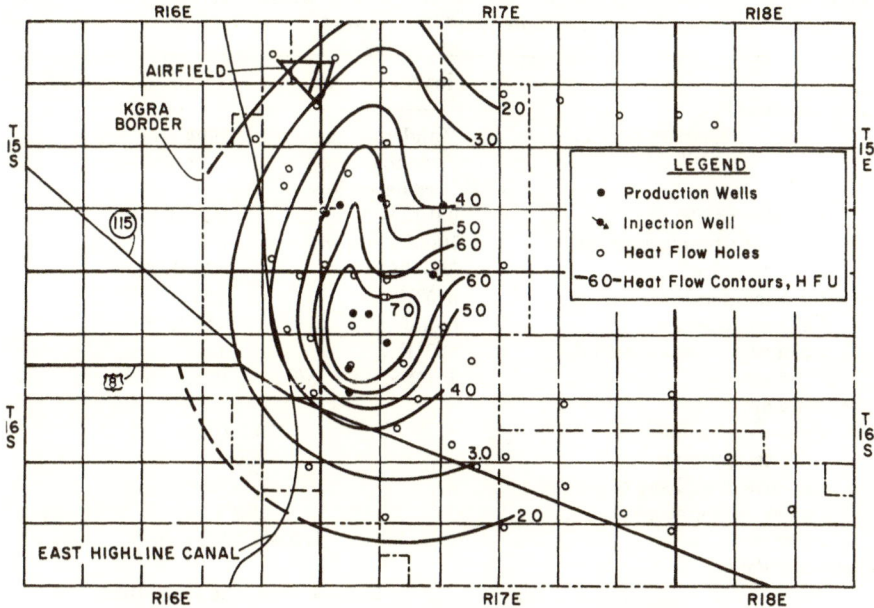

FIGURE 12.16—Heat flow contours, East Mesa KGRA [BuRec, 1977].

circle the earth. The valley is oriented northwest-southeast and is bounded by the San Andreas fault zone on the northeast and by the San Jacinto fault zone on the southwest. The most prominent feature of the valley is the Salton Sea, 48×16 km (30×10 mi) in size and lying at about 72 m (235 ft) below sea level. During the last 2000 years and until about 300 years ago, the area now occupied by the Salton Sea was the site of the larger Lake Cahuilla. The irrigated agricultural land found today in the Imperial Valley is confined roughly within the ancient shoreline of Lake Cahuilla; the East Mesa KGRA lies at the eastern boundary of this ancient shoreline [Saint, 1976]. The Salton Sea was formed when the Colorado River flooded the area during 1905/1906; its level is maintained by the runoff from irrigation in the valley.

The structure of the valley is characterized by predominantly transform faults with lateral movement, crustal spreading, high heat flow, faulting, and volcanic activity. The crystalline basement of the valley in the East Mesa area is at a depth of about 3.7 km (12,100 ft) and is overlain with water-saturated sediments of the later Cenozoic era (Quaternary and late Tertiary periods), having been deposited by the Colorado River and other streams that flowed through the Colorado River drainage basin during the last 4 million years. The depth to the top of the hydrothermal (convective) reservoir is estimated to be 1 km (3281 ft) [Williams et al., 1977]. Although there are no surface thermal manifestations in the East Mesa area, the subsurface hydrothermal reservoir is quite extensive, con-

Table 12.16—*Preliminary design specifications for the Magmamax dual binary plant, East Mesa, California* [a]

	Isobutane expander	Propane expander
Turbine data:		
Type.....	Tandem-compound, double-flow, radial-inflow	Single-cylinder, radial-inflow
Rated capacity... ..	9.0 MW	2.2 MW
Speed (turbine/generator). . . .	7000/3600 rpm	([b])
Inlet pressure.....	500 lbf/in²	474 lbf/in²
Inlet temperature.... ..	345°F	205°F
Exhaust pressure..	60 lbf/in²	142 lbf/in²
Exhaust temperature...... ...	230°F	102°F
Flow rate. .	1.031 × 10⁶ lbm/h	274 × 10³ lbm/h
Geofluid data:		
Pressure after pumps..		270 lbf/in²
Inlet temperature......		360°F
Pressure after heat exchangers....		117 lbf/in²
Outlet temperature........		180°F
Flow rate........		1.444 × 10⁶ lbm/h
Condenser data:		
Type.......		Surface type, shell-and-tube
Pressure.......		59 lbf/in²
Cooling water temperature..... .		62°F
Outlet water temperature...		79.5°F
Water flow rate..... . .		([b])
Heat rejection system data:		
Type.............		Phased cooling, storage ponds with sprays
Number of ponds......		2
Design wet-bulb temperature.....		58°F

[a] Year of startup: 1979.
[b] Not available.

---

taining perhaps 56 km³ (45×10⁶ acre · ft) of water-bearing sediment at a temperature of 180°C (356°F). The reservoir is sealed and insulated from the surface by a layer of impermeable silts and clay [Combs and Muffler, 1973; Saint, 1976]. The reservoir cap is of variable thickness, maximum being about 700 m (2300 ft).

Geophysical surveys of the East Mesa KGRA have been conducted by a number of people [Biehler, 1971; Biehler and Combs, 1972; Biehler et al., 1964; Combs, 1971; Combs, 1972; Douze, 1971]. Some results of these studies are shown in figures 12.17–12.19. Figure 12.17 gives contours of constant temperature gradient obtained by means of 37 bore holes ranging in depth from 30 to 450 m (98 to 1476 ft). Gradients as high as 20°C/100 m (11°F/100 ft), or about six times the normal average temperature gradient, are present near the center of the KGRA. Figure 12.18 is a residual Bouguer gravity map revealing an area of about 250 ha (618 acres) in which there is at least a 4 mgal or 0.04 mm/s² closure. It has

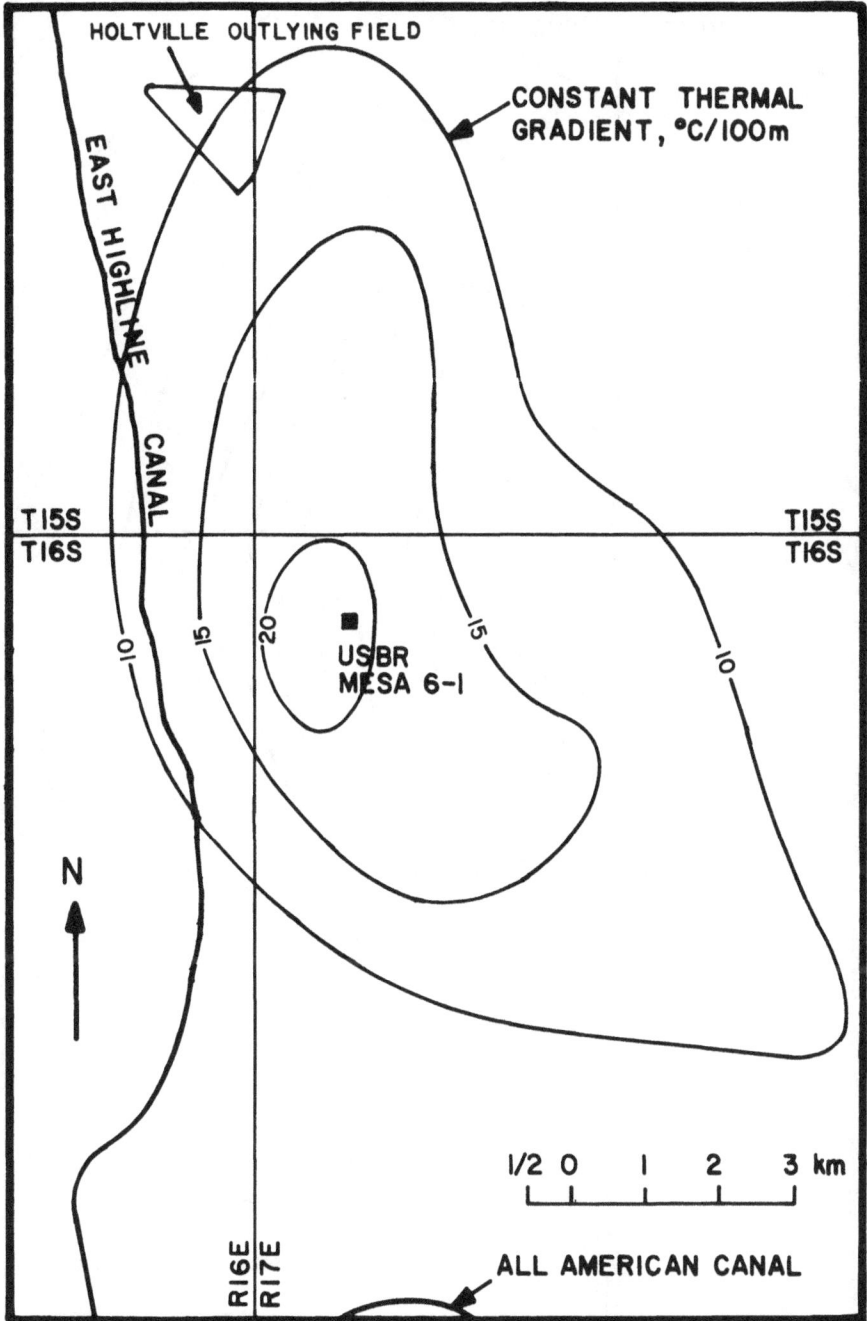

FIGURE 12.17—Thermal gradient map, East Mesa KGRA [Combs, 1971].

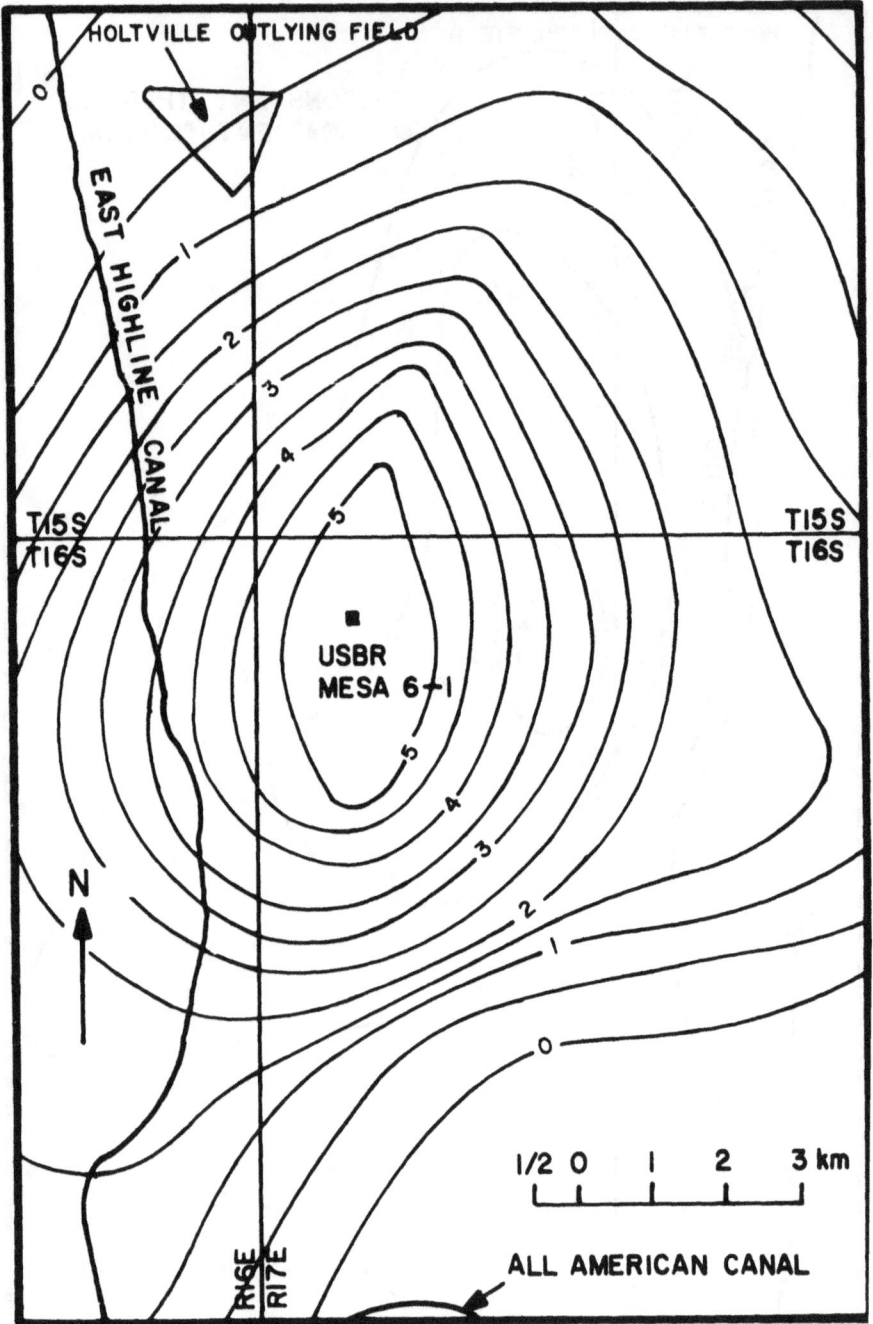

FIGURE 12.18—Residual Bouguer gravity anomaly map, East Mesa KGRA. Contour values in mgal (1 mgal=10 μm/s^2) [Biehler, 1971].

been suggested that this gravity high is an indication of either an increase in density of the sediments caused by cementation and thermal alteration by circulating hot brines, or the emplacement of higher-density igneous rocks [Biehler and Combs, 1972].

Seismic ground noise may be an indicator of a geothermal reservoir when reinforcing evidence from other geophysical measurements exists. Figure 12.19 shows a ground-noise map of the East Mesa KGRA, where

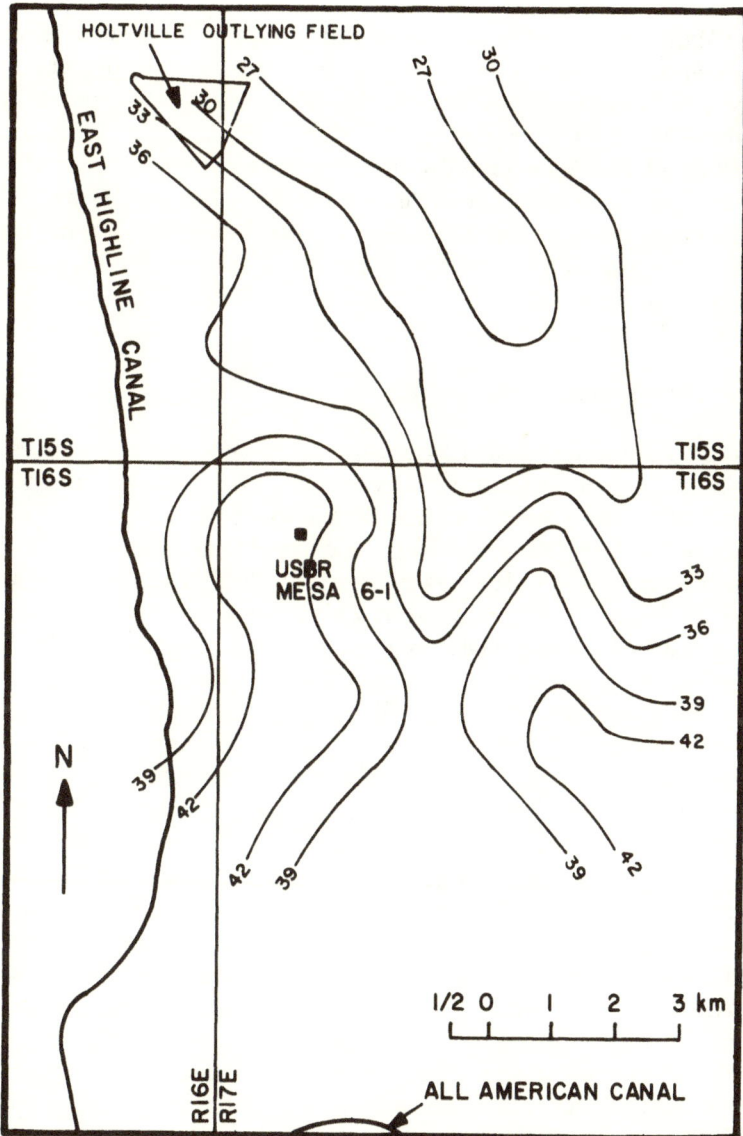

FIGURE 12.19—Seismic ground-noise map, East Mesa KGRA. Contour values in dB relative to 1 (nm/s)^2/Hz in passband 3–5 Hz [Douze, 1971].

contours are given in decibels relative to 1 (mm/s)^2 per hertz in the pass-band 3–5 Hz. Since the East Mesa region constitutes a "hidden reservoir"—that is, there are no surface manifestations—its discovery marked a major triumph for the science of geothermal exploration, even though the original indication of a possible thermal anomaly came about accidentally through oil and gas drilling in the area [Koenig, 1977].

12.3.2   Well programs

The Magmamax dual binary plant will be supplied with pressurized geofluid from three wells. Down-hole pumps will be used to prevent flashing and help inhibit the deposition of scale. Each well is expected to produce about 218 Mg/h (481,000 lbm/h) of brine at wellhead condition of 182°C (360°F) and 1.35 MPa (196 lbf/in²). A booster pump at the power plant will raise the pressure to 1.86 MPa (270 lbf/in²). The geothermal fluid will circulate through a number of heat exchangers before being reinjected into the reservoir through two reinjection wells. The spent fluid will enter the reinjection pump at 82°C (180°F) and 0.81 MPa (117 lbf/in²); the pump will raise the pressure to 1.3 MPa.

12.3.3   Energy conversion system

A highly simplified flow diagram for the plant is shown in figure 12.20; the planned layout design as of December 20, 1976, is shown in figure 12.21. Preliminary design specifications for the plant are given in table 12.16. There are two parallel power loops, one using isobutane (i-C_4H_{10}) and one using propane (C_3H_8) as the working fluid. The two loops are connected by means of the recuperator, which uses the superheated isobutane leaving the turbine to preheat and vaporize the propane. On the basis of the preliminary specifications quoted for the plant, the system will be capable of a resource utilization efficiency of about 52%, assuming a sink temperature of 27°C (80°F). Specific brine consumption will be about 58.5 kg/kW · h (129 lbm/kW · h). The objective of the plant's developers is to reduce consumption to less than 45.4 kg/kW · h (100 lbm/kW · h) [Hinrichs and Falk, 1978].

The isobutane turbine was built by the York Division of Borg-Warner Corporation to the specifications of J. Hilbert Anderson [Anderson, 1973]. The machine is of the double-flow type with each side being a three-stage radial inflow turbine. The unit is essentially a compressor that has been redesigned for turbine duty. The turbine is expected to operate at an isentropic efficiency of about 77%. The propane turbine was built by Mafi-Trench and is typical of machines of this type used for low-temperature applications; of the radial inflow type, its preliminary design specifications indicate an expected isentropic efficiency of about 86% [Mafi, 1978].

About 10.5 ha (26 acres) will be dedicated to the storage ponds required for the phased cooling system. The two ponds will hold roughly 1.7×10^5

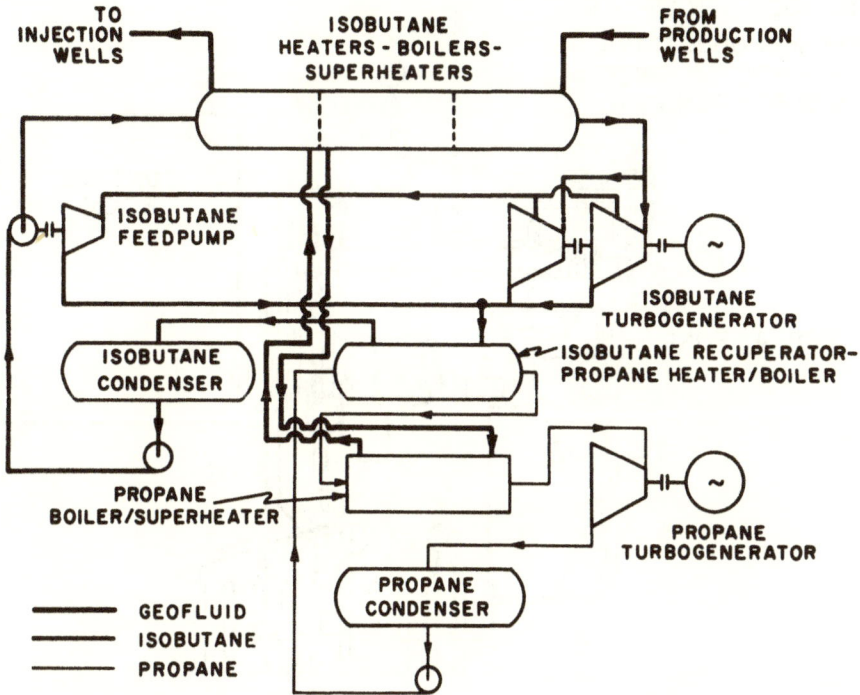

FIGURE 12.20—Simplified flow diagram for the Magmamax Process (U.S. Pat. No. 3757516) of the dual binary plant at East Mesa [Anderson, 1976a].

m³ (138 acre · ft) and 9.0×10^4 m³ (73 acre · ft) of water, respectively, for a storage time of about 16 hours [Hinrichs, 1978]. The combined use of phased storage and spray cooling is expected to produce cooling water at a temperature only 2.2°C (4°F) above the design wet-bulb temperature of 14.4°C (58°F).

12.4 Republic Geothermal at East Mesa, California

12.4.1 Setting

Republic Geothermal is currently developing a portion of the East Mesa geothermal field with the intention of building a 48-MW(e) double-flash power plant to be operating in 1980. Republic's plant will be located about 5 km (3 mi) north of the Magmamax dual binary plant (see figure 12.16). Republic holds leases to 2713 ha (6705 acres); as can be seen from figure 12.16, Republic's holdings lie in a region of the Known Geothermal Resource Area (KGRA) in which the thermal anomaly ranges from a high of about 6 HFU to less than 4 HFU. Otherwise the geology of the area is practically identical to the portion of the KGRA described in section 12.3.1.

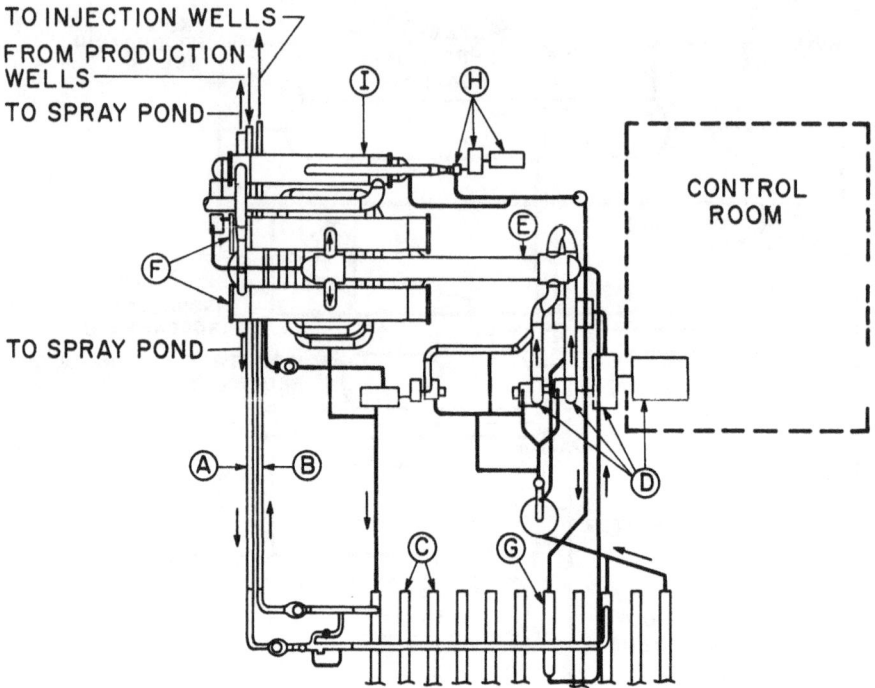

FIGURE 12.21—Plant layout diagram for Magmamax binary plant, East Mesa. Key:
A=brine inlet; B=brine outlet; C=isobutane heaters, boilers, superheaters (typ.);
D=isobutane turbine, gearbox, generator; E=recuperator; F=isobutane condensers; G=propane superheater; H=propane turbine, gearbox, generator; I=
propane condenser [Anderson, 1976b; T. C. Hinrichs, personal communication].

12.4.2 Well programs and productivity

A recent analysis of flow tests was carried out for well EM 16-29, which
extends to a depth of 2420 m (7940 ft) with the following casing program
[Elliott, 1978; J. H. Barkman, personal communication]:

> 384 mm (15.124 in) I.D. from wellhead to 329 m (1079 ft)
> 276 mm (10.880 in) I.D. from 283 to 1408 m (929 to 4619 ft)
> 201 mm (7.921 in) I.D. from 1355 to 2420 m (4446 to 7940 ft).

The analysis indicated that the reservoir pressure was 14.80 MPa (2146
lbf/in²) at 1524 m (5000 ft), and the draw-down coefficient was 21
kPa/(kg/s), or 1.38 lbf/in²/(lbm/s). True reservoir temperature (obtained by a weighted average over the production interval) was about
168°C (334°F). The maximum observed flow rate was limited by the size
of certain elements of the surface equipment to about 45 kg/s (357,000
lbm/h). The wells will be operated not in the self-flowing mode (as in
these tests), but in a pumped mode using down-hole pumps of either line-
shaft or electric, submersible design. Each well is expected to deliver

about 113 kg/s (900,000 lbm/h) of fluid under pumped conditions with the pump set at approximately 305 m (1000 ft).

Republic received the first loan guaranty made by the government under the Geothermal Loan Guaranty Program (GLGP). Issued in May 1977, the guaranty provides $9 million to drill a number of additional production and reinjection wells at the East Mesa site. Each well is expected to be capable of producing more than 2 MW of electrical power [Silverman, 1977; ERDA News, 1977].

12.4.3   Energy conversion system

A single-flash plant of 10 MW output is planned, the first of its type in the United States. This commercial plant should be completed in 1979 [Holt, 1977b] and will be the first element of a larger, 48-MW(e) double-flash plant presently being designed. This plant will include the above-mentioned 10-MW turbine-generator and is expected to begin operating in 1980. An artist's conception of the proposed plant is shown in figure 12.22.

12.5   Southern California Edison/Union Oil at Brawley, California

A 10-MW separated-steam (or "single-flash") geothermal power plant is being designed for the Brawley field in the Imperial Valley. The plant will be operated by Southern California Edison using steam supplied by Union Oil Company. The Brawley field lies about 38 km (24 mi) northwest of East Mesa with a reservoir characterized by higher fluid temperatures and salinities than at East Mesa. Six geothermal wells have been sunk; the deepest extends to a depth of about 4 km (13,000 ft). At a depth of 2.4 km (7870 ft), a maximum temperature of 262°C (504°F) has been reported. Mean reservoir temperature is believed to be 253°C (478°F) [Muffler, 1979]; dissolved solids amount to roughly 10,000 ppm.

The energy conversion system will be supplied by Mitsubishi Heavy Industries, Ltd., and consist of that company's portable Modular-10 turbine-generator package [Aikawa et al., 1978], a unit designed for simplicity of installation and maintenance. It is patterned after the plant at Onuma, Japan (see section 6.4.3). The turbine, generator, and all auxiliaries are located at ground level; the exhaust is led from the upper part of the turbine casing to a shell-and-tube condenser. Table 12.17 lists specifications for the turbine.

12.6   Double-boiling binary plant at Raft River, Idaho

A 5-MW (gross) binary plant is being designed by the Idaho National Engineering Laboratory for operation at the Raft River KGRA in Idaho [Ingvarsson and Madsen, 1976]. This plant will use geothermal fluid at the relatively low temperature of 143°C (290°F) and will employ isobutane as the cycle working fluid. A simplified process flow diagram is

FIGURE 12.22—Artist's conception of Republic Geothermal 48-MW double-flash plant, East Mesa. [Courtesy of Republic Geothermal, J. H. Barkman, personal communication.]

Table 12.17—*Turbine specifications for 10-MW flash-steam plant at Brawley, California* [a]

Type..............................	Single-cylinder, single-flow, 5-stage impulse
Rated output......................	10,000 kW
Maximum output...................	11,765 kW
Speed.............................	3600 rev/min
Steam inlet pressure...............	95.5 lbf/in²
Steam inlet temperature............	324°F
Noncondensable gas content.........	~2.0% by wt of steam
Exhaust pressure....................	6.0 in Hg
Steam flow rate....................	188,500 lbm/h
Last-stage blade height.............	11.22 in
Maximum allowable pressure.........	228 lbf/in²

[a] Source: Aikawa et al., 1978.

shown in figure 12.23. Optimization studies show that the system should be designed with isobutane as the working fluid and with two boilers, one at 116°C (240°F) or 2.63 MPa (382 lbf/in²) and one at 82°C (180°F) or 1.40 MPa (203 lbf/in²). Net output of the plant will be 3.35 MW(e). Table 12.18 lists cycle conditions and state properties for the nominal design case; the state points refer to those shown in figure 12.23. The plant

Table 12.18—*Cycle properties for nominal design case of Raft River isobutane binary power plant* [a]

State point [b]	Temperature		Mass Flow Rate	
	°C	°F	Mg/h	10⁶ lbm/h
1............................	41	105	424	0. 934
2............................	82	180	278	0. 613
3............................	116	240	278	0. 613
4............................	116	240	278	0. 613
5............................	54	130	278	0. 613
6............................	53	128	424	0. 934
7............................	38	101	424	0. 934
8............................	82	180	146	0. 321
9............................	82	180	146	0. 321
10............................	51	123	146	0. 321
11............................	143	290	472	1. 04
12............................	121	250	472	1. 04
13............................	106	222	472	1. 04
14............................	88	190	472	1. 04
15............................	62	144	472	1. 04
16............................	24	75	3048	6. 72
17............................	35	95	3048	6. 72

[a] Source: Ingvarsson and Madsen, 1976.
[b] These numbers refer to figure 12.23.

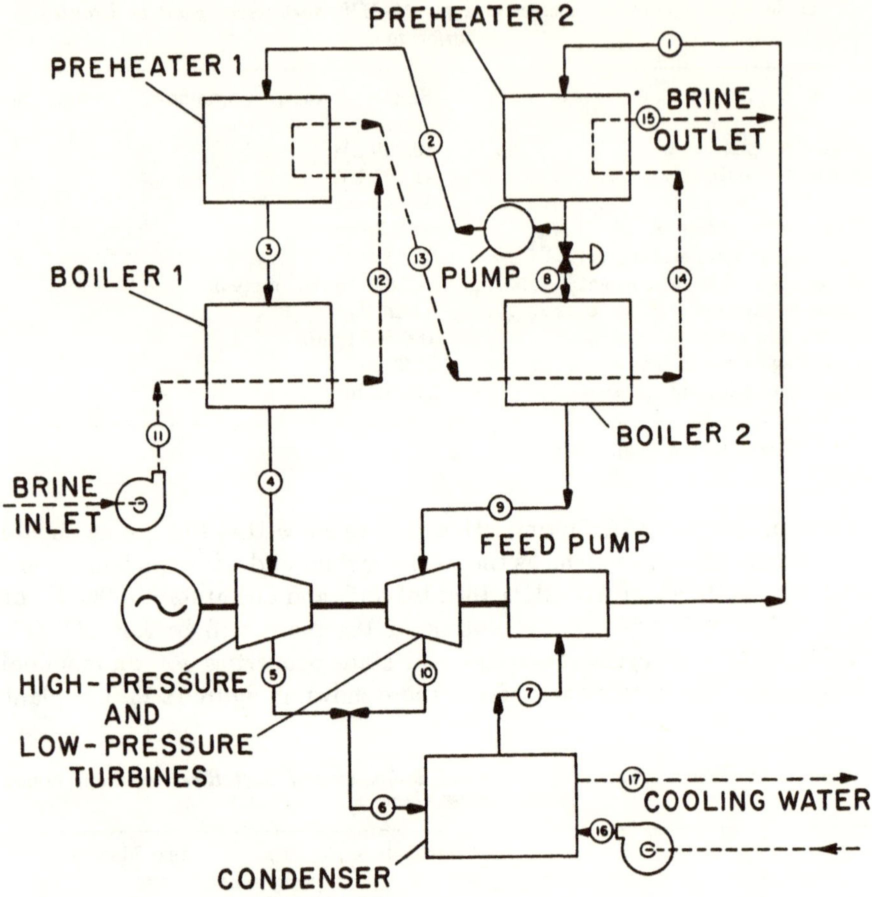

FIGURE 12.23—Process flow diagram for Raft River double-boiling isobutane binary cycle power plant [Ingvarsson and Madsen, 1976].

will require about 141 kg/kW · h (310 lbm/kW · h) and have a resource utilization efficiency $\eta_u = 32\%$, based on a wellhead availability of 14.88 kJ/kg (34.62 Btu/lbm) and a plant output of 4.72 kJ/kg (10.99 Btu/lbm). Although the plant will serve primarily as a test bed for low-temperature geothermal power plants, the electricity produced will be fed into the grid of the Raft River Electrical Cooperative. The cost of electricity is estimated to be 31.15 mill/kW · h [Ingvarsson and Madsen, 1976]. The plant is expected to begin operating in January 1980.

12.7   Hawaii geothermal project at Puna, Hawaii

A separated-steam (or "single-flash") plant of 5 MW capacity will be installed at the Kapoho reservoir near Cape Kumukahi in the Puna (Puulena) region of the Big Island of Hawaii in 1980. The plant will consist of a single wellhead unit. The geothermal area lies in the east rift zone at the easternmost tip of the island [Furumoto, 1978]. Six wells have been drilled in the area, but only one of these was successful, well HGP-A, drilled to a depth of 1871 m (6140 ft). Reservoir temperature is 358°C (676°F) measured at the bottom of the hole in a zone of conductive heat flow. A convective zone characterized by near-isothermal conditions appears to lie between 1220 and 1769 m (4003 and 5804 ft) with a temperature of 290°C (554°F) [Muffler, 1979]. The dryness fraction of the two-phase geothermal mixture ranges from 52%–64% at the wellhead [Chen et al., 1978]. Results of flow tests on this well have been highly encouraging, and it has been estimated that the Kapoho geothermal reservoir may be capable of supporting 50,000 MW·years [Chen and Grabbe, 1978]: a more conservative figure of 1230 MW·years is given by the U.S. Geological Survey (U.S.G.S.) [Muffler, 1979].

The greatest demand for electricity in Hawaii is on the island of Oahu, whereas the greatest potential for geothermal power production is on the Big Island. Nevertheless, a geothermal development group was formed in 1977 to promote this resource in an attempt to reduce Hawaii's dependence on imported fuel oil. The group consists of the State Department of Planning and Economic Development (DPED), the University of Hawaii's Hawaii Geothermal Project (HGP), and the County of Hawaii. In addition the Hawaiian Electric Company (Honolulu) and the Hawaii Electric Light Company (Hilo) are participating as consultants [Chen and Grabbe, 1978].

12.8 Demonstration plant at Valles Caldera, New Mexico

The U.S. Department of Energy, through the Division of Geothermal Energy, is helping support the design and development of a 50-MW flash-steam geothermal power plant in the Jemez Mountains of north-central New Mexico. The plant site is in the western arm of the Rocky Mountains at an area known as Baca No. 1 within the Valles Caldera KGRA, about 56 km (35 mi) west of Los Alamos and 120 km (75 mi) north of Albuquerque. The caldera was formed about one million years ago by an enormous volcanic eruption that resulted in a huge depression as the volcano settled into its empty magma chamber [Pettitt, 1978]. Five hot springs are located in the caldera, with temperatures ranging from 25° to 87°C (77° to 189° F). These springs are classified as acid-sulfate, and there is extensive hydrothermal alteration; associated with the springs are gas seeps.

The geothermal reservoir is believed to be liquid dominated with an overlying vapor-dominated zone. There are 17 wells in the southwest quadrant of the caldera, and bottom-hole temperatures range from 250° to 290°C (480° to 554°F) with a mean reservoir temperature of 273°C (523°F). The wells vary in depth from 1525 to 2745 m (5004 to 9006 ft). The top of the reservoir is estimated to lie about 1 km (3281 ft) below the surface, and the mean volume of the reservoir is about 125 km³ (30 mi³) [Muffler, 1979]. The overlying rocks, for drilling purposes, are classified as hard. The U.S.G.S. reported that the Valles Caldera KGRA has the potential to support 2700 MW of electrical power production for 30 years [Muffler, 1979]. At this stage it is known that the plant will utilize the two-phase geofluid in a separated-steam (or so-called "single-flash") power system. The plant flow diagram will likely follow the general layout shown earlier in figure 1.5.

The geothermal field has been developed by Union Oil Company, which will supply steam to Public Service Company of New Mexico. Each production well is believed capable of providing 91 Mg/h (200,000 lbm/h) of geothermal fluid with a quality of 35% at a wellhead pressure of 965 kPa (140 lbf/in²). Roughly 15 wells will be needed to supply the 50-MW plant. The intent of the demonstration project is to show that geothermal power plants using liquid-dominated resources can be built and operated in the United States on an economically competitive, commercial basis. If all goes as planned, electricity will begin flowing in 1982 [GEM, 1978].

12.9   Potential growth of geothermal power in the United States

The most recent assessment by the U.S.G.S. [Muffler, 1979] estimates that the identified hydrothermal convection systems with temperatures in excess of 90°C (194°F) in the United States (excluding National Parks) are capable of generating about 23,000 MW of electricity for a period of 30 years, i.e., the equivalent of 23 large nuclear power plants. When undiscovered systems are included, the estimate jumps to 95,000–150,000 MW for 30 years. A comprehensive list of known geothermal sites together with their characteristics may be found in U.S.G.S. Circular 790 [Muffler, 1979].

The thermal energy contained in geopressured-geothermal reservoirs may ultimately deliver between 23,000 and 240,000 MW of electricity for 30 years. This estimate does not include any electricity generated from the separate combustion of dissolved methane gas. The technology for converting the energy in hot, dry rocks into electricity is not yet established; the U.S.G.S. could not assess the potential of hot, dry-rock systems, although it is estimated that the energy associated with such systems is several thousand times that associated with either hydrothermal or geopressured reservoirs [Muffler, 1979].

12.9.1   Imperial Valley, California

The Imperial Valley of southern California holds a huge reserve of geothermal energy. A recent conservative estimate of the potential of this area suggests that 8700 MW of geothermal electricity capacity may be possible assuming 20–30-year plant lifetimes [Younker and Kasameyer, 1978]. A number of power plants of various designs are either under construction or in the advanced stages of planning. Table 12.19 lists particulars for these

Table 12.19—*Geothermal power plants under construction or planned for the Imperial Valley, California*

Site	Utility or Plant Owner	Field Developer	Year of Startup	Capacity MW	Plant Type
East Mesa	Imperial Magma	Magma Power	1979	11. 2	Magmamax dual binary
East Mesa	San Diego G & E	Republic Geothermal	1980, 1982	48. 0	Single- and double-flash
Brawley	So. Cal. Edison	Union Oil	1980	10. 0	Single-flash
Niland	So. Cal. Edison	Union Oil	1982	10. 0	Single-flash
Niland	San Diego G & E	Magma Electric	1982	50	Single- or double-flash
Heber	So. Cal. Edison	Chevron	1982	45	Double-flash
Heber	San Diego G & E	Chevron	(a)	50	Binary
Westmorland	(b)	Republic Geothermal	(a)	50	Double-flash

a In planning.   b Not available.

plants. The first three have been described in sections 12.3–12.5; of the plants now in planning, two are slated for the Heber geothermal field. One of these will be a 50-MW "double-flash" plant to be operated by Southern California Edison, and the other will be a 45-MW binary plant for San Diego Gas & Electric.

The Heber KGRA is located about 32 km (20 mi) west-southwest of East Mesa. Reservoir temperature is about 180°C (356°F); 11 wells, ranging in depth from 0.9 to 3.3 km (2950 to 10,800 ft), have been drilled to date [Muffler, 1979]. The total amount of dissolved solids in the Heber geofluid is between 14,000 and 25,000 ppm, or about one order of magnitude greater than the fluid at East Mesa. Also, Heber is located in the midst of valuable, irrigated agricultural land and will necessitate careful handling of the geofluid—from the production wells through the power plants and ultimately to reinjection wells for disposal. The depth to the crystalline basement in the Heber area, estimated at 7 km (4.3 mi), is greater than at any other known site in the Imperial Valley [Geonomics, 1976]. The convecting reservoir consists of deltaic sediments considered relatively soft for drilling purposes. The U.S.G.S. estimates that Heber is capable of supporting 650 MW of electrical generating capacity for 30 years [Muffler, 1979].

Preliminary specifications for the SDG&E 45-MW binary plant are given in table 12.20. (See also figure 1.8.) Although the working fluid has not yet been determined, it is expected to be a mixture of isobutane and isopentane. It has been suggested that the composition of the mixture be changed over the lifetime of the plant in order to maintain as near an optimum operating condition as possible in the event the reservoir temperature should decline in the course of exploitation. Resource utilization efficiency would be about 41% based on present design specifications and allowing for 15,000 ppm of total dissolved solids in the brine. The plant would require a specific brine flow of about 68 kg/kW·h (149 lbm/kW·h) [SDG&E, 1977].

Republic Geothermal is planning to build a 50-MW true "double-flash" plant at Westmorland, about 35 km (22 mi) north of Heber. Reservoir temperatures have been quoted ranging from 217°C (423°F) [Muffler, 1979] to 260°C (500°F) [Anderson, 1977]. Six geothermal test wells have been drilled with a maximum depth of about 2.6 km (8530 ft). The geofluid carries between 20,000 and 70,000 ppm of dissolved solids, and flow rates of up to 263 Mg/h (580,000 lbm/h) of two-phase fluid have been reported [Anderson, 1977].

A 50-MW flash-steam plant is being contemplated by San Diego Gas & Electric for the Salton Sea KGRA. This geothermal site, sometimes referred to as Niland, is situated at the southeastern shore of the Salton Sea, about 18 km (11 mi) north of Westmorland. The geofluid is quite hot, but hypersaline; temperature is 323°C (613°F) and salinity ranges from 250,000 to 300,000 ppm. Chlorides of sodium, calcium, and potassium may

Table 12.20—*Preliminary technical specifications for proposed 45-MW binary plant at Heber*

Turbine data:	
Rated capacity.	45 MW
Maximum capacity.............	65 MW
Secondary working fluid........	Isobutane-isopentane mixture
Inlet pressure	500 lbf/in²
Inlet temperature . .	295° F
Exhaust pressure....	72.3 lbf/in²
Exhaust temperature............	165.6° F
Mass flow rate........	7,687,000 lbm/h
Goethermal fluid data:	
Inlet pressure....	153 lbf/in²
Inlet temperature	360° F
Outlet pressure	133 lbf/in²
Outlet temperature	153° F
Mass flow rate......	6,722,000 lbm/h
Condenser data:	
Type...	Single-pass, counterflow, shell-and-tube
Cooling water inlet temperature...	90° F
Cooling water outlet temperature..	108.2° F
Cooling water flow rate....... ...	67,783,000 lbm/h
Heat rejection system data:	
Type...	Crossflow, mechanically-induced-draft water cooling tower
Number of cells............	10
Design wet-bulb temperature.. .	80° F
Water pump power......	5000 kW

be economically recovered from the brines, in addition to production of electricity by means of appropriate energy conversion schemes. Before a large-scale commercial power station can be installed, methods must be perfected for the proper handling of the extremely corrosive brine. The field has been developed by Magma Power which has contracted with the engineering firm of Morrison-Knudsen of Boise, Idaho, to carry out engineering and construction management for the plant at a cost of $50 million. Electricity may begin to flow in 1982 if construction proceeds according to plan [Smith, 1979].

12.9.2   Western United States

Table 12.21 lists the geothermal plants outside California that are now under construction or being planned. The first three have been discussed in sections 12.6–12.8. The plants shown for Roosevelt Hot Springs, Utah, Desert Peak, Nevada, and the hybrid coal/geothermal plant proposed by the City of Burbank, California, are not definite but are in advanced planning stages.

The Roosevelt Hot Springs area is located in Beaver County near Milford, Utah, at the junction of the Escalante Valley (a north-south trending

Table 12.21—*Proposed U.S. geothermal power plants outside California*

Site	Utility or Plant Owner	Field Developer	Year of Startup	Capacity MW	Plant type
Raft River, ID	EG&G	EG&G	1980	5	Double-boiling binary
Puna, HI	[a] HELCO	Univ. of Hawaii	1980	5	Single-flash
Valles Caldera, NM	Public Service Co. of New Mexico	Union Oil	1982	50	Single-flash
Roosevelt Hot Springs, UT	Rogers Engineering Co	Phillips	[b]	55	Double-flash
Desert Peak, NV	Sierra Pacific	Phillips	[b]	20–50	Single- or double-flash
[c]	[d] City of Burbank	[c]	[b]	250–750	Hybrid coal-geothermal

[a] Hawaii Electric Light Company.
[b] In planning.
[c] Not available.
[d] In conjunction with a consortium of neighboring cities in the Los Angeles area.

graben) and the Mineral Range (a horst block running parallel to the east side of the valley). The Mineral Range is characterized by rugged topography and is about 48 km (30 mi) long and from 96 to 16 km (6 to 10 mi) wide. The Escalante Valley is filled with sediments to a thickness of roughly 1.5 km (0.9 mi) in the center [Lenzer et al., 1977a, 1977b].

Many faults are present throughout the area, several of which have an important influence on the hydrology. The hydrothermal reservoir itself is fracture dominated. The top of the reservoir lies at a depth of about 900 m (2970 ft) over most of the area. A thermal gradient profile was obtained from 39 test holes; 5 of these at the center of the anomaly registered gradients in excess of 40°C/100 m (22°F/100 ft), or about 12 times the normal gradient. Test holes ranged in depth from 60 to 610 m (197 to 2000 ft). The region within which the thermal gradient exceeds 10°C/100 m (5.5°F/100 ft) covers an area roughly 8×13 km (5×8 mi). The predominant rock types are Precambrian granite and Pliocene or Pleistocene volcanics. The overlaying rock is judged medium-to-hard for drilling purposes [Williams et al., 1977]. The U.S.G.S. reports that the reservoir should be able to support 970 MW for 30 years [Muffler, 1979]. So far, a total of seven wells are each capable of producing, on the average, about 450 Mg/h (992×10^3 lbm/h) of liquid and steam mixture at 260°C (500°F). The geofluid contains less than 8000 ppm of dissolved solids, and the reservoir pressure is nearly hydrostatic.

Negotiations are underway among the Utah Power and Light Company (UP&L), Phillips Petroleum Company, and Rogers International, a subsidiary of Rogers Engineering Company, concerning the financing, design, construction, and operation of a 52-MW geothermal power plant at Roosevelt Hot Springs. It is expected that Phillips Petroleum, developer of the geothermal field, will supply steam to Rogers, which will be responsible for financing the design and construction of the plant and for operating it. UP&L will buy the power and have an option to acquire the plant after it has evaluated the long-term reliability of the plant and the resource. No details are currently available concerning the energy conversion system, although the plant is expected to be of the separated-steam/flash-steam type.

The City of Burbank and a number of neighboring cities in the Los Angeles metropolitan area are seriously considering a hybrid power plant utilizing both fossil and geothermal energy resources. Although plant location has not yet been decided, four possible areas have been studied: East Mesa, Coso Hot Springs, and Long Valley, California; and Roosevelt Hot Springs, Utah [Burbank, 1977]. The best site from an overall economic standpoint appears to be Roosevelt. Design is at present in the hands of an architect/engineering firm, which is considering a plant of 750 MW capacity.

Thermodynamic analysis of several possible hybrid fossil/geothermal energy conversion systems has been carried out at Brown University

[Khalifa et al., 1978a]. Candidate systems include the so-called geothermal-preheat system (figure 12.24) ; the fossil-superheat system (figure
12.25) ; and the compound-hybrid system (figure 12.26).

The Pacific Sierra Research Corporation has conducted economic studies
that compare the cost of constructing and operating a hybrid plant of the
geothermal-preheat type at each of the four sites mentioned above
[Thomas, 1977]. The calculated capital costs and the cost of electricity for
a 750-MW hybrid plant at these respective sites are as follows: $289/kW

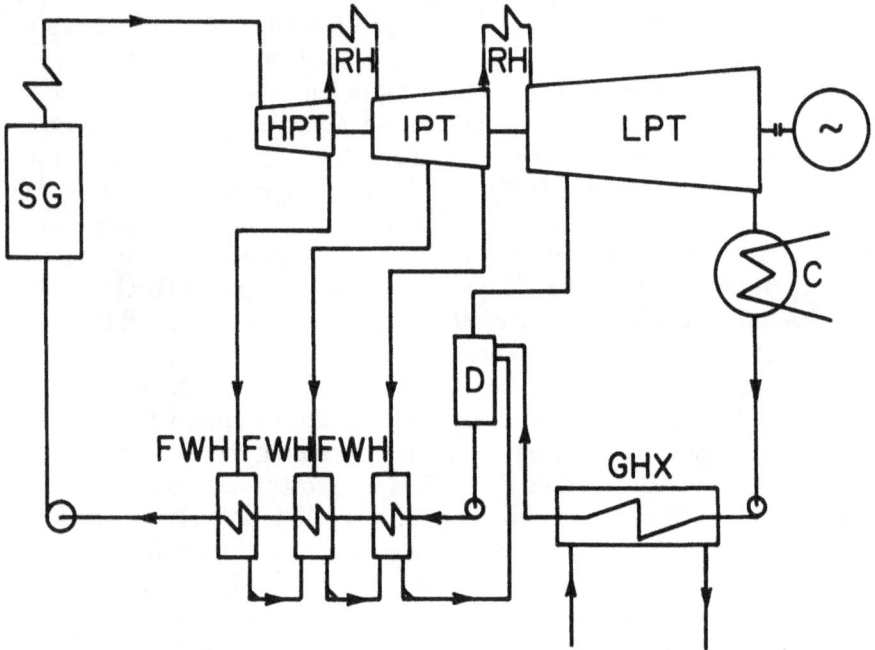

FIGURE 12.24—Simplified flow diagram for geothermal-preheat fossil/geothermal
power plant [Khalifa et al., 1978b].

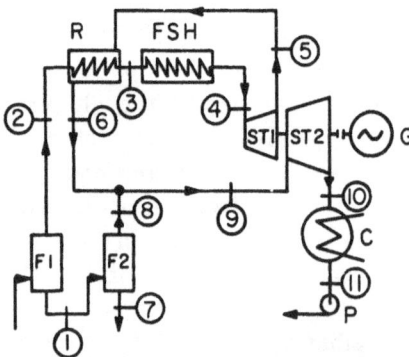

FIGURE 12.25—Simplified flow diagram for
fossil-superheat hybrid geothermal/fossil
power plant [DiPippo et al., 1978].

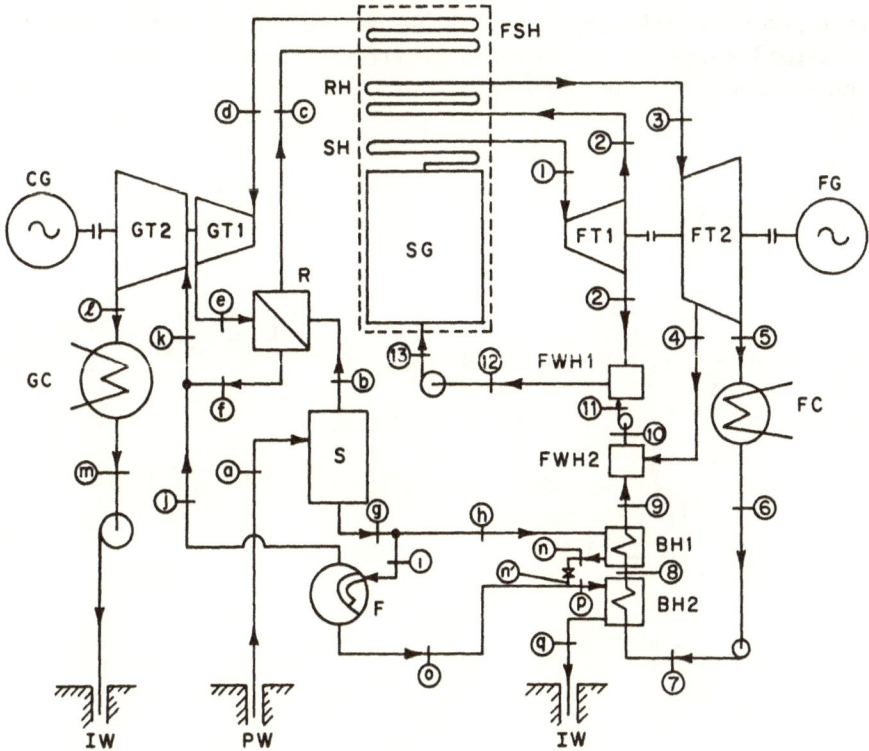

FIGURE 12.26—Compound hybrid fossil/geothermal power plant, which combines features of the geothermal-preheat and fossil-superheat systems [DiPippo and Avelar, 1979; Khalifa et al., 1978a].

and 20.4 mill/kW · h for Roosevelt Hot Springs; $493/kW and 25.3 mill/kW · h for Coso Hot Springs; $533/kW and 28.3 mill/kW · h for East Mesa; and $521/kW and 30.5 mill/kW · h for Long Valley. All figures are in 1977 dollars. Owing to the thermodynamic advantage of the hybrid plant over individual fossil and geothermal plants [DiPippo et al., 1977], a geothermal-preheat hybrid plant of 750 MW net output at Roosevelt would save 50 Mg/h (110×10³ lbm/h) of coal compared to an all-coal plant of the same capacity [Thomas, 1977].

12.9.3   Gulf Coast

Beneath the Gulf Coast of the United States, principally along the Louisiana and northeastern Texas shorelines, lies a vast reservoir of brine at moderate temperatures and enormous pressure that contains various concentrations of dissolved methane [Bebout, 1976; Britton, 1979; Holt, 1977a]. The potential exists for three different modes of energy extraction: (1) direct use of methane gas after separation at high pressure, (2) hydraulic turbogenerator driven by high-pressure liquid, and (3) binary-cycle power generation by means of a low boiling point fluid receiving heat from the hot brine. A plant that could perform these energy conversions is shown schematically in figure 12.27.

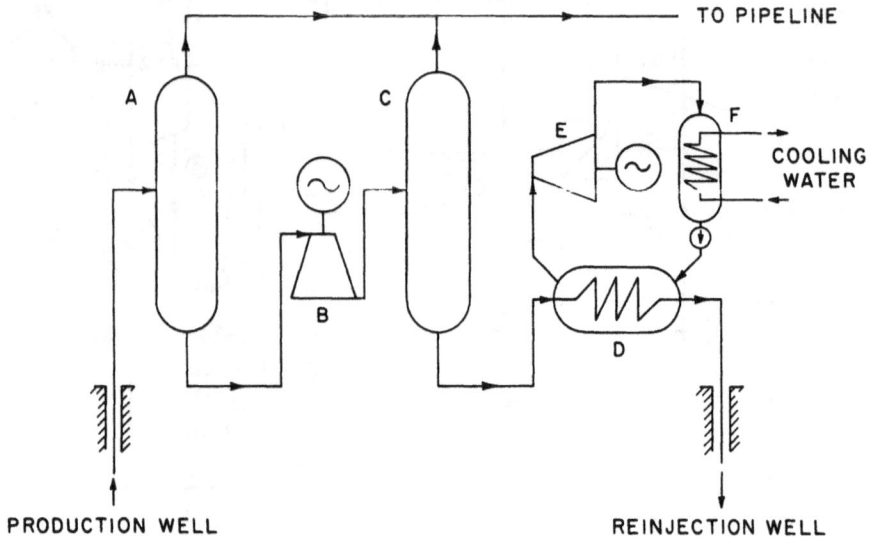

FIGURE 12.27—Proposed system for three-way utilization of geopressured geothermal energy resources. Key: A=high-pressure methane separator; B=hydraulic turbine; C=intermediate-pressure methane separator; D=heat exchanger; E=hydrocarbon or halocarbon turbine; F=condenser [after Britton, 1978; Holt, 1977a].

The U.S. Department of Energy is actively supporting the drilling of deep exploratory wells in the geopressured reservoirs of Brazoria County, south of Houston. While it is possible to estimate the total energy contained in the Gulf Coast geopressured zones (and this amount is comparable to the total hydrothermal resources of the rest of the United States), it is more difficult to assess what fraction of this energy can eventually be brought to the surface for exploitation. The U.S.G.S. says that the electricity that could be generated for 8% recovery amounts to between 23,000 and 240,000 MW for a period of 30 years [Muffler, 1979].

12.9.4   East Coast

The geothermal exploration program along the East Coast of the United States from New Jersey to Georgia is aimed at delineating a hot water reservoir that might be exploited for space heating or other industrial, commercial, or residential purposes. The Department of Energy is conducting an extensive drilling program involving about 60 shallow wells 300 m

(1000 ft) deep to gauge the thermal gradient and allow selection of sites for deep exploratory wells. The first of these will be drilled near Crisfield, Maryland, and extend to a depth of about 1525 m (5000 ft).

Since the highest temperatures expected to be encountered are about 90°C (194°F), it is unlikely that East Coast geothermal energy will be used for generation of electricity. It should be recalled, however, that binary-cycle power plants can operate with geofluids as cool as these. The Soviet pilot plant at Paratunka, for example, functioned with brine at 81.5°C (178.7°F) (see section 11.3). The pioneering work in the discovery of the East Coast geothermal potential was done by J. K. Costain at the Virginia Polytechnic Institute and State University [Costain et al., 1978].

12.9.5 Summary

The installed electrical generating capacity of the geothermal resources in the United States has grown dramatically since the first 11-MW unit began operating at The Geysers in 1960. This pattern of growth is traced in figure 12.28, which includes projections of expected new generating capacity up to the year 1984. It seems certain that by 1985 there will be 2000 MW of geothermal-based electrical generating capacity in this country, representing a savings of about 25 million barrels of oil equivalent per year. Furthermore, considering only those states in which geothermal power plants are expected to be operating at that time (California, Hawaii, Nevada, New Mexico, and Utah), geothermal energy may account for as much as 3% of the total electricity generated.

REFERENCES [a]

Aikawa, K., Fukuda, S., and Tahara, M., 1978. "MHI's Recent Achievements in Field of Geothermal Power Generation," *MHI Technical Review,* Vol. 15, No. 3, Ser. 43, pp. 195–207, Mitsubishi Heavy Industries, Ltd., Tokyo.

Anderson, D. N., 1977. "Recent Developments of Geothermal Resources in the United States: A Survey of the Ten Most Promising New Areas," Geothermal Resources Council Spec. Short Course No. 6, Dec. 12–13, Houston.

Anderson, D. N. and Hall, B. A., eds., 1973. "Geothermal Exploration in the First Quarter Century," *Geothermal Resources Council Spec. Rep. No. 3,* Davis, CA, p. 172.

Anderson, J. H., 1973. "The Vapor-Turbine Cycle for Geothermal Power Generation," in *Geothermal Energy: Resources, Production, Stimulation,* P. Kruger and C. Otte, eds., Stanford Univ. Press, Stanford, CA, pp. 163–175.

Anderson, J. H., 1976a, "Process Flow for Magmamax Plant—Magma Electric Co., East Mesa, CA," J. Hilbert Anderson, Inc., Dwg. No. 13–C–540, York, PA.

Anderson, J. H., 1976b. "Plant Layout, Magma Electric Co., East Mesa, CA, Plant," J. Hilbert Anderson, Inc., Dwg. No. 13–E–383, York, PA.

Austin, A. L., Lundberg, A. W., Owen, L. B., and Tardiff, G. E., 1977. "The LLL Geothermal Energy Program Status Report: January 1976–January 1977," Lawrence Livermore Laboratory UCRL–50046–76, Livermore, CA, pp. 51–126.

---

[a] See note on p. 24.

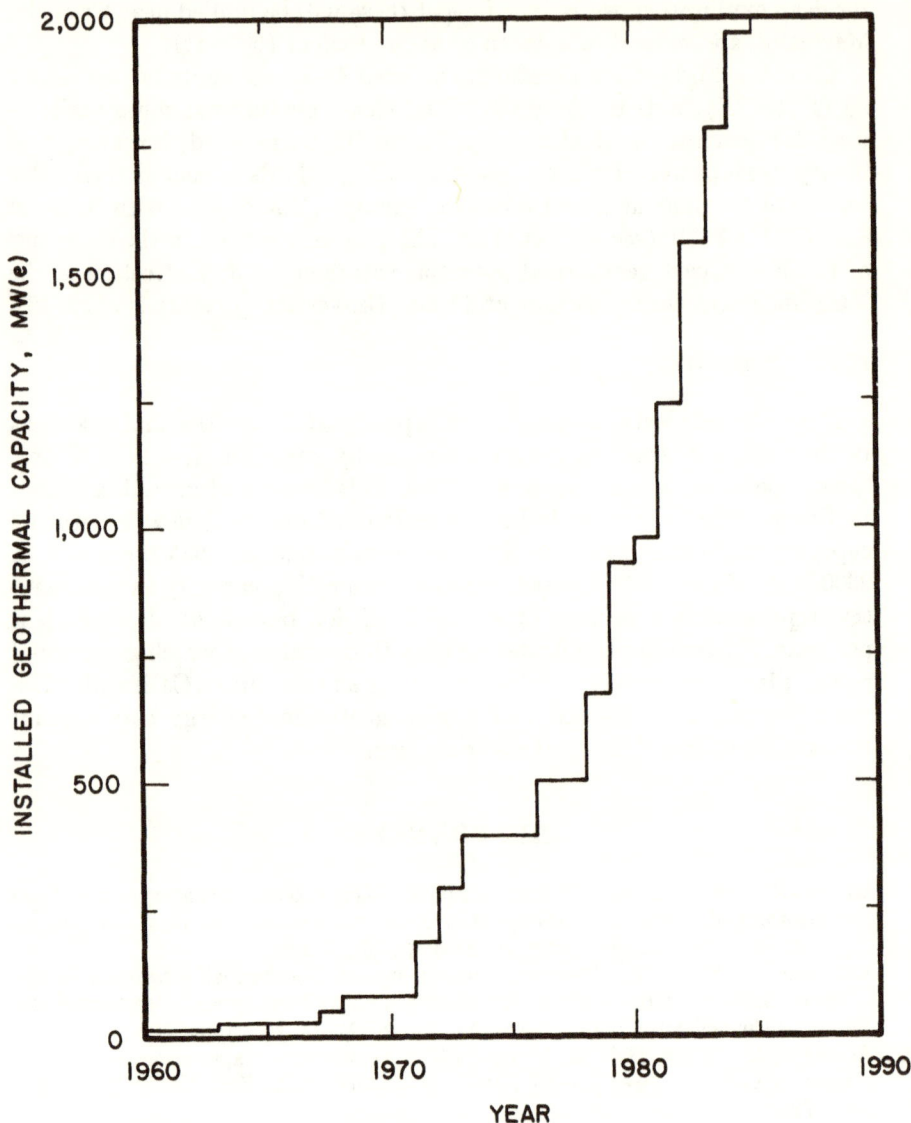

FIGURE 12.28—Growth of installed geothermal electric generating capacity in the United States, projected to 1985.

Barton, D. B., 1970. "Current Status of Geothermal Power Plants at The Geysers, Sonoma County, California, U.S.A.," *Pisa, 1970,* Vol. 2, pp. 1552–1559.

Bebout, D. G., 1976. "Subsurface Technique for Locating and Evaluating Geopressured Geothermal Reservoirs Along the Texas Gulf Coast," *Proc. 2nd Geopressured Geothermal Energy Conf.,* Vol. 2, Univ. of Texas, Austin.

Biehler, S., 1971. "Gravity Studies in the Imperial Valley," in *Cooperative Geological-Geophysical-Geochemical Investigations of Geothermal Resources in the Imperial Valley Area of California,* Educ. Res. Serv., Univ. of California at Riverside, pp. 29–41.

Biehler, S. and Combs, J., 1972, "Correlation of Gravity and Geothermal Anomalies in the Imperial Valley, Southern California," *Geol. Soc. Amer. Abstracts with Programs*, Vol. 4, No. 3, p. 128.

Biehler, S., Kovach, R. L., and Allen, C. R., 1964. "Geophysical Framework of Northern End of Gulf of California Structural Province," in *Marine Geology of the Gulf of California, A Symposium*, T. H. Van Andel and G. G. Shor, Jr., eds., Amer. Assoc. Petrol. Geol. Mem. 3, pp. 126–143.

Britton, P., 1979. "Geothermal Goes East," *Popular Science*, Vol. 214, No. 2, pp 66–69.

Bruce, A. W., 1961. "Experience Generating Geothermal Power at The Geysers Power Plant, Sonoma County, California," *Rome, 1961*, Vol. 3, pp. 284–296.

Bruce, A. W., 1970. "Engineering Aspects of a Geothermal Power Plant," *Pisa, 1970*, Vol. 2, pp. 1516–1520.

Burbank, City of, 1977. "Site-Specific Analysis of Hybrid Geothermal/Fossil Power Plants," Rep. prep. for ERDA under Contract No. E (0–4–1311).

BuRec, 1977. "Geothermal Resource Investigations, East Mesa Test Site, Imperial Valley, California: Status Report," U.S. Dept. of Interior, Bureau of Reclamation.

Chen, B. H. and Grabbe, E. M., 1978. "Planning for Geothermal Development in Hawaii," *Geothermal Resources Council Trans.*, Vol. 2, Sect. 1, pp. 95–98.

Chen, B. H., Kihara, D. H., Yuen, P. C., and Takahashi, P. K., 1978. "Well Test Results from HGP-A," *Geothermal Resources Council Trans.*, Vol. 2, Sec. 1, pp. 99–102.

Combs, J., 1971. "Heat Flow and Geothermal Resource Estimates for the Imperial Valley," in *Cooperative Geological-Geophysical-Geochemical Investigations of Geothermal Resources in the Imperial Valley Area of California*, Educ. Res. Serv., Univ. of California at Riverside, pp. 5–27.

Combs, J., 1972. "Preliminary Heat Flow Values and Temperature Distributions Associated with the Mesa and Dunes Geothermal Anomalies, Imperial Valley, Southern California," *EOS*, Vol. 53, No. 4, pp. 515–516.

Combs, J. and Muffler, L. J. P., 1973. "Exploration for Geothermal Resources," in *Geothermal Energy: Resources, Production, Stimulation*, P. Kruger and C. Otte, eds., Stanford Univ. Press, Stanford, CA, pp. 95–128.

Costain, J. K., Glover, J., III, and Sinha, A. K., 1978. "Geothermal Resource Potential of the Eastern United States," Geothermal Resources Council Spec. Short Course No. 7, *Geothermal Energy: A National Opportunity*, Washington.

Dan, F. J., Hersam, D. E., Kho, S K., and Krumland, L. R., 1975. "Development of a Typical Generating Unit at The Geysers Geothermal Project, A Case Study," *San Francisco, 1975*, Vol. 3, pp. 1949–1958.

DiPippo, R. and Avelar, E.M., 1979. "Compound Hybrid Geothermal-Fossil Power Plants," *Geothermal Resources Council Trans.*, Vol. 3, pp. 165–168.

DiPippo, R., Khalifa, H. E., Correia, R. J., and Kestin, J., 1978. "Fossil Superheating in Geothermal Steam Power Plants," *Proc. 13th Intersociety Energy Conv. Engin. Conf.*, Vol. 2, pp. 1095–1101.

DiPippo, R., Khalifa, H. E., and Kestin, J., 1977. "Hybrid Fossil-Geothermal Power Plants," ASME Paper No. 77–WA/Ener-2, ASME Winter Annual Meeting, Atlanta, Ga.

Donnelly, J. M., Goff, F. E., Thompson, J. M., and Hearn, B. C., Jr., 1976. "Implications of Thermal Water Chemistry in The Geysers, Clear Lake Area," *Proc. Geothermal Environ. Sem. '76*, Lake County, CA, pp. 99–103.

Douze, E. J., 1971. "Seismic Ground-Noise Survey of the Mesa Anomaly," Teledyne-Geotech, Dallas.

Dutcher, J. L. and Moir, L. J., 1976. "Geothermal Steam Pricing at The Geysers, Lake and Sonoma Counties, California," *Proc. 11th Intersociety Energy Conv. Engin. Conf.*, Vol. 1, pp. 786–789.

Elliott, D. G., 1978. "Analysis of EM 16–29 Well Flow Data," Letter Rep. No. 2, ERDA Interagency Agree. No. EG–77–A–36–1016, Jet Propulsion Laboratory, Pasadena.

ERDA News, 1977. "ERDA Okays First Geothermal Loan for Republic Geothermal," *Geothermal Energy Magazine*, Vol. 5, No. 6, p. 21.

Finney, J. P., 1972. "The Geysers Geothermal Power Plant," *Chem. Eng. Prog.*, Vol. 68, pp. 83–86.

Furumoto, A. S., 1978. "The Relationship of a Geothermal Reservoir to the Geological Structure of the East Rift of Kilauea Volcano, Hawaii," *Geothermal Resources Council Trans.*, Vol. 2, Sec. 1, pp. 199–201.

GEM, 1977a. "Shell and NCPA Sign Geothermal Pact," *Geothermal Energy Magazine*, Vol. 5, No. 9, p. 42.

GEM, 1977b. "PG&E Unit 13 Currently under Construction," *Geothermal Energy Magazine*. Vol. 6. No. 2. p. 49.

GEM, 1978. "Geothermal Plant Proposed in N.M.," *Geothermal Energy Magazine*, Vol. 6, No. 2, p 49.

Geonomics, 1976, "Geotechnical Environmental Aspects of Geothermal Power Generation at Heber, Imperial Valley, California," submitted to EPRI for Holt/Procon, October.

Hartley, R. P., 1978. *Pollution Control Guidance for Geothermal Energy Development*, EPA Rep. No. 600/7–78–101, Ind. Env. Res. Lab., Cincinnati.

Hinrichs, T. C., 1973. "Magmamax Binary Pilot Plant," in *Minutes 8th CATMEC Meeting*, Brown Univ. Rep. No. CATMEC/10, DOE No. COO/4051–17, Providence, RI, pp. 18–19.

Hinrichs, T. C. and Falk, H. W., Jr., 1978. "The East Mesa 'Magmamax Process' Power Generation Plant," *Geothermal Energy Magazine*, Vol. 6, No. 1, pp. 44–46.

Holt, B., 1977a. "Geopressured Resource: A Sleeping Giant," *Geothermal Energy Magazine*, Vol. 5, No. 11, pp. 30–32. (Reprinted from *Hydrocarbon Processing*, July 1977.)

Holt, B., 1977b. "Geothermal Utilization Projects in the Western United States," Geothermal Resources Council Spec. Short Course No. 6, Dec. 12–13, Houston.

Ingvarsson, I. J. and Madsen, W. W., 1976. "Determination of the 5 MW Gross Nominal Design Case Binary Cycle for Power Generation at Raft River, Idaho," INEL, EG&G Idaho, Rep No. TREE–1039.

Khalifa, H. E., DiPippo, R., and Kestin, J., 1978a. "Hybrid Fossil-Geothermal Power Plants," *Proc. 5th Energy Tech. Conf.*, pp. 960–970.

Khalifa, H. E., DiPippo, R., and Kestin, J., 1978b. "Geothermal Preheating in Fossil-Fired Steam Power Plants," *Proc. 13th Intersociety Energy Conv. Engin. Conf.*, Vol. 2, pp. 1068–1073.

Koenig, J. B., 1977. "Exploration for Geothermal Resources," Geothermal Resource Council Spec. Short Course No. 6, Dec. 12–13, Houston.

Laszlo, J., 1976. "Application of the Stretford Process for H_2S Abatement at The Geysers Geothermal Power Plant," *Proc. 11th Intersociety Energy Conv. Engin. Conf.*, Vol. 1, pp. 724–730.

Lengquist, R. and Hirschfeld, F., 1976. "Geothermal Power: The Sleeper in the Energy Race," *Mech. Engineering*, Vol. 98, No. 12, pp. 25–31.

Lenzer, R. C., Crosby, G. W., and Berge, C. W., 1977a. "Geothermal Exploration of Roosevelt KGRA, Utah," Geothermal Resources Council Spec. Short Course No. 6, Dec. 12–13, Houston.

Lenzer, R. C., Crosby, G. W., and Berge, C. W., 1977b. "Recent Developments at the Roosevelt Hot Springs KGRA," Geothermal Resources Council Spec. Short Course No. 6, Dec. 12–13, Houston.

Lindsay, D. R., 1977. "A New Steam Sales Contract at The Geysers," Geothermal Resources Council Spec. Short Course No. 6, Dec. 12–13, Houston.

Mafi, S., 1978. "Turboexpanders in a Geothermal Power Recovery Cycle, *Turbo-machinery Int.*, Vol. 19, No. 3, pp. 41–42.

Mahoney, D. J. and Bangert, A. C., 1977. "Economic Impact of Geothermal Development, Sonoma and Lake Counties, California," Pacific Gas and Electric Co., San Francisco.

Matthew, P., 1975. "Geothermal Operating Experience, Geysers Power Plant," *San Francisco, 1975*, Vol. 3, pp. 2049–2054.

Muffler, L. J. P., ed., 1979. *Assessment of Geothermal Resources of the United States—1978*, Geological Survey Circular 790, U.S. Dept. of the Interior.

Pettitt, R. A., 1978 "Hot Dry Rock: A New Potential for Energy," *Geothermal Energy Magazine*, Vol. 6, No. 11, pp. 11–19.

PG&E, 1975. "Amended Environmental Data Statement, Geysers Unit 13," Pacific Gas and Electric Co., San Francisco.

Ramachandran, G., 1977. "Economic Analyses of Geothermal Energy Development in California, Vol. 1," Stanford Res. Inst., SAN–115 P 108–1 (Vol. 1), Menlo Park, CA.

Reed, M. J. and Campbell, G. E., 1975. "Environmental Impact of Development in The Geysers Geothermal Field, U.S.A." *San Francisco, 1975*, Vol. 2, pp. 1399–1410.

Saint, P. K., 1976. "Geothermal Guide to Mexicali-Imperial Rift Valley," Geothermal Energy Assoc., W. Covina, CA.

SDG&E, 1977. "Expression of Interest for a Geothermal Demonstration Power Plant," presented to U.S. ERDA by San Diego Gas and Electric Co., San Diego, June 20.

Semrau, K. T., 1976. "Control of Hydrogen Sulfide from Geothermal Power Production," *Proc. Geothermal Environm. Sem. '76*, Oct. 27–29, Lake County, CA, pp. 185–189.

Siegfried, H. N, 1925 "The Geysers," in Geothermal Exploration in the First Quarter Century, D. N. Anderson and B. A. Hall, eds., *Geothermal Resources Council Spec. Rep. No. 3*, Davis, CA, 1973, pp. 58–88.

Silverman, M., 1977. "Geothermal Loan Guarantee Program," paper presented at *Geothermal: State of the Art*, 1977 Annual Meeting Geothermal Resources Council, May 9–11, San Diego.

Smith, R. A., ed, 1979. *Geothermal Report*, Vol. 8, No. 4, Feb. 15, p. 3.

Thomas, F. J., 1977. "Cost Factors Related to Fossil-Geothermal Power Plants," in *Minutes 6th CATMEC Meeting*, Brown Univ. Rep. No. CATMEC/5, DOE No. COO/4051–9, Providence, RI, pp. 16–17 and Appendix M.

Truesdell, A. H. and White, D. E., 1973. "Production of Superheated Steam from Vapor-Dominated Reservoirs," *Geothermics*, Vol. 2, p. 154.

Weres, O., 1976. "Environmental Implications of the Exploitation of Geothermal Brines," *Proc. Geothermal Environm. Sem. '76*, Oct. 27–29, Lake County, CA, pp. 115–123.

Weres, O., Tsao, K., and Wood, B., 1977. "Resource, Technology, and Environment at The Geysers," Rep. No. LBL–5231, Lawrence Berkeley Laboratory, Berkeley.

White, D. E., Muffler, L. J. P., and Truesdell, A. H., 1971. "Vapor-Dominated Hydrothermal Systems Compared with Hot-Water Systems," *Econ. Geol.*, Vol. 66, p. 75.

Williams, F., Cohen, A., Pfundstein, R., and Pond, S., 1977. "Site-Specific Analysis of Geothermal Development—Data Files of Prospective Sites," Mitre/Metrek Rep. No. MTR–7586, Vol. 3.

Witherspoon, P. A., 1977. "Geothermal Reservoir Engineering," in *Minutes 4th CATMEC Meeting*, Brown Univ. Rep. No. CATMEC/3, DOE No. COO/4051–4, Providence, RI, pp. 1–10.

Younker, L. and Kasameyer, P., 1978. "A Revised Estimate of Recoverable Thermal Energy in the Salton Sea Geothermal Resource Area," Lawrence Livermore Laboratory Rep. No. UCRL–52450, Livermore, CA.

CHAPTER 13

COUNTRIES PLANNING GEOTHERMAL POWER PLANTS

13.1 Overall survey

The number of countries engaged in geothermal exploration, development, or exploitation for all purposes or those interested in putting their geothermal resources to use is estimated to be at least 65. These include Australia, Austria, Bahamas, Barbados, Belgium, Bhutan, Bolivia, Brazil, Canada, Chile, China, Colombia, Congo, Costa Rica, Dominican Republic, Ecuador, Egypt, El Salvador, Ethiopia, Fiji, France, Germany, Ghana, Greece, Guatemala, Guinea, Guyana, Haiti, Honduras, Hungary, Iceland, India, Indonesia, Iran, Israel, Italy, Jamaica, Japan, Kenya, Kuwait, Malaysia, Mexico, New Zealand, Nicaragua, Panama, Philippines, Portugal (Azores), Saudi Arabia, Spain (Canary Islands), Sri Lanka, Switzerland, Taiwan, Tanzania, Trinidad and Tobago, Turkey, Uganda, Union of Soviet Socialist Republics, United Arab Emirates, United Kingdom, United States, Venezuela, Yugoslavia, Zaire, and Zambia. This chapter discusses those countries on the threshold of exploiting their geothermal resources for electricity generation.

13.2 Azores (Portugal)

The islands of the Azores lie on the Mid-Atlantic Ridge, a spreading tectonic plate boundary. Of the nine islands that comprise the group, the largest and most heavily populated is São Miguel. In 1970 a well drilled on the northern flank of the Agua de Pau Volcano encountered fluids in excess of 200°C (392°F) at depths greater than 550 m (1805 ft). The full depth of the well was 981 m (3219 ft) [Meucke et al., 1974]. A 3-MW wellhead power unit is being designed for São Miguel, and could be in operation as soon as 1979 [K. Aikawa, personal communication].

13.3 Chile

The El Tatio geothermal field has been the subject of considerable exploration and drilling. The site is located in northern Chile, in Antofagasta Province, in a region consisting of a volcanic desert plateau at an elevation of over 4000 m (13,100 ft) with Quaternary volcanic mountains rising to nearly 6000 m (19,700 ft). Owing to the extreme remoteness and near-inaccessibility of the field, exploration is proceeding slowly. Furthermore, because the region is essentially arid, any geothermal development is likely to include the production of fresh water [Koenig, 1973]. It has been estimated that about 18 MW of electricity could be generated from the existing 13 wells [Lahsen and Trujillo, 1975]. These range in depth from 600 to 1820 m (1970 to 5970 ft) and have encountered geofluids at temperatures from 180° to 265°C (356° to 509°F). A small pilot plant is in operation and plans are underway to construct a 15-MW plant in the near future [Ellis and Mahon, 1977].

13.4 Costa Rica

The geothermal development program in Costa Rica is directed by the Instituto Costarrince de Electricidad (ICE) and has concentrated on Guanacaste Province in the northwestern part of the country. The geothermal area extends for 30 km (19 mi) along the flank of a chain of active volcanoes [Furgerson and Afonso, 1977]. An integrated program involving heat flow, temperature gradient, geochemical, electrical, and hydrological investigations is being carried out in the southwestern portion of the Cordillera de Quanacaste.

Particular attention is being given to the areas of Las Hornillas de Miravalles, Las Pailas, and Borinquen, where some drilling has been conducted [Blackwell et al., 1977]. A total of 35 exploratory wells have been sunk in the area; 24 have been to depths of 50 m (164 ft) or less, and 9 have exceeded 90 m (295 ft). Active development is underway and a 40-MW geothermal power plant is scheduled by ICE to be installed by 1984/1985. At Las Hornillas de Miravalles geochemical studies have revealed the possibility of a deep, chlorinated aquifer with reservoir base temperatures as high as 240°C (464°F) [Gardner and Corrales, 1977]. Deep drilling is underway with the objective of achieving a total depth of 4000 m (13,000 ft), which should allow for the completion of 4 wells since the aquifer is estimated to lie at a depth of between 800 and 1200 m (2625 and 3937 ft) [J. T. Kuwada, personal communication].

13.5 Guatemala

The national electric company of Guatemala, INDE, is aiming for 100 MW of installed geothermal capacity by the early 1980's [Meidav et al., 1977]. Three areas—Moyuta, Amatitlán, and Zunil—have been under exploration with technical assistance from Japan. There were high hopes for the field at Moyuta, about 25 km (16 mi) northwest of the successful project at Ahuachapán across the border in El Salvador. Shallow wells revealed gradients of about 0.25°C/m (0.14°F/ft), but two wells produced low temperatures, and the site has been abandoned [Dominco, 1977; J. T. Kuwada, personal communication]. Attention is still being given to the other two sites. Amatitlán may someday support 50–100 MW, whereas Zunil appears to be a rather small area with limited prospects.

13.6 Honduras

In 1977 the National Electric Authority of Honduras (ENEE) began a program of geothermal exploration that focused on two areas: Pavana, in the southernmost part of the country near Choluteca, and San Ignacio, which is located northwest of the capital city of Tegucigalpa [Meidav et al., 1977]. At the present time the exploration program is temporarily in abeyance [J. T. Kuwada, personal communication]. By 1982, ENEE hopes to have 50 MW of geothermal power on-line, with an additional 50 MW by 1984/1985 [Meidav et al., 1977].

13.7 Indonesia

Indonesia's location at the junction of three tectonic plates (with associated vulcanism and earthquake activity) together with its average annual rainfall of 2000 mm (79 in) create a potentially valuable source of geothermal (hydrothermal) power. Exploration for geothermal energy began in 1926 and extensive geophysical, geological, and geochemical surveys have been conducted by various teams of scientists from France, Japan, New Zealand, the United States, and the United Nations (UNESCO). A summary of these studies has been published recently [Radja, 1975]. Among the many promising thermal areas, the one at which a geothermal power plant is likely to appear first is Kawah Kamojang. A 250-kW wellhead power generating unit has been purchased for $100,000 and has recently become operational there [GEM, 1978]. Fumaroles abound at this thermal site, the first to be discovered and explored in the Indonesian archipelago.

The geothermal field of Kawah Kamojang, along with two other solfatara and fumarolic areas, Kawah Manuk and G. Wayang, is located in the southern Priangan thermal anomaly on the western side of the island of Java, about 42 km (26 mi) southeast of the city of Bandung. The area is marked by acid intrusions of quartz diorite and extrusions (dacites). Siliceous sinter deposited by several hot springs leads to the conclusion that high temperatures exist at depth [Radja, 1975]. The area of the geothermal field is estimated to be about 1400 ha (3500 acres) [Hochstein, 1975]; the area enclosed within a 10 $\Omega \cdot$m-contour of a resistivity mapping is about 600 ha (1500 acres) [Radja, 1975]. Figure 13.1 shows the Kamojang area, including the location of resistivity traverses, resistivity sounding profiles, and temperature gradient holes [Hochstein, 1975]. Figure 13.2 is a more detailed view and includes the location of exploratory shallow and deep wells [Hochstein, 1975].

Temperature gradients as high as 0.35°C/m (0.19°F/ft), or about 12 times normal, have been measured in the Pangkalan portion of the field.

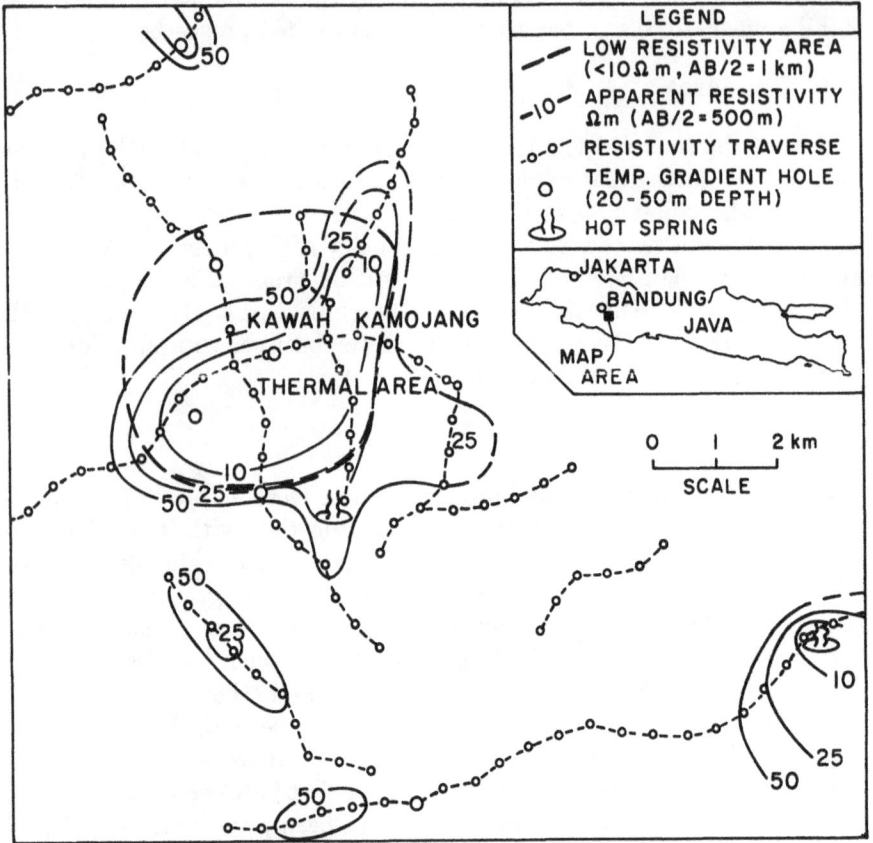

FIGURE 13.1—Exploration of the Kawah Kamojang geothermal field, Indonesia [Hochstein, 1975].

One of the earliest wells drilled is still producing; steam flows at 130°C (266°F) and 12.4 Mg/h (27.3 klbm/h). This particular well (No. 3) was drilled in 1926 [Stehn, 1929; Kartokusumo et al., 1975]. The steam contains 2%–4% by weight of noncondensable gases: CO_2 makes up about 96.5% by volume, H_2S about 1.85%, with CH_4, H_2, and N_2 making up the bulk of the residuals [Kartokusumo et al., 1975]. The findings indicate that Kawah Kamojang is a vapor-dominated field, perhaps a dry steam field, to shallow depths of at least 130 m (425 ft). Maximum temperature at these depths is about 238°C (460°F). Hochstein states that the reservoir is quite extensive and is 200–500 m (656–1640 ft) thick, with a power potential of between 100 and 250 MW for 50 years [Hochstein, 1975]. Furthermore, he concludes that vapor-dominated hydrothermal systems in young volcanic rocks may be much more common than previously believed.

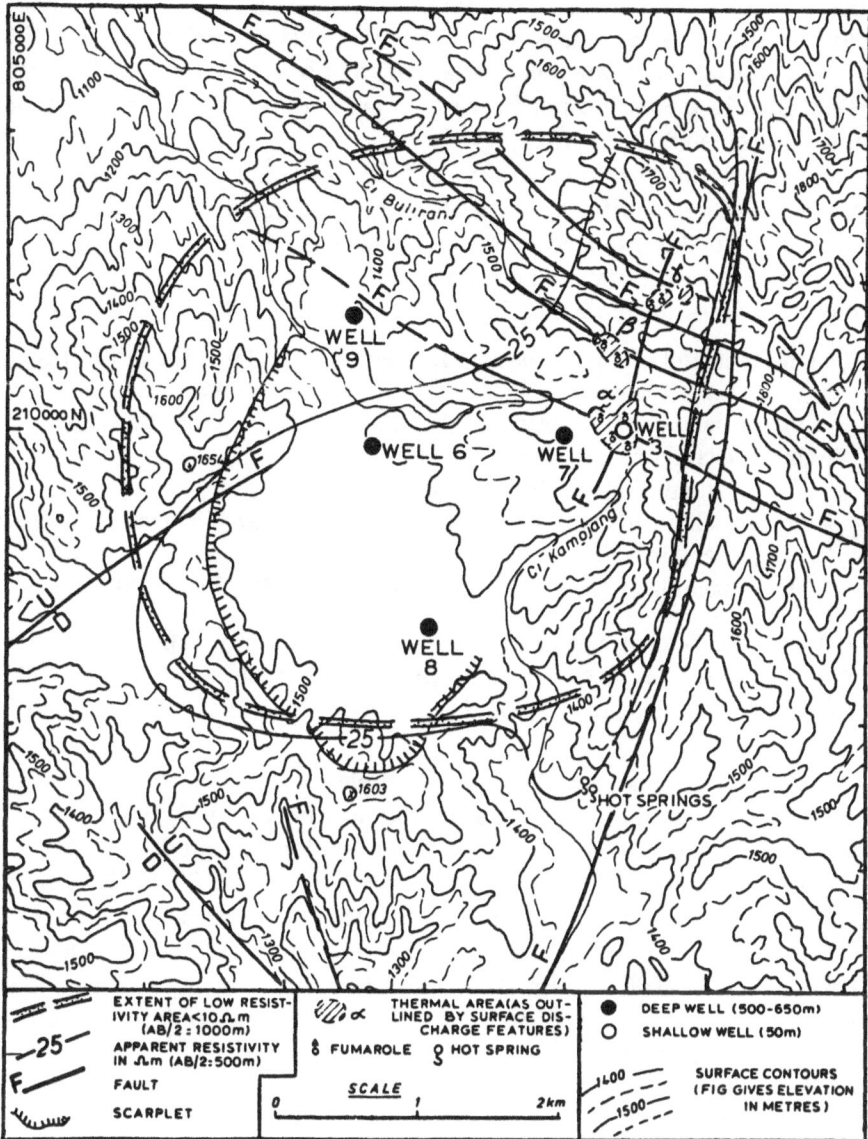

FIGURE 13.2—Location of exploratory wells and results of resistivity survey at Kawah Kamojang, Indonesia [Hochstein, 1975].

Power production at Kamojang is expected to begin during 1979. Geothermal Power Company of New York has supplied a 250-kW, noncondensing, wellhead turbogenerator to PERTAMINA, the state oil and natural gas mining company [GR, 1978; GEM, 1978]. The self-contained unit consists of a turbine, generator, controls, gearbox, and exhaust silencer/diffuser mounted on a platform; GPC calls the unit a "Mono-blok" system. The package cost about $400/kW. The power generated will be used during the development phase of the project. After a 3-MW unit is installed in the second phase, the plan is to build a 30-MW condensing unit [Basoeki and Radja, 1978].

Many other thermal areas are evident throughout the Indonesian archi-pelago. Surface manifestations such as hot springs, boiling mud pools, solfataras, and/or fumaroles are present at the sites listed below. Some surveys have been conducted at a few of these, and the reader is referred to other sources for more details [Akil, 1975; Muffler, 1975; Radja, 1975].

On the island of Java: Danau (Banten), Dieng, Ijen, Kawah Derajat, Kromong-Careme.

On the island of Sumatra: Toba, Padang Highlands, Pasumah.

On the island of Borneo: Kalimantan.

On the island of Halmahera: North Halmahera.

On the island of Sulawesi: Minahasa, Gorontalo, Central Sulawesi, South Sulawesi.

On the islands of Nusa Tenggara: Waikokor, Wai Pesith, Magekoba.

Of these, the field at Dieng has been quite extensively explored. It is a liquid-dominated reservoir with a bottom-hole temperature of 173°C (343°F) at a depth of 139 m (456 ft). A total of five wells have been drilled to date.

Aerial reconnaissance has provided ample evidence of the large store of geothermal energy in the Indonesian archipelago. While there are large supplies of petroleum within Indonesia, it is often difficult to transport it to the places where power is needed. Furthermore, PERTAMINA may decide to exploit its reserves of petroleum as a valuable export commodity. Demand for electric power is expected to reach 5100 MW in 1990, whereas it was less than 1000 MW in 1975. Thus geothermal energy, with its low cost and indigenous advantages, figures to play an important role in meet-ing the growing demand for electrical power in Indonesia.

13.8 Kenya

Six wells have been drilled in the Olkaria geothermal region in the Rift Valley province of Kenya in east Africa. Although the majority of the wells encountered conditions of low permeability, the two best wells yielded roughly 30–40 t/h (66–88 × 10³ lbm/h) of liquid-vapor mixture. The reser-voir lies at 700–800 m (2297–2625 ft), and the fluid temperature is 245°C (473°F). Temperatures as high as 300°C (572°F) have been reported at a depth of 1650 m (5414 ft) [Ellis and Mahon, 1977]. A 15-MW geo-thermal power unit is being designed for this resource [K. Aikawa, per-sonal communication].

13.9 Nicaragua

The national electric authority of Nicaragua (ENALUF) has predicted that 100 MW of geothermal power will be installed in Nicaragua by the early 1980's, with about 150–220 MW installed by 1985, 300–400 MW by 2000, and as much as 800 MW by the year 2020 [Meidav et al., 1977]. The most likely candidate site for the first geothermal power plant is the Momotombo field, which was investigated from 1969 to 1971, along with the San Jacinto-Tisate area, both explored under a program sponsored by the U.S. Agency for International Development. Work at the sites was delayed several years by the Managua earthquake of December 23, 1972 [Muffler, 1975], and has suffered another setback because of political problems which erupted in 1978.

The geothermal field at Momotombo is located on the lower slopes of the Momotombo Volcano, on the edge of Lake Managua. A total of 25 wells have been drilled in the field. Some of these wells show drastic temperature inversions, as much as $-1.5°C/m$ ($-0.8°F/ft$), indicating the presence of colder fluid at depth. Some flow rates from a few of the wells have been reported [Girelli, 1977]: Momotombo well No. 3, 85 Mg/h (187 klbm/h); well No. 9, 56 Mg/h (123 klbm/h); well No. 12, 40 Mg/h (88 klbm/h). Construction was scheduled to begin in 1979 on a geothermal power plant at Momotombo, but the size of the unit has not been decided. The site is believed capable of supporting 100 MW, but a smaller 30-MW unit may be installed initially until confidence in the field is thoroughly established [Girelli, 1977].

13.10 Panama

Panama presently has a total installed electric capacity of 237 MW with projections of 534 MW by 1984. Although hydroelectric plants constitute a significant fraction of Panama's generating capacity, it is believed that a 75-MW powerplant, either conventional thermal or geothermal, will be needed by 1985. The most promising geothermal site in Panama is at Cerro Pando. It is too early, however, to assess the quality and the potential of this area in light of the minimal amount of exploratory work completed at this time [Ho, 1977; Meidav et al, 1977].

REFERENCES [a]

Akil, I., 1975. "Development of Geothermal Resources in Indonesia," *San Francisco, 1975,* Vol. 1, pp. 11–15.

Basoeki, M. and Radja, V. T., 1978. "Recent Development of 30 MW Kamojang Geothermal Power Project, West Java, Indonesia," *Geothermal Resources Council Trans.,* Vol. 2, Sec. 1, pp. 35–38.

---

[a] See note on p. 24.

Blackwell, D. D., Granados, G. E., and Koenig, J. B., 1977. "Heat Flow and Geothermal Gradient Exploration of Geothermal Areas in the Cordillera de Guanacaste Costa Rica of," *Geothermal Resources Council Trans.*, Vol. 1, pp. 17–18.

Dominco, E., 1977. "Guatemala," *Geothermal Resources: Survey of an Emerging Industry*, Geothermal Resources Council Spec. Short Course No. 6, Houston.

Ellis, A. J. and Mahon, W. A. J., 1977. *Chemistry and Geothermal Systems*, Academic Press, New York.

Furgerson, R. B. and Afonso L., P. S., 1977. "Electrical Investigations in the Guanacaste Geothermal Area (Costa Rica)," *Geothermal Resources Council Trans.*, Vol. 1, pp. 99–100.

Gardner, M. C. and Corrales, R., 1977. "Geochemical and Hydrological Investigations of the Guanacaste Geothermal Project, Costa Rica," *Geothermal Resources Council Trans.*, Vol. 1, pp. 101–102.

GEM, 1978. "Indonesia Uses U.S. Technology to Obtain Electricity from the Earth," *Geothermal Energy Magazine*, Vol. 6, No. 11, pp. 20–21.

Girelli, M., 1977. "Nicaragua," *Geothermal Resources: Survey of an Emerging Industry*, Geothermal Resources Council Spec. Short Course No. 6, Houston.

GR, 1978. *Geothermal Report*, R. A. Smith, ed., Vol. 7, No. 11, June 1, p. 4.

Ho, A., 1977. "Panama," *Geothermal Resources: Survey of an Emerging Industry*, Geothermal Resources Council Spec. Short Course No. 6, Houston.

Hochstein, M. P., 1975. "Geophysical Exploration of the Kawah Kamojang Geothermal Field, West Java," *San Francisco, 1975*, Vol. 2, pp. 1049–1058.

Kartokusumo, W., Mahon, W. A. J., and Seal, K. E., 1975. "Geochemistry of the Kawah Kamojang Geothermal System, Indonesia," *San Francisco, 1975*, Vol. 1, pp. 757–759.

Koenig, J. B., 1973. "Worldwide Status of Geothermal Resources Development," in *Geothermal Energy: Resources, Production, Stimulation*, P. Kruger and C. Otte, eds., Stanford Univ. Press, Stanford, CA, pp. 15–58.

Lahsen, A. and Trujillo, P., 1975. "The Geothermal Field of El Tatio, Chile," *San Francisco, 1975*, Vol. 1, pp. 157–177.

Meidav, T., Sanyal, S., and Facca, G., 1977. "An Update of World Geothermal Energy Development," *Geothermal Energy Magazine*, Vol. 5, No. 5, pp. 30–34.

Meucke, G. K., Ade-Hall, J. M., Aumento, F., MacDonald, A., Reynolds, P. H., Hyndman, R. D., Quintino, J., Opdyke, N., and Lowrie, W., 1974. "Deep Drilling in an Active Geothermal Area in the Azores," *Nature*, Vol. 252, pp. 281–285.

Muffler, L. J. P., 1975. "Summary of Section I: Present Status of Resources Development," *San Francisco, 1975*, Vol. 1, pp. xxxiii–xliv.

Radja, V. T., 1975. "Overview of Geothermal Energy Studies in Indonesia," *San Francisco, 1975*, Vol. 1, pp. 233–240.

Stehn, C. E., 1929. "Kawah Kamojang," Fourth Pacific Science Congress, Java, Indonesia.

Index

www.ingramcontent.com/pod-product-compliance
Lightning Source LLC
Chambersburg PA
CBHW021026210326
41598CB00016B/925